AF767624

Selected Titles in This Series

244 **Leovigildo Alonso Tarrío, Ana Jeremías López, and Joseph Lipman,** Studies in duality on noetherian formal schemes and non-noetherian ordinary schemes, 1999

243 **Tsit-Yuan Lam and Andy R. Magid, Editors,** Algebra, K-theory, groups, and education, 1999

242 **Bernhelm Booss-Bavnbek and Krzysztof Wojciechowski, Editors,** Geometric aspects of partial differential equations, 1999

241 **Piotr Pragacz, Michał Szurek, and Jarosław Wiśniewski, Editors,** Algebraic geometry: Hirzebruch 70, 1999

240 **Angel Carocca, Víctor González-Aguilera, and Rubí E. Rodríguez, Editors,** Complex geometry of groups, 1999

239 **Jean-Pierre Meyer, Jack Morava, and W. Stephen Wilson, Editors,** Homotopy invariant algebraic structures, 1999

238 **Gui-Qiang Chen and Emmanuele DiBenedetto, Editors,** Nonlinear partial differential equations, 1999

237 **Thomas Branson, Editor,** Spectral problems in geometry and arithmetic, 1999

236 **Bruce C. Berndt and Fritz Gesztesy, Editors,** Continued fractions: From analytic number theory to constructive approximation, 1999

235 **Walter A. Carnielli and Itala M. L. D'Ottaviano, Editors,** Advances in contemporary logic and computer science, 1999

234 **Theodore P. Hill and Christian Houdré, Editors,** Advances in stochastic inequalities, 1999

233 **Hanna Nencka, Editor,** Low dimensional topology, 1999

232 **Krzysztof Jarosz, Editor,** Function spaces, 1999

231 **Michael Farber, Wolfgang Lück, and Shmuel Weinberger, Editors,** Tel Aviv topology conference: Rothenberg Festschrift, 1999

230 **Ezra Getzler and Mikhail Kapranov, Editors,** Higher category theory, 1998

229 **Edward L. Green and Birge Huisgen-Zimmermann, Editors,** Trends in the representation theory of finite dimensional algebras, 1998

228 **Liming Ge, Huaxin Lin, Zhong-Jin Ruan, Dianzhou Zhang, and Shuang Zhang, Editors,** Operator algebras and operator theory, 1999

227 **John McCleary, Editor,** Higher homotopy structures in topology and mathematical physics, 1999

226 **Luis A. Caffarelli and Mario Milman, Editors,** Monge Ampère equation: Applications to geometry and optimization, 1999

225 **Ronald C. Mullin and Gary L. Mullen, Editors,** Finite fields: Theory, applications, and algorithms, 1999

224 **Sang Geun Hahn, Hyo Chul Myung, and Efim Zelmanov, Editors,** Recent progress in algebra, 1999

223 **Bernard Chazelle, Jacob E. Goodman, and Richard Pollack, Editors,** Advances in discrete and computational geometry, 1999

222 **Kang-Tae Kim and Steven G. Krantz, Editors,** Complex geometric analysis in Pohang, 1999

221 **J. Robert Dorroh, Giséle Ruiz Goldstein, Jerome A. Goldstein, and Michael Mudi Tom, Editors,** Applied analysis, 1999

220 **Mark Mahowald and Stewart Priddy, Editors,** Homotopy theory via algebraic geometry and group representations, 1998

For a complete list of titles in this series, visit the AMS Bookstore at **www.ams.org/bookstore/**.

Studies in Duality
on
Noetherian Formal Schemes
and
Non-Noetherian Ordinary Schemes

CONTEMPORARY
MATHEMATICS

244

Studies in Duality on Noetherian Formal Schemes and Non-Noetherian Ordinary Schemes

Leovigildo Alonso Tarrío
Ana Jeremías López
Joseph Lipman

American Mathematical Society
Providence, Rhode Island

1991 *Mathematics Subject Classification.* Primary 14F99;
Secondary 13D99, 14B15, 32C37.

Library of Congress Cataloging-in-Publication Data

Alonso Tarrío, Leovigildo, 1962–
Studies in duality on noetherian formal schemes and non-noetherian ordinary schemes / Leovigildo Alonso Tarrío, Ana Jeremías López, Joseph Lipman.
p. cm. — (Contemporary mathematics, ISSN 0271-4132 ; 244)
Includes bibliographical references and index.
ISBN 0-8218-1942-9 (alk. paper)
1. Duality theory (Mathematics) 2. Schemes (Algebraic geometry) I. Jeremías López, Ana, 1962– . II. Lipman, Joseph. III. Title. IV. Series: Contemporary mathematics (American Mathematical Society) ; v. 244.
QA564.A52 1999
515′.782–dc21 99-42685
CIP

Visit the AMS home page at URL: http://www.ams.org/

10 9 8 7 6 5 4 3 2 1 04 03 02 01 00 99

Contents

Preface

This volume comprises three essentially independent, but related, papers treating the foundations of Grothendieck Duality on noetherian formal schemes and on not-necessarily noetherian ordinary schemes. Here, briefly, is what is done and what is left undone.

Grothendieck Duality starts with the existence of a right adjoint for the (suitably restricted) derived direct image functor associated to a proper map, and the compatibility of such a right adjoint with flat base change. Our treatment in the first paper is the first for arbitrary noetherian formal schemes. (The classical case of ordinary schemes is the one where the topology on the structure sheaf is discrete.) It is indicated how the main results synthesize several duality-related topics such as local duality, formal duality, dualizing complexes, and residue theorems. The overall approach is abstract, in the style of Verdier and Deligne. Enlivening concrete interpretations—often involving differential forms—are left to another occasion.

It should be noted that the proof of the base-change theorem on formal schemes given in §7 of the first paper uses the special case of base-change on ordinary schemes, under weaker assumptions than those which support published proofs of the latter. There is at least outlined in the third paper a method for proving a sufficiently general base change theorem on ordinary schemes, even without noetherian hypotheses. Moreover, while details are not given, it seems that the method could be modified so as to apply directly to formal schemes.

Grothendieck Duality continues with the construction of a *pseudofunctor* agreeing with the above right adjoint on the category of proper maps, and with the usual inverse-image functor on the category of étale maps (cf. [**De,** §3, pp. 303–318]). A first step in this construction is showing that the construction of the said right adjoint is "local on the source." Following Verdier, we can deduce this from flat base change (§8.3, page 88). But we do not yet have a pseudofunctor for, say, separated pseudo-finite-type maps of formal schemes (see §1.2.2, page 7), because at present we lack a compactification theorem for such maps, analogous to the well-known one of Nagata for separated finite-type maps of ordinary noetherian schemes (factoring them as proper $\circ$ open immersion).

The role played by quasi-coherent sheaves on ordinary schemes is taken over on formal schemes by limits of coherent sheaves (which are quasi-coherent, though the converse doesn't always hold), see Theorem 4.1, page 41. The most general duality theorems are stated for quasi-coherent *torsion* sheaves—sheaves with sections annihilated locally by an *open* ideal (see §6, beginning on page 58). The expected statements also continue to hold for coherent sheaves on formal schemes (see e.g., Theorem 8.4, page 89), the transition from torsion sheaves being effected via Proposition 6.2.1, a special case of Greenlees-May duality.

Greenlees-May duality on noetherian formal schemes (first proved in [**GM**] for modules over rings) is the subject of the second paper. This is a canonical duality between the right-derived torsion functor and the left-derived completion functor associated to a coherent ideal on a formal scheme—a far-reaching generalization of classical local duality.

Though noetherian hypotheses play an essential role in the formal context, they can be largely eliminated for ordinary schemes, and that is the subject of the third paper. An underlying strategy illustrated there is to reduce as much as possible of the complexity of duality theory, both in definitions of maps and in proofs of theorems, to a purely formal category-theoretic context. Thus we use resolutions (injective or flat) of complexes only to establish a few basic formal relations—the axioms, so to speak; and after that the deduction of commutativity of various functorial diagrams becomes a central issue, one which was unduly slighted in the standard treatments (see e.g., p. 117 in [**Ha**]).[1] The categorical context is generated by five of the famous "six operations" (derived tensor and Hom, inverse and direct image, and a right adjoint for the latter—compactly-supported direct images not being considered here). It is very rich in relations. Dealing directly with the variety of possible commutativities, at first amusing, then tedious, appears before long to outgrow human capacities. This sort of situation is treated in the literature under the rubric "coherence in categories"; but the tools developed in that area seem as yet unable to offer much in the present complex setup.

Could logicians or computer scientists help?

References

[Co] B. Conrad, *Base Change and Grothendieck Duality,* Lecture Notes in Math., Springer-Verlag, New York, to appear.

[De] P. Deligne, Cohomologie à supports propres, in *Théorie des topos et Cohomologie Étale des Schémes (SGA 4) Tome 3,* Lecture Notes in Math., no. **305**, Springer-Verlag, New York, 1973, pp. 250–461.

[GM] J. P. C. Greenlees and J. P. May, Derived functors of I-adic completion and local homology *J. Algebra* **149** (1992), 438–453.

[Ha] R. Hartshorne, *Residues and Duality,* Lecture Notes in Math., no. **20**, Springer-Verlag, New York, 1966.

[1] In [**Co**], his reworking and clarification of [**Ha**], Conrad does come to grips with this issue.

Part 1

Duality and Flat Base Change on Formal Schemes

Contemporary Mathematics
Volume **244**, 1999

Duality and Flat Base Change on Formal Schemes

Leovigildo Alonso Tarrío, Ana Jeremías López, and Joseph Lipman

Abstract. We give several related versions of global Grothendieck Duality for unbounded complexes on noetherian formal schemes. The proofs, based on a non-trivial adaptation of Deligne's method for the special case of ordinary schemes, are reasonably self-contained, modulo the Special Adjoint Functor Theorem. An alternative approach, inspired by Neeman and based on recent results about "Brown Representability," is indicated as well. A section on applications and examples illustrates how our results synthesize a number of different duality-related topics (local duality, formal duality, residue theorems, dualizing complexes, ...).

A flat-base-change theorem for pseudo-proper maps leads in particular to sheafified versions of duality for bounded-below complexes with quasi-coherent homology. Thanks to Greenlees-May duality, the results take a specially nice form for proper maps and bounded-below complexes with coherent homology.

Contents

1. Preliminaries and main theorems.

First we need some notation and terminology. Let X be a ringed space, i.e., a topological space together with a sheaf of commutative rings $\mathcal{O}_X$. Let $\mathcal{A}(X)$ be the

First two authors partially supported by Xunta de Galicia research project XUGA20701A96 and Spain's DGES grant PB97-0530. They also thank the Mathematics Department of Purdue University for its hospitality, help and support.

Third author partially supported by the National Security Agency.

category of $\mathcal{O}_X$-modules, and $\mathcal{A}_{\mathrm{qc}}(X)$ (resp. $\mathcal{A}_{\mathrm{c}}(X)$, resp. $\mathcal{A}_{\vec{\mathrm{c}}}(X)$) the full subcategory of $\mathcal{A}(X)$ whose objects are the quasi-coherent (resp. coherent, resp. $\varinjlim$'s of coherent) $\mathcal{O}_X$-modules.[1] Let $\mathbf{K}(X)$ be the homotopy category of $\mathcal{A}(X)$-complexes, and let $\mathbf{D}(X)$ be the corresponding derived category, obtained from $\mathbf{K}(X)$ by adjoining an inverse for every quasi-isomorphism (= homotopy class of maps of complexes inducing homology isomorphisms).

For any full subcategory $\mathcal{A}_{\dots}(X)$ of $\mathcal{A}(X)$, denote by $\mathbf{D}_{\dots}(X)$ the full subcategory of $\mathbf{D}(X)$ whose objects are those complexes whose homology sheaves all lie in $\mathcal{A}_{\dots}(X)$, and by $\mathbf{D}^+_{\dots}(X)$ (resp. $\mathbf{D}^-_{\dots}(X)$) the full subcategory of $\mathbf{D}_{\dots}(X)$ whose objects are those complexes $\mathcal{F} \in \mathbf{D}_{\dots}(X)$ such that the homology $H^m(\mathcal{F})$ vanishes for all $m \ll 0$ (resp. $m \gg 0$).

The full subcategory $\mathcal{A}_{\dots}(X)$ of $\mathcal{A}(X)$ is *plump* if it contains 0 and for every exact sequence $\mathcal{M}_1 \to \mathcal{M}_2 \to \mathcal{M} \to \mathcal{M}_3 \to \mathcal{M}_4$ in $\mathcal{A}(X)$ with $\mathcal{M}_1$, $\mathcal{M}_2$, $\mathcal{M}_3$ and $\mathcal{M}_4$ in $\mathcal{A}_{\dots}(X)$, $\mathcal{M}$ is in $\mathcal{A}_{\dots}(X)$ too. If $\mathcal{A}_{\dots}(X)$ is plump then it is abelian, and has a derived category $\mathbf{D}(\mathcal{A}_{\dots}(X))$. For example, $\mathcal{A}_{\mathrm{c}}(X)$ is plump [**GD**, p. 113, (5.3.5)]. If $\mathfrak{X}$ is a locally noetherian formal scheme,[2] then $\mathcal{A}_{\vec{\mathrm{c}}}(\mathfrak{X}) \subset \mathcal{A}_{\mathrm{qc}}(\mathfrak{X})$ (Corollary 3.1.5)—with equality when $\mathfrak{X}$ is an ordinary scheme, i.e., when $\mathcal{O}_{\mathfrak{X}}$ has discrete topology [**GD**, p. 319, (6.9.9)]—and both of these are plump subcategories of $\mathcal{A}(\mathfrak{X})$, see Proposition 3.2.2.

Let $\mathbf{K}_1$, $\mathbf{K}_2$ be triangulated categories with respective translation functors T_1, T_2 [**H1**, p. 20]. A (covariant) Δ*-functor* is a pair (F, Θ) consisting of an additive functor $F\colon \mathbf{K}_1 \to \mathbf{K}_2$ together with an isomorphism of functors $\Theta : FT_1 \xrightarrow{\sim} T_2F$ such that for every triangle $A \xrightarrow{u} B \xrightarrow{v} C \xrightarrow{w} T_1A$ in $\mathbf{K}_1$, the diagram

$$FA \xrightarrow{Fu} FB \xrightarrow{Fv} FC \xrightarrow{\Theta\circ Fw} T_2FA$$

is a triangle in $\mathbf{K}_2$. Explicit reference to Θ is often suppressed—but one should keep it in mind. (For example, if $\mathcal{A}_{\dots}(X) \subset \mathcal{A}(X)$ is plump, then each of $\mathbf{D}_{\dots}(X)$ and $\mathbf{D}^{\pm}_{\dots}(X)$ carries a unique triangulation for which the translation is the restriction of that on $\mathbf{D}(X)$ and such that inclusion into $\mathbf{D}(X)$ together with Θ:=identity is a Δ-functor; in other words, they are all *triangulated subcategories* of $\mathbf{D}(X)$. See e.g., Proposition 3.2.4 for the usefulness of this remark.) Compositions of Δ-functors, and morphisms between Δ-functors, are defined in the natural way.[3] A Δ-functor $(G, \Psi)\colon \mathbf{K}_2 \to \mathbf{K}_1$ is a *right* Δ*-adjoint* of (F, Θ) if G is a right adjoint of F and the resulting functorial map $FG \to \mathbf{1}$ (or equivalently, $\mathbf{1} \to GF$) is a morphism of Δ-functors.

We use $\mathbf{R}$ to denote right-derived functors, constructed e.g., via K-injective resolutions (which exist for all $\mathcal{A}(X)$-complexes [**Sp**, p. 138, Thm. 4.5]).[4] For a map $f\colon X \to Y$ of ringed spaces (i.e., a continuous map $f\colon X \to Y$ together with a ring-homomorphism $\mathcal{O}_Y \to f_*\mathcal{O}_X$), $\mathbf{L}f^*$ denotes the left-derived functor of f^*,

[1] "$\varinjlim$" always denotes a direct limit over a small ordered index set in which any two elements have an upper bound. More general direct limits will be referred to as *colimits*.

[2] Basic properties of formal schemes can be found in [**GD**, Chap. 1, §10].

[3] See also [**De**, §0, §1] for the multivariate case, where signs come into play—and Δ-functors are called "exact functors."

[4] A complex F in an abelian category $\mathcal{A}$ is K-injective if for each exact $\mathcal{A}$-complex G the abelian-group complex $\mathrm{Hom}^\bullet_{\mathcal{A}}(G, F)$ is again exact. In particular, any bounded-below complex of injectives is K-injective. If every $\mathcal{A}$-complex E admits a K-injective resolution $E \to I(E)$ (i.e., a quasi-isomorphism into a K-injective complex $I(E)$), then every additive functor $\Gamma\colon \mathcal{A} \to \mathcal{A}'$ ($\mathcal{A}'$ abelian) has a right-derived functor $\mathbf{R}\Gamma\colon \mathbf{D}(\mathcal{A}) \to \mathbf{D}(\mathcal{A}')$ which satisfies $\mathbf{R}\Gamma(E) = \Gamma(I(E))$. For example, $\mathbf{R}\mathrm{Hom}^\bullet_{\mathcal{A}}(E_1, E_2) = \mathrm{Hom}^\bullet_{\mathcal{A}}(E_1, I(E_2))$.

constructed via K-flat resolutions [**Sp**, p. 147, 6.7]. Each derived functor in this paper comes equipped, implicitly, with a Θ making it into a Δ-functor (modulo obvious modifications for contravariance), cf. [**L4**, Example (2.2.4)].[5] Conscientious readers may verify that such morphisms between derived functors as occur in this paper are in fact morphisms of Δ-functors.

1.1. Our **first main result,** global Grothendieck Duality for a map $f\colon \mathfrak{X} \to \mathfrak{Y}$ of quasi-compact formal schemes with $\mathfrak{X}$ noetherian, is that, $\mathbf{D}(\mathcal{A}_{\vec{c}}(\mathfrak{X}))$ being the derived category of $\mathcal{A}_{\vec{c}}(\mathfrak{X})$ and $\boldsymbol{j}\colon \mathbf{D}(\mathcal{A}_{\vec{c}}(\mathfrak{X})) \to \mathbf{D}(\mathfrak{X})$ being the natural functor, *the Δ-functor $\mathbf{R}f_*\circ\boldsymbol{j}$ has a right Δ-adjoint.*

A more elaborate—but readily shown equivalent—statement is:

THEOREM 1. *Let $f\colon \mathfrak{X} \to \mathfrak{Y}$ be a map of quasi-compact formal schemes, with $\mathfrak{X}$ noetherian, and let $\boldsymbol{j}\colon \mathbf{D}(\mathcal{A}_{\vec{c}}(\mathfrak{X})) \to \mathbf{D}(\mathfrak{X})$ be the natural functor. Then there exists a Δ-functor $f^\times\colon \mathbf{D}(\mathfrak{Y}) \to \mathbf{D}(\mathcal{A}_{\vec{c}}(\mathfrak{X}))$ together with a morphism of Δ-functors $\tau : \mathbf{R}f_*\boldsymbol{j}f^\times \to \mathbf{1}$ such that for all $\mathcal{G} \in \mathbf{D}(\mathcal{A}_{\vec{c}}(\mathfrak{X}))$ and $\mathcal{F} \in \mathbf{D}(\mathfrak{Y})$, the composed map (in the derived category of abelian groups)*

$$\mathbf{R}\mathrm{Hom}^\bullet_{\mathcal{A}_{\vec{c}}(\mathfrak{X})}(\mathcal{G}, f^\times\mathcal{F}) \xrightarrow{\text{natural}} \mathbf{R}\mathrm{Hom}^\bullet_{\mathcal{A}(\mathfrak{Y})}(\mathbf{R}f_*\mathcal{G}, \mathbf{R}f_*f^\times\mathcal{F}) \xrightarrow{\text{via } \tau} \mathbf{R}\mathrm{Hom}^\bullet_{\mathcal{A}(\mathfrak{Y})}(\mathbf{R}f_*\mathcal{G}, \mathcal{F})$$

is an isomorphism.

Here we think of the $\mathcal{A}_{\vec{c}}(\mathfrak{X})$-complexes $\mathcal{G}$ and $f^\times\mathcal{F}$ as objects in both $\mathbf{D}(\mathcal{A}_{\vec{c}}(\mathfrak{X}))$ and $\mathbf{D}(\mathfrak{X})$. But as far as we know, the natural map $\mathrm{Hom}_{\mathbf{D}(\mathcal{A}_{\vec{c}}(\mathfrak{X}))} \to \mathrm{Hom}_{\mathbf{D}(\mathfrak{X})}$ need not always be an isomorphism. It *is* when $\mathfrak{X}$ is *properly algebraic,* i.e., the J-adic completion of a proper B-scheme with B a noetherian ring and J a B-ideal: then $\boldsymbol{j}$ induces an equivalence of categories $\mathbf{D}(\mathcal{A}_{\vec{c}}(\mathfrak{X})) \to \mathbf{D}_{\vec{c}}(\mathfrak{X})$, see Corollary 3.3.4. So for properly algebraic $\mathfrak{X}$, we can replace $\mathbf{D}(\mathcal{A}_{\vec{c}}(\mathfrak{X}))$ in Theorem 1 by $\mathbf{D}_{\vec{c}}(\mathfrak{X})$, and let $\mathcal{G}$ be any $\mathcal{A}(\mathfrak{X})$-complex with $\mathcal{A}_{\vec{c}}(\mathfrak{X})$-homology.

We prove Theorem 1 (= Theorem 4.1) in §4, adapting the argument of Deligne in [**H1**, Appendix] (see also [**De**, §1.1.12]) to the category $\mathcal{A}_{\vec{c}}(\mathfrak{X})$, which presents itself as an appropriate generalization to formal schemes of the category of quasi-coherent sheaves on an ordinary noetherian scheme. For this adaptation what is needed, mainly, is the plumpness of $\mathcal{A}_{\vec{c}}(\mathfrak{X})$ in $\mathcal{A}(\mathfrak{X})$, a non-obvious fact mentioned above. In addition, we need some facts on "boundedness" of certain derived functors in order to extend the argument to unbounded complexes. (See section 3.4, which makes use of techniques from [**Sp**].)[6]

In Deligne's approach the "Special Adjoint Functor Theorem" is used to get right adjoints for certain functors on $\mathcal{A}_{\mathrm{qc}}(X)$, and then these right adjoints are applied to injective resolutions of complexes ... There is now a neater approach to duality on a quasi-compact separated ordinary scheme X, due to Neeman [**N1**], in which "Brown Representability" shows directly that a Δ-functor F on $\mathbf{D}(\mathcal{A}_{\mathrm{qc}}(X))$ has a right adjoint if and only if F commutes with coproducts. Both approaches need a small set of category-generators: coherent sheaves for $\mathcal{A}_{\mathrm{qc}}(X)$ in Deligne's,

---

[5] We do not know, for instance, whether $\mathbf{L}f^*$—which is defined only up to isomorphism—can always be chosen so as to commute with translation, i.e., so that $\Theta =$ Identity will do.

[6] A Δ-functor ϕ is *bounded above* if there is an integer b such that for any n and any complex $\mathcal{E}$ such that $H^i\mathcal{E} = 0$ for all $i \le n$ it holds that $H^j(\phi\mathcal{E}) = 0$ for all $j < n + b$. *Bounded below* and *bounded* (above and below) are defined analogously. Boundedness (way-outness) is what makes the very useful "way-out Lemma" [**H1**, p. 68, 7.1] applicable.

and perfect complexes for $\mathbf{D}(\mathcal{A}_{\mathrm{qc}}(X))$ in Neeman's. Lack of knowledge about perfect complexes over formal schemes discouraged us from pursuing Neeman's strategy. Recently however (after this paper was essentially written), Franke showed in [**Fe**] that Brown Representability holds for the derived category of an arbitrary Grothendieck category $\mathcal{A}$.[7] Consequently Theorem 1 also follows from the fact that $\mathcal{A}_{\vec{c}}(\mathfrak{X})$ is a Grothendieck category (straightforward to see once we know it—by plumpness in $\mathcal{A}(\mathfrak{X})$—to be abelian) together with the fact that $\mathbf{R}f_* \circ \boldsymbol{j}$ commutes with coproducts (Proposition 3.5.2).

1.2. Two other, probably more useful, generalizations—from ordinary schemes to formal schemes—of global Grothendieck Duality are stated below in Theorem 2 and treated in detail in §6. To describe them, and related results, we need some preliminaries about *torsion functors.*

1.2.1. Once again let $(X, \mathcal{O}_X)$ be a ringed space. For any $\mathcal{O}_X$-ideal $\mathcal{J}$, set

$$\varGamma_{\mathcal{J}}\mathcal{M} := \varinjlim_{n>0} \mathcal{H}om_{\mathcal{O}_X}(\mathcal{O}_X/\mathcal{J}^n, \mathcal{M}) \qquad (\mathcal{M} \in \mathcal{A}(X)),$$

and regard $\varGamma_{\mathcal{J}}$ as a subfunctor of the identity functor on $\mathcal{O}_X$-modules. If $\mathcal{N} \subset \mathcal{M}$ then $\varGamma_{\mathcal{J}}\mathcal{N} = \varGamma_{\mathcal{J}}\mathcal{M} \cap \mathcal{N}$; and it follows formally that the functor $\varGamma_{\mathcal{J}}$ is idempotent ($\varGamma_{\mathcal{J}}\varGamma_{\mathcal{J}}\mathcal{M} = \varGamma_{\mathcal{J}}\mathcal{M}$) and left exact [**St**, p. 138, Proposition 1.7].

Set $\mathcal{A}_{\mathcal{J}}(X) := \varGamma_{\mathcal{J}}(\mathcal{A}(X))$, the full subcategory of $\mathcal{A}(X)$ whose objects are the *$\mathcal{J}$-torsion sheaves,* i.e., the $\mathcal{O}_X$-modules $\mathcal{M}$ such that $\varGamma_{\mathcal{J}}\mathcal{M} = \mathcal{M}$. Since $\varGamma_{\mathcal{J}}$ is an idempotent subfunctor of the identity functor, therefore it is right-adjoint to the inclusion $i = i_{\mathcal{J}} \colon \mathcal{A}_{\mathcal{J}}(X) \hookrightarrow \mathcal{A}(X)$. Moreover, $\mathcal{A}_{\mathcal{J}}(X)$ is closed under $\mathcal{A}(X)$-colimits: if F is any functor into $\mathcal{A}_{\mathcal{J}}(X)$ such that iF has a colimit $\mathcal{M} \in \mathcal{A}(X)$, then, since i and $\varGamma_{\mathcal{J}}$ are adjoint, the corresponding functorial map from iF to the constant functor with value $\mathcal{M}$ factors via a functorial map from iF to the constant functor with value $\varGamma_{\mathcal{J}}\mathcal{M}$, and from the definition of colimits it follows that the monomorphism $\varGamma_{\mathcal{J}}\mathcal{M} \hookrightarrow \mathcal{M}$ has a right inverse, so that it is an isomorphism, and thus $\mathcal{M} \in \mathcal{A}_{\mathcal{J}}(X)$. In particular, if the domain of a functor G into $\mathcal{A}_{\mathcal{J}}(X)$ is a small category, then iG does have a colimit, which is also a colimit of G; and so $\mathcal{A}_{\mathcal{J}}(X)$ has small colimits, i.e., it is small-cocomplete.

Submodules and quotient modules of $\mathcal{J}$-torsion sheaves are $\mathcal{J}$-torsion sheaves. If $\mathcal{J}$ is *finitely-generated* (locally) and if $\mathcal{N} \subset \mathcal{M}$ are $\mathcal{O}_X$-modules such that $\mathcal{N}$ and $\mathcal{M}/\mathcal{N}$ are $\mathcal{J}$-torsion sheaves then $\mathcal{M}$ is a $\mathcal{J}$-torsion sheaf too; and hence $\mathcal{A}_{\mathcal{J}}(X)$ is plump in $\mathcal{A}(X)$.[8] In this case, the stalk of $\varGamma_{\mathcal{J}}\mathcal{M}$ at $x \in X$ is

$$(\varGamma_{\mathcal{J}}\mathcal{M})_x = \varinjlim_{n>0} \mathrm{Hom}_{\mathcal{O}_{X,x}}(\mathcal{O}_{X,x}/\mathcal{J}_x^n, \mathcal{M}_x).$$

Let X be a locally noetherian scheme and $Z \subset X$ a closed subset, the support of $\mathcal{O}_X/\mathcal{J}$ for some quasi-coherent $\mathcal{O}_X$-ideal $\mathcal{J}$. The functor $\varGamma'_Z := \varGamma_{\mathcal{J}}$ does not depend on the quasi-coherent ideal $\mathcal{J}$ determining Z. It is a subfunctor of the left-exact functor $\varGamma_Z$ which associates to each $\mathcal{O}_X$-module $\mathcal{M}$ its subsheaf of sections supported in Z. If $\mathcal{M}$ is quasi-coherent, then $\varGamma'_Z(\mathcal{M}) = \varGamma_Z(\mathcal{M})$.

---

[7] So does the closely-related existence of K-injective resolutions for all $\mathcal{A}$-complexes. (See also [**AJS**, §5].)

[8] Thus the subcategory $\mathcal{A}_{\mathcal{J}}(X)$ is a *hereditary torsion class* in $\mathcal{A}(X)$, in the sense of Dickson, see [**St**, pp. 139–141].

More generally, for any complex $\mathcal{E} \in \mathbf{D}_{\mathrm{qc}}(X)$, the $\mathbf{D}(X)$-map $\mathbf{R}\varGamma'_Z\mathcal{E} \to \mathbf{R}\varGamma_Z\mathcal{E}$ induced by the inclusion $\varGamma'_Z \hookrightarrow \varGamma_Z$ is an isomorphism [**AJL**, p. 25, Corollary (3.2.4)]; so for such $\mathcal{E}$ we usually identify $\mathbf{R}\varGamma'_Z\mathcal{E}$ and $\mathbf{R}\varGamma_Z\mathcal{E}$.

Set $\mathcal{A}_Z(X) := \mathcal{A}_{\mathcal{J}}(X)$, the plump subcategory of $\mathcal{A}(X)$ whose objects are the *Z-torsion sheaves,* that is, the $\mathcal{O}_X$-modules $\mathcal{M}$ such that $\varGamma'_Z\mathcal{M} = \mathcal{M}$; and set $\mathcal{A}_{\mathrm{qc}Z}(X) := \mathcal{A}_{\mathrm{qc}}(X) \cap \mathcal{A}_Z(X)$, the plump subcategory of $\mathcal{A}(X)$ whose objects are the quasi-coherent $\mathcal{O}_X$-modules supported in Z.

For a locally noetherian formal scheme $\mathfrak{X}$ with ideal of definition $\mathcal{J}$, set $\varGamma'_{\mathfrak{X}} := \varGamma_{\mathcal{J}}$, a left-exact functor depending only on the sheaf of topological rings $\mathcal{O}_{\mathfrak{X}}$, not on the choice of $\mathcal{J}$—for $\mathcal{M} \in \mathcal{A}(\mathfrak{X})$, $\varGamma'_{\mathfrak{X}}\mathcal{M} \subset \mathcal{M}$ is the submodule whose sections are those of $\mathcal{M}$ annihilated locally by an open ideal. Say that $\mathcal{M}$ is a *torsion sheaf* if $\varGamma'_{\mathfrak{X}}\mathcal{M} = \mathcal{M}$. Let $\mathcal{A}_{\mathrm{t}}(\mathfrak{X}) := \mathcal{A}_{\mathcal{J}}(\mathfrak{X})$ be the plump subcategory of $\mathcal{A}(\mathfrak{X})$ whose objects are all the torsion sheaves; and set $\mathcal{A}_{\mathrm{qct}}(\mathfrak{X}) := \mathcal{A}_{\mathrm{qc}}(\mathfrak{X}) \cap \mathcal{A}_{\mathrm{t}}(\mathfrak{X})$, the full (in fact plump, see Corollary 5.1.3) subcategory of $\mathcal{A}(\mathfrak{X})$ whose objects are the quasi-coherent torsion sheaves. It holds that $\mathcal{A}_{\mathrm{qct}}(\mathfrak{X}) \subset \mathcal{A}_{\vec{\mathrm{c}}}(\mathfrak{X})$, see Corollary 5.1.4. If $\mathfrak{X}$ is an ordinary locally noetherian scheme (i.e., $\mathcal{J} = 0$), then $\mathcal{A}_{\mathrm{t}}(\mathfrak{X}) = \mathcal{A}(\mathfrak{X})$ and $\mathcal{A}_{\mathrm{qct}}(\mathfrak{X}) = \mathcal{A}_{\mathrm{qc}}(\mathfrak{X}) = \mathcal{A}_{\vec{\mathrm{c}}}(\mathfrak{X})$.

1.2.2. For any map $f\colon \mathfrak{X} \to \mathfrak{Y}$ of locally noetherian formal schemes there are ideals of definition $\mathcal{I} \subset \mathcal{O}_{\mathfrak{Y}}$ and $\mathcal{J} \subset \mathcal{O}_{\mathfrak{X}}$ such that $\mathcal{I}\mathcal{O}_{\mathfrak{X}} \subset \mathcal{J}$ [**GD**, p. 416, (10.6.10)]; and correspondingly there is a map of ordinary schemes (= formal schemes having (0) as ideal of definition) $(\mathfrak{X}, \mathcal{O}_{\mathfrak{X}}/\mathcal{J}) \to (\mathfrak{Y}, \mathcal{O}_{\mathfrak{Y}}/\mathcal{I})$ [**GD**, p. 410, (10.5.6)]. We say that f is *separated* (resp. *affine,* resp. *pseudo-proper,* resp. *pseudo-finite,* resp. *of pseudo-finite type*) if for some—and hence any—such $\mathcal{I}, \mathcal{J}$ the corresponding scheme-map is separated (resp. affine, resp. proper, resp. finite, resp. of finite type), see [**GD**, §§10.15–10.16, p. 444 *ff.*], keeping in mind [**GD**, p. 416, (10.6.10)(ii)].[9] Any affine map is separated. Any pseudo-proper map is separated and of pseudo-finite type. The map f is pseudo-finite $\Leftrightarrow$ it is pseudo-proper and affine $\Leftrightarrow$ it is pseudo-proper and has finite fibers [**EGA**, p. 136, (4.4.2)].

We say that f is *adic* if for some—and hence any—ideal of definition $\mathcal{I} \subset \mathcal{O}_{\mathfrak{Y}}$, $\mathcal{I}\mathcal{O}_{\mathfrak{X}}$ is an ideal of definition of $\mathfrak{X}$ [**GD**, p. 436, (10.12.1)]. We say that f is *proper* (resp. *finite,* resp. *of finite type*) if f is pseudo-proper (resp. pseudo-finite, resp. of pseudo-finite type) and adic, see [**EGA**, p. 119, (3.4.1)], [**EGA**, p. 148, (4.8.11)] and [**GD**, p. 440, (10.13.3)].

1.2.3. Here is our **second main result**, Torsion Duality for formal schemes. (See Theorem 6.1 and Corollary 6.1.4 for more elaborate statements.) In the assertion, $\widetilde{\mathbf{D}}_{\mathrm{qc}}(\mathfrak{X}) := \mathbf{R}\varGamma'^{-1}_{\mathfrak{X}}(\mathbf{D}_{\mathrm{qct}}(\mathfrak{X}))$ is the least Δ-subcategory of $\mathbf{D}(\mathfrak{X})$ containing both $\mathbf{D}_{\mathrm{qc}}(\mathfrak{X})$ and $\mathbf{R}\varGamma'^{-1}_{\mathfrak{X}}(0)$ (Definition 5.2.9, Remarks 5.2.10, (1) and (2)). For example, when $\mathfrak{X}$ is an ordinary scheme then $\widetilde{\mathbf{D}}_{\mathrm{qc}}(\mathfrak{X}) = \mathbf{D}_{\mathrm{qc}}(\mathfrak{X})$.

[9] In [**Y**, Definition 1.14], pseudo-finite-type maps are called "maps of formally finite type." The proof of Prop. 1.4 in [**Y**] (with $A' = A$) yields the following characterization of pseudo-finite-type maps of affine formal schemes (cf. [**GD**, p. 439, Prop. (10.13.1)]): The map $f\colon \mathrm{Spf}(B) \to \mathrm{Spf}(A)$ corresponding to a continuous homomorphism $h\colon A \to B$ of noetherian adic rings is of pseudo-finite type $\Leftrightarrow$ for any ideal of definition I of A, there exists an A-algebra of finite type A', an A'-ideal $I' \supset IA'$, and an A-algebra homomorphism $A' \to B$ inducing an adic surjective map $\widehat{A'} \twoheadrightarrow B$ where $\widehat{A'}$ is the I'-adic completion of A'.

THEOREM 2. *Let $f\colon \mathcal{X} \to \mathcal{Y}$ be a map of noetherian formal schemes. Assume either that f is separated or that $\mathcal{X}$ has finite Krull dimension, or else restrict to bounded-below complexes.*

(a) *The restriction of $\mathbf{R}f_*\colon \mathbf{D}(\mathcal{X}) \to \mathbf{D}(\mathcal{Y})$ takes $\mathbf{D}_{\mathrm{qct}}(\mathcal{X})$ to $\mathbf{D}_{\mathrm{qct}}(\mathcal{Y})$, and it has a right Δ-adjoint $f_{\mathrm{t}}^{\times}\colon \mathbf{D}(\mathcal{Y}) \to \mathbf{D}_{\mathrm{qct}}(\mathcal{X})$.*

(b) *The restriction of $\mathbf{R}f_*\mathbf{R}\varGamma'_{\mathcal{X}}$ takes $\widetilde{\mathbf{D}}_{\mathrm{qc}}(\mathcal{X})$ to $\mathbf{D}_{\mathrm{qct}}(\mathcal{Y}) \subset \widetilde{\mathbf{D}}_{\mathrm{qc}}(\mathcal{Y})$, and it has a right Δ-adjoint $f^{\#}\colon \mathbf{D}(\mathcal{Y}) \to \widetilde{\mathbf{D}}_{\mathrm{qc}}(\mathcal{X})$.*

REMARKS 1.2.4. (1) The "homology localization" functor

$$\mathbf{\Lambda}_{\mathcal{X}}(-) := \mathbf{R}\mathcal{H}om^{\bullet}(\mathbf{R}\varGamma'_{\mathcal{X}}\mathcal{O}_{\mathcal{X}}, -)$$

is right-adjoint to $\mathbf{R}\varGamma'_{\mathcal{X}}$, and $\mathbf{\Lambda}_{\mathcal{X}}^{-1}(0) = \mathbf{R}\varGamma'^{-1}_{\mathcal{X}}(0)$ (Remarks 6.3.1). The Δ-functors $f^{\#}$ and $f_{\mathrm{t}}^{\times}$ are connected thus (Corollaries 6.1.4 and 6.1.5(a)):

$$f^{\#} = \mathbf{\Lambda}_{\mathcal{X}} f_{\mathrm{t}}^{\times}, \qquad f_{\mathrm{t}}^{\times} = \mathbf{R}\varGamma'_{\mathcal{X}} f^{\#}.$$

(2) In the footnote on page 69 it is indicated that $\mathbf{R}\varGamma'^{-1}_{\mathcal{X}}(0)$ admits a "Bousfield colocalization" in $\mathbf{D}(\mathcal{X})$, with associated "cohomology colocalization" functor $\mathbf{R}\varGamma'_{\mathcal{X}}$; and in Remark 6.3.1(3), Theorem 2 is interpreted as duality with coefficients in the corresponding quotient $\widetilde{\mathbf{D}}_{\mathrm{qc}}(\mathcal{X})/\mathbf{R}\varGamma'^{-1}_{\mathcal{X}}(0) \cong \mathbf{D}_{\mathrm{qc}}(\mathcal{X})/(\mathbf{D}_{\mathrm{qc}}(\mathcal{X}) \cap \mathbf{R}\varGamma'^{-1}_{\mathcal{X}}(0))$.

(3) The proof of Theorem 2 is similar to that of Theorem 1, at least when the formal scheme $\mathcal{X}$ is separated (i.e., the unique formal-scheme map $\mathcal{X} \to \mathrm{Spec}(\mathbb{Z})$ is separated) or finite-dimensional, in which case there is an *equivalence of categories* $\mathbf{D}(\mathcal{A}_{\mathrm{qct}}(\mathcal{X})) \to \mathbf{D}_{\mathrm{qct}}(\mathcal{X})$ (Proposition 5.3.1). (As mentioned before, we know the corresponding result with "$\vec{\mathrm{c}}$" in place of "qct" only for *properly algebraic* formal schemes.) In addition, replacing separatedness of $\mathcal{X}$ by separatedness of f takes a technical pasting argument.

(4) For an ordinary scheme X (having (0) as ideal of definition), $\varGamma'_X$ is just the identity functor of $\mathcal{A}(X)$, and $\mathbf{D}_{\mathrm{qct}}(X) = \mathbf{D}_{\mathrm{qc}}(X)$. In this case, Theorems 1 and 2 both reduce to the usual global (non-sheafified) version of Grothendieck Duality. In §2 we will describe how Theorem 2 generalizes and ties together various strands in the literature on local, formal, and global duality. In particular, the behavior of Theorem 2 vis-à-vis variable f gives compatibility of local and global duality, at least on an abstract level—i.e., without the involvement of differentials, residues, etc. (See Corollary 6.1.6.)

1.3. As in the classic paper [**V**] of Verdier, the **culminating results** devolve from flat-base-change isomorphisms, established here for the functors $f_{\mathrm{t}}^{\times}$ and $f^{\#}$ of Theorem 2, with f *pseudo-proper*—in which case we denote $f_{\mathrm{t}}^{\times}$ by $f^{!}$.

THEOREM 3. *Let $\mathcal{X}$, $\mathcal{Y}$ and $\mathcal{U}$ be noetherian formal schemes, let $f\colon \mathcal{X} \to \mathcal{Y}$ be a pseudo-proper map, and let $u\colon \mathcal{U} \to \mathcal{Y}$ be flat, so that in the natural diagram*

$$\begin{array}{ccc} \mathcal{X} \times_{\mathcal{Y}} \mathcal{U} =: \mathcal{V} & \xrightarrow{\ v\ } & \mathcal{X} \\ {\scriptstyle g}\big\downarrow & & \big\downarrow{\scriptstyle f} \\ \mathcal{U} & \xrightarrow[\ u\ ]{} & \mathcal{Y} \end{array}$$

the formal scheme $\mathcal{V}$ is noetherian, g is pseudo-proper, and v is flat (*Proposition 7.1*).

Then for all $\mathcal{F} \in \widetilde{\mathbf{D}}^+_{\mathrm{qc}}(\mathcal{Y}) := \widetilde{\mathbf{D}}_{\mathrm{qc}}(\mathcal{Y}) \cap \mathbf{D}^+(\mathcal{Y})$ *the base-change map* $\beta_{\mathcal{F}}$ *of Definition 7.3 is an isomorphism*

$$\beta_{\mathcal{F}} \colon \mathbf{R}\varGamma'_{\mathcal{V}} v^* f^! \mathcal{F} \xrightarrow{\ \sim\ } g^! \mathbf{R}\varGamma'_{\mathcal{U}} u^* \mathcal{F} \underset{6.1.5(\mathrm{b})}{\cong} g^! u^* \mathcal{F}.$$

In particular, if u *is* adic *then we have a functorial isomorphism* $v^* f^! \mathcal{F} \xrightarrow{\sim} g^! u^* \mathcal{F}$.

This theorem is proved in §7 (Theorem 7.4). The functor $\mathbf{R}\varGamma'_{\mathcal{V}}$ has a right adjoint $\mathbf{\Lambda}_{\mathcal{V}}$, see (5.2.10.1). Theorem 3 leads quickly to the corresponding result for $f^{\#}$ (see Theorem 8.1 and Corollary 8.3.3):

Theorem 4. *Under the preceding conditions, let*

$$\beta^{\#}_{\mathcal{F}} \colon v^* f^{\#} \mathcal{F} \to g^{\#} u^* \mathcal{F} \qquad (\mathcal{F} \in \widetilde{\mathbf{D}}^+_{\mathrm{qc}}(\mathcal{Y}))$$

be the map adjoint to the natural composition

$$\mathbf{R}g_* \mathbf{R}\varGamma'_{\mathcal{V}} v^* f^{\#} \mathcal{F} \underset{\text{Thm. 3}}{\xrightarrow{\ \sim\ }} \mathbf{R}g_* g^! u^* \mathcal{F} \to u^* \mathcal{F}.$$

Then the map $\mathbf{\Lambda}_{\mathcal{V}}(\beta^{\#}_{\mathcal{F}})$ *is an* isomorphism

$$\mathbf{\Lambda}_{\mathcal{V}}(\beta^{\#}_{\mathcal{F}}) \colon \mathbf{\Lambda}_{\mathcal{V}} v^* f^{\#} \mathcal{F} \xrightarrow{\ \sim\ } \mathbf{\Lambda}_{\mathcal{V}} g^{\#} u^* \mathcal{F} \underset{6.1.5(\mathrm{a})}{\cong} g^{\#} u^* \mathcal{F}.$$

Moreover, if u *is an open immersion, or if* $\mathcal{F} \in \mathbf{D}^+_{\mathrm{c}}(\mathcal{Y})$, *then* $\beta^{\#}_{\mathcal{F}}$ *itself is an isomorphism.*

The special case of Theorems 3 and 4 when u is an open immersion is equivalent to what may be properly referred to as Grothendieck Duality (unqualified by the prefix "global"), namely the following *sheafified* version of Theorem 2 (see Theorem 8.2):

Theorem 5. *Let* $\mathcal{X}$ *and* $\mathcal{Y}$ *be noetherian formal schemes and let* $f \colon \mathcal{X} \to \mathcal{Y}$ *be a pseudo-proper map. Then the following natural compositions are* isomorphisms:

$$\begin{aligned} \mathbf{R}f_* \mathbf{R}\mathcal{H}om^{\bullet}_{\mathcal{X}}(\mathcal{G}, f^{\#}\mathcal{F}) &\to \mathbf{R}\mathcal{H}om^{\bullet}_{\mathcal{Y}}(\mathbf{R}f_*\mathbf{R}\varGamma'_{\mathcal{X}}\mathcal{G}, \mathbf{R}f_*\mathbf{R}\varGamma'_{\mathcal{X}} f^{\#}\mathcal{F}) \\ &\to \mathbf{R}\mathcal{H}om^{\bullet}_{\mathcal{Y}}(\mathbf{R}f_*\mathbf{R}\varGamma'_{\mathcal{X}}\mathcal{G}, \mathcal{F}) \quad (\mathcal{G} \in \widetilde{\mathbf{D}}_{\mathrm{qc}}(\mathcal{X}),\ \mathcal{F} \in \widetilde{\mathbf{D}}^+_{\mathrm{qc}}(\mathcal{Y})); \end{aligned}$$

$$\begin{aligned} \mathbf{R}f_* \mathbf{R}\mathcal{H}om^{\bullet}_{\mathcal{X}}(\mathcal{G}, f^{!}\mathcal{F}) \to \mathbf{R}\mathcal{H}om^{\bullet}_{\mathcal{Y}}(\mathbf{R}f_*\mathcal{G}, \mathbf{R}f_* f^{!}\mathcal{F}) &\to \mathbf{R}\mathcal{H}om^{\bullet}_{\mathcal{Y}}(\mathbf{R}f_*\mathcal{G}, \mathcal{F}) \\ &(\mathcal{G} \in \mathbf{D}_{\mathrm{qct}}(\mathcal{X}),\ \mathcal{F} \in \widetilde{\mathbf{D}}^+_{\mathrm{qc}}(\mathcal{Y})). \end{aligned}$$

Finally, if f is *proper* and $\mathcal{F} \in \mathbf{D}^+_{\mathrm{c}}(\mathcal{Y})$, then $f^{\#} \colon \mathbf{D}^+_{\mathrm{c}}(\mathcal{Y}) \to \mathbf{D}^+_{\mathrm{c}}(\mathcal{X})$ *is right-adjoint to* $\mathbf{R}f_* \colon \mathbf{D}^+_{\mathrm{c}}(\mathcal{X}) \to \mathbf{D}^+_{\mathrm{c}}(\mathcal{Y})$, *and* $\mathbf{R}\varGamma'_{\mathcal{X}}$ *in Theorem* 5 *can be deleted,* see Theorem 8.4.

In this—and several other results about complexes with coherent homology—an essential ingredient is Proposition 6.2.1, deduced here from Greenlees-May duality for ordinary affine schemes, see [**AJL**]:

Let $\mathcal{X}$ *be a locally noetherian formal scheme, and let* $\mathcal{E} \in \mathbf{D}(\mathcal{X})$. *Then for all* $\mathcal{F} \in \mathbf{D}_{\mathrm{c}}(\mathcal{X})$ *the natural map* $\mathbf{R}\varGamma'_{\mathcal{X}}\mathcal{E} \to \mathcal{E}$ *induces an isomorphism*

$$\mathbf{R}\mathcal{H}om^{\bullet}(\mathcal{E}, \mathcal{F}) \xrightarrow{\ \sim\ } \mathbf{R}\mathcal{H}om^{\bullet}(\mathbf{R}\varGamma'_{\mathcal{X}}\mathcal{E}, \mathcal{F}).$$

In closing this introductory section, we wish to express our appreciation for illuminating interchanges with Amnon Neeman and Amnon Yekutieli.

2. Applications and examples.

Again, Theorem 2 generalizes global Grothendieck Duality on ordinary schemes. This section illustrates further how Theorem 2 provides a common home for a number of different duality-related results (local duality, formal duality, residue theorems, dualizing complexes ...). For a quick example, see Remark 2.3.8.

Section 2.1 reviews several forms of local duality. In section 2.2 we sheafify these results, and connect them to Theorem 2. In particular, Proposition 2.1.6 is an abstract version of the Local Duality theorem of [**HüK**, p. 73, Theorem 3.4]; and Theorem 2.2.3 (Pseudo-finite Duality) globalizes it to formal schemes.

Section 2.3 relates Theorems 1 and 2 to the central "Residue Theorems" in [**L1**] and [**HüS**] (but does not subsume those results).

Section 2.4 indicates how both the Formal Duality theorem of [**H2**, p. 48, Proposition (5.2)] and the Local-Global Duality theorem in [**L3**, p. 188] can be deduced from Theorem 2.

Section 2.5, building on work of Yekutieli [**Y**, §5], treats *dualizing complexes* on formal schemes, and their associated dualizing functors. For a pseudo-proper map f, the functor $f^{\#}$ of Theorem 2 lifts dualizing complexes to dualizing complexes (Proposition 2.5.11). For any map $f\colon \mathcal{X} \to \mathcal{Y}$ of noetherian formal schemes, there is natural isomorphism

$$\mathbf{R}\mathcal{H}om^{\bullet}_{\mathcal{X}}(\mathbf{L}f^{*}\mathcal{F}, f^{\#}\mathcal{G}) \xrightarrow{\sim} f^{\#}\mathbf{R}\mathcal{H}om^{\bullet}_{\mathcal{Y}}(\mathcal{F},\mathcal{G}) \qquad (\mathcal{F} \in \mathbf{D}_{\mathrm{c}}^{-}(\mathcal{Y}),\ \mathcal{G} \in \mathbf{D}^{+}(\mathcal{Y})),$$

(Proposition 2.5.13). For pseudo-proper f, if $\mathcal{Y}$ has a dualizing complex $\mathcal{R}$, so that $f^{\#}\mathcal{R}$ is a dualizing complex on $\mathcal{X}$, and if $\mathcal{D}^{\mathcal{Y}} := \mathbf{R}\mathcal{H}om^{\bullet}(-,\mathcal{R})$ and $\mathcal{D}^{\mathcal{X}}$ are the corresponding dualizing functors, one deduces a natural isomorphism (well-known for ordinary schemes)

$$f^{\#}\mathcal{E} \cong \mathcal{D}^{\mathcal{X}}\mathbf{L}f^{*}\mathcal{D}^{\mathcal{Y}}\mathcal{E} \qquad (\mathcal{E} \in \mathbf{D}_{\mathrm{c}}^{+}(\mathcal{Y})),$$

see Proposition 2.5.12.

There are corresponding results for $f_{\mathrm{t}}^{\times}$ as well.

2.1. (Local Duality.) All rings are commutative, unless otherwise specified.

Let $\varphi\colon R \to S$ be a ring homomorphism with S noetherian, let J be an S-ideal, and let Γ_J be the functor taking any S-module to its submodule of elements which are annihilated by some power of J. Let E and E' be complexes in $\mathbf{D}(S)$, the derived category of S-modules, and let $F \in \mathbf{D}(R)$. With $\underline{\underline{\otimes}}$ denoting derived tensor product in $\mathbf{D}(S)$ (defined via K-flat resolutions [**Sp**, p. 147, Proposition 6.5]), there is a natural isomorphism $E \underline{\underline{\otimes}} \mathbf{R}\Gamma_J E' \xrightarrow{\sim} \mathbf{R}\Gamma_J(E \underline{\underline{\otimes}} E')$, see e.g., [**AJL**, p. 20, Corollary(3.1.2)]. Also, viewing $\mathbf{R}\mathrm{Hom}^{\bullet}_R(E', F)$ as a functor from $\mathbf{D}(S)^{\mathrm{op}} \times \mathbf{D}(R)$ to $\mathbf{D}(S)$, one has a canonical $\mathbf{D}(S)$-isomorphism

$$\mathbf{R}\mathrm{Hom}^{\bullet}_R(E \underline{\underline{\otimes}} E', F) \xrightarrow{\sim} \mathbf{R}\mathrm{Hom}^{\bullet}_S(E, \mathbf{R}\mathrm{Hom}^{\bullet}_R(E', F)),$$

see [**Sp**, p. 147; 6.6]. Thus, with $\varphi_J^{\#}\colon \mathbf{D}(R) \to \mathbf{D}(S)$ the functor given by

$$\varphi_J^{\#}(-) := \mathbf{R}\mathrm{Hom}^{\bullet}_R(\mathbf{R}\Gamma_J S, -) \cong \mathbf{R}\mathrm{Hom}^{\bullet}_S(\mathbf{R}\Gamma_J S, \mathbf{R}\mathrm{Hom}^{\bullet}_R(S, -)),$$

there is a composed isomorphism

$$\mathbf{R}\mathrm{Hom}^{\bullet}_S(E, \varphi_J^{\#}F) \xrightarrow{\sim} \mathbf{R}\mathrm{Hom}^{\bullet}_R(E \underline{\underline{\otimes}} \mathbf{R}\Gamma_J S, F) \xrightarrow{\sim} \mathbf{R}\mathrm{Hom}^{\bullet}_R(\mathbf{R}\Gamma_J E, F).$$

Application of homology H^0 yields the (rather trivial) *local duality isomorphism*

$$\text{(2.1.1)} \qquad \operatorname{Hom}_{\mathbf{D}(S)}(E, \varphi_J^{\#}F) \xrightarrow{\sim} \operatorname{Hom}_{\mathbf{D}(R)}(\mathbf{R}\Gamma_J E, F).$$

"Non-trivial" versions of (2.1.1) include more information about $\varphi_J^{\#}$. For example, Greenlees-May duality [**AJL**, p. 4, $(0.3)_{\text{aff}}$] gives a canonical isomorphism

$$\text{(2.1.2)} \qquad \varphi_J^{\#}F \cong \mathbf{L}\Lambda_J \mathbf{R}\operatorname{Hom}^{\bullet}_R(S, F),$$

where Λ_J is the *J-adic completion functor,* and $\mathbf{L}$ denotes "left-derived." In particular, if R is noetherian, S is a finite R-module, and $F \in \mathbf{D}_{\mathrm{c}}(R)$ (i.e., each homology module of F is finitely generated), then as in [**AJL**, p. 6, Proposition (0.4.1)],

$$\text{(2.1.3)} \qquad \varphi_J^{\#}F = \mathbf{R}\operatorname{Hom}^{\bullet}_R(S, F) \otimes_S \hat{S} \qquad (\hat{S} = J\text{-adic completion of } S).$$

More particularly, for $S = R$ and $\varphi = \mathrm{id}$ (the identity map) we get

$$\mathrm{id}_J^{\#}F = F \otimes_R \hat{R} \qquad (F \in \mathbf{D}_{\mathrm{c}}(R)).$$

Hence, *classical local duality* [**H1**, p. 278 (modulo Matlis dualization)] is just (2.1.1) when R is local, $\varphi = \mathrm{id}$, J is the maximal ideal of R, and F is a normalized dualizing complex—so that, as in Corollary 5.2.3, and by [**H1**, p. 276, Proposition 6.1],

$$\operatorname{Hom}_{\mathbf{D}(R)}(\mathbf{R}\Gamma_J E, F) = \operatorname{Hom}_{\mathbf{D}(R)}(\mathbf{R}\Gamma_J E, \mathbf{R}\Gamma_J F) = \operatorname{Hom}_{\mathbf{D}(R)}(\mathbf{R}\Gamma_J E, I)$$

where I is an R-injective hull of the residue field R/J. (See also Lemma 2.5.7.)

For another example, let $S = R[[\mathbf{t}]]$ where $\mathbf{t} := (t_1, \dots, t_d)$ is a sequence of variables, and set $J := \mathbf{t}S$. The standard calculation (via Koszul complexes) gives an isomorphism $\mathbf{R}\Gamma_J S \cong \nu[-d]$ where ν is the free R-submodule of the localization $S_{t_1 \dots t_d}$ generated by those monomials $t_1^{n_1} \dots t_d^{n_d}$ with all exponents $n_i < 0$, the S-module structure being induced by that of $S_{t_1 \dots t_d}/S \supset \nu$. The *relative canonical module* $\omega_{R[[\mathbf{t}]]/R} := \operatorname{Hom}_R(\nu, R)$ is a *free, rank one, S-module.* There result, for finitely-generated R-modules F, functorial isomorphisms

$$\text{(2.1.4)} \qquad \varphi^{\#}_{\mathbf{t}R[[\mathbf{t}]]}F \cong \operatorname{Hom}_R(\nu[-d], F) \cong \omega_{R[[\mathbf{t}]]/R}[d] \otimes_R F \cong R[[\mathbf{t}]] \otimes_R F[d];$$

and when R is noetherian, the usual way-out argument [**H1**, p. 69, (ii)] yields the same for any $F \in \mathbf{D}^+_{\mathrm{c}}(R)$.

Next, we give a commutative-algebra analogue of Theorem 2 in §1, in the form of a "torsion" variant of the duality isomorphism (2.1.1). Proposition 2.2.1 will clarify the relation between the algebraic and formal-scheme contexts.

With $\varphi\colon R \to S$ and J an S-ideal as before, let $\mathcal{A}_J(S)$ be the category of J-torsion S-modules, i.e., S-modules M such that

$$M = \Gamma_J M := \{\, m \in M \mid J^n m = 0 \text{ for some } n > 0 \,\}.$$

The derived category of $\mathcal{A}_J(S)$ is equivalent to the full subcategory $\mathbf{D}_J(S)$ of $\mathbf{D}(S)$ with objects those S-complexes E whose homology lies in $\mathcal{A}_J(S)$, (or equivalently, such that the natural map $\mathbf{R}\Gamma_J E \to E$ is an isomorphism), and the functor $\mathbf{R}\Gamma_J$ is right-adjoint to the inclusion $\mathbf{D}_J(S) \hookrightarrow \mathbf{D}(S)$ (cf. Proposition 5.2.1 and its proof).

Hence the functor $\varphi_J^\times : \mathbf{D}(R) \to \mathbf{D}_J(S)$ defined by

$$\varphi_J^\times(-) := \mathbf{R}\Gamma_J \mathbf{R}\mathrm{Hom}^\bullet_R(S,-) \cong \mathbf{R}\Gamma_J S \underline{\underline{\otimes}} \mathbf{R}\mathrm{Hom}^\bullet_R(S,-)$$

is right-adjoint to the natural composition $\mathbf{D}_J(S) \hookrightarrow \mathbf{D}(S) \to \mathbf{D}(R)$: in fact, for $E \in \mathbf{D}_J(S)$ and $F \in \mathbf{D}(R)$ there are natural isomorphisms

$$\mathbf{R}\mathrm{Hom}^\bullet_S(E, \varphi_J^\times F) \xrightarrow{\sim} \mathbf{R}\mathrm{Hom}^\bullet_S(E, \mathbf{R}\mathrm{Hom}^\bullet_R(S,F)) \xrightarrow{\sim} \mathbf{R}\mathrm{Hom}^\bullet_R(E,F). \tag{2.1.5}$$

Here is another interpretation of $\varphi_J^\times F$. For S-modules A and R-modules B set

$$\mathrm{Hom}_{R,J}(A,B) := \Gamma_J \mathrm{Hom}_R(A,B),$$

the S-module of R-homomorphisms α vanishing on $J^n A$ for some n (depending on α), i.e., *continuous* when A is J-adically topologized and B is discrete. If E is a K-flat S-complex and F is a K-injective R-complex, then $\mathrm{Hom}^\bullet_R(E,F)$ is a K-injective S-complex; and it follows for all $E \in \mathbf{D}(S)$ and $F \in \mathbf{D}(R)$ that

$$\mathbf{R}\mathrm{Hom}^\bullet_{R,J}(E,F) \cong \mathbf{R}\Gamma_J \mathbf{R}\mathrm{Hom}^\bullet_R(E,F).$$

Thus,

$$\varphi_J^\times F = \mathbf{R}\mathrm{Hom}^\bullet_{R,J}(S,F).$$

A *torsion version of local duality* is the isomorphism, derived from (2.1.5):

$$\mathrm{Hom}_{\mathbf{D}_J(S)}(E, \mathbf{R}\mathrm{Hom}^\bullet_{R,J}(S,F)) \xrightarrow{\sim} \mathrm{Hom}_{\mathbf{D}(R)}(E,F) \qquad (E \in \mathbf{D}_J(S),\ F \in \mathbf{D}(R)).$$

Apropos of Remark 1.2.4(1), the functors $\varphi_J^\times$ and $\varphi_J^\#$ are related by

$$\begin{aligned} \mathbf{L}\Lambda_J \mathbf{R}\mathrm{Hom}^\bullet_R(S,F) &\underset{(2.1.2)}{\cong} \varphi_J^\# F \cong \mathbf{L}\Lambda_J \varphi_J^\times F, \\ \mathbf{R}\Gamma_J \mathbf{R}\mathrm{Hom}^\bullet_R(S,F) &= \varphi_J^\times F \cong \mathbf{R}\Gamma_J \varphi_J^\# F. \end{aligned}$$

The first relation is the case $E = \mathbf{R}\Gamma_J S$ of (2.1.5), followed by Greenlees-May duality. The second results, e.g., from the sequence of natural isomorphisms, holding for $G \in \mathbf{D}_J(S)$, $E \in \mathbf{D}(S)$, and $F \in \mathbf{D}(R)$:

$$\begin{aligned} \mathrm{Hom}_{\mathbf{D}(S)}(G, \mathbf{R}\Gamma_J \mathbf{R}\mathrm{Hom}^\bullet_R(E,F)) &\cong \mathrm{Hom}_{\mathbf{D}(S)}(G, \mathbf{R}\mathrm{Hom}^\bullet_R(E,F)) \\ &\cong \mathrm{Hom}_{\mathbf{D}(R)}(\mathbf{R}\Gamma_J S \underline{\underline{\otimes}}_S G \underline{\underline{\otimes}}_S E, F) \\ &\cong \mathrm{Hom}_{\mathbf{D}(S)}(G, \mathbf{R}\mathrm{Hom}^\bullet_R(\mathbf{R}\Gamma_J E, F)) \\ &\cong \mathrm{Hom}_{\mathbf{D}(S)}(G, \mathbf{R}\Gamma_J \mathbf{R}\mathrm{Hom}^\bullet_R(\mathbf{R}\Gamma_J E, F)), \end{aligned}$$

which entail that the natural map is an isomorphism

$$\mathbf{R}\Gamma_J \mathbf{R}\mathrm{Hom}^\bullet_R(E,F) \xrightarrow{\sim} \mathbf{R}\Gamma_J \mathbf{R}\mathrm{Hom}^\bullet_R(\mathbf{R}\Gamma_J E, F).$$

Local Duality theorems are often formulated, as in (c) of the following, in terms of modules and local cohomology ($\mathrm{H}^\bullet_J := \mathrm{H}^\bullet \mathbf{R}\Gamma_J$) rather than derived categories.

PROPOSITION 2.1.6. *Let $\varphi : R \to S$ be a homomorphism of noetherian rings, let J be an S-ideal, and suppose that there exists a sequence $\mathbf{u} = (u_1, \dots, u_d)$ in J such that $S/\mathbf{u}S$ is R-finite. Then for any R-finite module F:*

(a) $\mathrm{H}^n \varphi_J^\# F = 0$ *for all* $n < -d$, *so that there is a natural* $\mathbf{D}(S)$*-map*

$$h \colon (\mathrm{H}^{-d} \varphi_J^\# F)[d] \to \varphi_J^\# F.$$

(b) *If* $\tau_F\colon \mathbf{R}\Gamma_J\varphi_J^{\#}F \to F$ *corresponds in* (2.1.1) *to the identity map of* $\varphi_J^{\#}F$,[10] *and* $\int = \int_{\varphi,J}^{d}(F)$ *is the composed map*

$$\mathbf{R}\Gamma_J(\mathrm{H}^{-d}\varphi_J^{\#}F)[d] \xrightarrow{\mathbf{R}\Gamma_J(h)} \mathbf{R}\Gamma_J\varphi_J^{\#}F \xrightarrow{\tau_F} F,$$

then $(\mathrm{H}^{-d}\varphi_J^{\#}F, \int)$ *represents the functor* $\mathrm{Hom}_{\mathbf{D}(R)}(\mathbf{R}\Gamma_J E[d], F)$ *of* S*-modules* E.

(c) *If* $J \subset \sqrt{\mathbf{u}S}$ *then there is a bifunctorial isomorphism* (*with* E, F *as before*):

$$\mathrm{Hom}_S(E, \mathrm{H}^{-d}\varphi_J^{\#}F) \xrightarrow{\sim} \mathrm{Hom}_R(\mathrm{H}_J^d E, F).$$

PROOF. If $\hat{\varphi}$ is the obvious map from R to the $\mathbf{u}$-adic completion $\hat{S}$ of S, then in $\mathbf{D}(S)$, $\varphi_J^{\#}F = \hat{\varphi}_J^{\#}F$ since $\mathbf{R}\Gamma_J S = \mathbf{R}\Gamma_J\hat{S}$. In proving (a), therefore, we may assume that S is $\mathbf{u}$-adically complete, so that φ factors as $R \xrightarrow{\psi} R[[\mathbf{t}]] \xrightarrow{\chi} S$ with $\mathbf{t} = (t_1, \dots, t_d)$ a sequence of indeterminates and S finite over $R[[\mathbf{t}]]$. (ψ is the natural map, and $\chi(t_i) = u_i$.) In view of the easily-verified relation $\varphi_J^{\#} = \chi_J^{\#} \circ \psi_{\mathbf{t}R[[\mathbf{t}]]}^{\#}$, (2.1.3) and (2.1.4) yield (a). Then (b) results from the natural isomorphisms

$$\mathrm{Hom}_S(E, \mathrm{H}^{-d}\varphi_J^{\#}F) \underset{\text{via } h}{\xrightarrow{\sim}} \mathrm{Hom}_{\mathbf{D}(S)}(E[d], \varphi_J^{\#}F) \underset{(2.1.1)}{\xrightarrow{\sim}} \mathrm{Hom}_{\mathbf{D}(R)}(\mathbf{R}\Gamma_J E[d], F).$$

Finally, (c) follows from (b) because $\mathrm{H}_J^i E = \mathrm{H}_{\mathbf{u}S}^i E = 0$ for all $i > d$ (as one sees from the usual calculation of $\mathrm{H}_{\mathbf{u}S}^i E$ via Koszul complexes), so that the natural map is an isomorphism $\mathrm{Hom}_{\mathbf{D}(R)}(\mathbf{R}\Gamma_J E[d], F) \xrightarrow{\sim} \mathrm{Hom}_R(\mathrm{H}_J^d E, F)$. □

2.1.7. Under the hypotheses of Proposition 2.1.6(c), the functor $\mathrm{Hom}_R(\mathrm{H}_J^d E, R)$ of S-modules E is representable. Under suitable extra conditions (for example, $\hat{S}$ a generic local complete intersection over $R[[\mathbf{t}]]$, Hübl and Kunz represent this functor by a *canonical* pair described explicitly via differential forms, residues, and certain trace maps [**HüK**, p. 73, Theorem 3.4]. For example, with $S = R[[\mathbf{t}]]$, $J = \mathbf{t}S$, and ν as in (2.1.4), the S-homomorphism from the module $\widehat{\Omega}_{S/R}^d$ of universally finite d-forms to the relative canonical module $\omega_{R[[\mathbf{t}]]/R} = \mathrm{Hom}_R(\nu, R)$ sending the form $dt_1 \dots dt_d$ to the R-homomorphism $\nu \to R$ which takes the monomial $t_1^{-1} \dots t_d^{-1}$ to 1 and all other monomials $t_1^{n_1} \dots t_d^{n_d}$ to 0, is clearly an isomorphism; and the resulting isomorphism $\widehat{\Omega}_{S/R}^d[d] \xrightarrow{\sim} \varphi_J^{\#}R$ *does not depend on the* d*-element sequence* $\mathbf{t}$ *generating* J—it corresponds under (2.1.1) to the *residue map*

$$\mathbf{R}\Gamma_J\widehat{\Omega}_{S/R}^d[d] = \mathrm{H}_J^d\widehat{\Omega}_{S/R}^d \to R$$

(see, e.g., [**L2**, §2.7]). Thus $\mathrm{Hom}_R(\mathrm{H}_J^d E, R)$ *is represented by* $\widehat{\Omega}_{S/R}^d$ *together with the residue map.* The general case reduces to this one via traces of differential forms.

2.2. (Formal sheafification of Local Duality). For $f\colon \mathcal{X} \to \mathcal{Y}$ as in Theorem 2 in §1, there is a right Δ-adjoint $f_{\mathrm{t}}^{\times}$ for the functor $\mathbf{R}f_*\colon \mathbf{D}_{\mathrm{qct}}(\mathcal{X}) \to \mathbf{D}_{\mathrm{qct}}(\mathcal{Y})$. Furthermore, with $\boldsymbol{j}\colon \mathbf{D}(\mathcal{A}_{\vec{\mathrm{c}}}(\mathcal{X})) \to \mathbf{D}_{\vec{\mathrm{c}}}(\mathcal{X})$ the canonical functor, we have

$$\mathbf{R}f_*\mathbf{R}\varGamma_{\mathcal{X}}'\boldsymbol{j}\mathbf{D}(\mathcal{A}_{\vec{\mathrm{c}}}(\mathcal{X})) \underset{(3.1.5)}{\subset} \mathbf{R}f_*\mathbf{R}\varGamma_{\mathcal{X}}'\mathbf{D}_{\mathrm{qc}}(\mathcal{X}) \underset{(5.2.1)}{\subset} \mathbf{R}f_*\mathbf{D}_{\mathrm{qct}}(\mathcal{X}) \underset{(5.2.6)}{\subset} \mathbf{D}_{\mathrm{qct}}(\mathcal{Y}) \underset{(3.1.7)}{\subset} \mathbf{D}_{\vec{\mathrm{c}}}(\mathcal{Y}).$$

It results from (5.2.10.1) and Proposition 3.2.3 that $\mathbf{R}f_*\mathbf{R}\varGamma_{\mathcal{X}}'\boldsymbol{j}\colon \mathbf{D}(\mathcal{A}_{\vec{\mathrm{c}}}(\mathcal{X})) \to \mathbf{D}_{\vec{\mathrm{c}}}(\mathcal{Y})$ has the right Δ-adjoint $\mathbf{R}Q_{\mathcal{X}}f^{\#} := \mathbf{R}Q_{\mathcal{X}}\mathbf{R}\mathcal{H}om^{\bullet}(\mathbf{R}\varGamma_{\mathcal{X}}'\mathcal{O}_{\mathcal{X}}, f_{\mathrm{t}}^{\times})$.

[10] τ_F may be thought of as "evaluation at 1": $\mathbf{R}\mathrm{Hom}_{R,J}^{\bullet}(S, F) \to F$.

If, moreover, $\mathcal{X}$ is *properly algebraic* (Definition 3.3.3)—in particular, if $\mathcal{X}$ is affine—then $\boldsymbol{j}$ is an equivalence of categories (Corollary 3.3.4), and so the functor $\mathbf{R}f_*\mathbf{R}\mathit{\Gamma}'_{\mathcal{X}}\colon \mathbf{D}_{\vec{c}}(\mathcal{X}) \to \mathbf{D}_{\vec{c}}(\mathcal{Y})$ has a right Δ-adjoint.

For *affine* f, these results are closely related to the Local Duality isomorphisms (2.1.5) and (2.1.1). Recall that an *adic ring* is a pair (R, I) with R a ring and I an R-ideal such that with respect to the I-adic topology R is Hausdorff and complete. The topology on R having been specified, the corresponding affine formal scheme is denoted $\mathrm{Spf}(R)$.

PROPOSITION 2.2.1. *Let $\varphi\colon (R, I) \to (S, J)$ be a continuous homomorphism of noetherian adic rings, and let $\mathcal{X} := \mathrm{Spf}(S) \xrightarrow{f} \mathrm{Spf}(R) =: \mathcal{Y}$ be the corresponding (affine) formal-scheme map. Let $\kappa_{\mathcal{X}}\colon \mathcal{X} \to X := \mathrm{Spec}(S)$, $\kappa_{\mathcal{Y}}\colon \mathcal{Y} \to Y := \mathrm{Spec}(R)$ be the completion maps, and let $^\sim = {}^{\sim_S}$ denote the standard exact functor from S-modules to quasi-coherent $\mathcal{O}_X$-modules. Then:*

(a) *The restriction of $\mathbf{R}f_*$ takes $\mathbf{D}_{\mathrm{qct}}(\mathcal{X})$ to $\mathbf{D}_{\mathrm{qct}}(\mathcal{Y})$, and this restricted functor has a right adjoint $f_{\mathrm{t}}^{\times}\colon \mathbf{D}_{\mathrm{qct}}(\mathcal{Y}) \to \mathbf{D}_{\mathrm{qct}}(\mathcal{X})$ given by*

$$f_{\mathrm{t}}^{\times}\mathcal{F} := \kappa_{\mathcal{X}}^*(\varphi_J^{\times}\mathbf{R}\Gamma(\mathcal{Y},\mathcal{F}))^{\sim} = \kappa_{\mathcal{X}}^*(\mathbf{R}\mathrm{Hom}^{\bullet}_{R,J}(S, \mathbf{R}\Gamma(\mathcal{Y},\mathcal{F})))^{\sim} \qquad (\mathcal{F} \in \mathbf{D}_{\mathrm{qct}}(\mathcal{Y})).$$

(b) *The restriction of $\mathbf{R}f_*\mathbf{R}\mathit{\Gamma}'_{\mathcal{X}}$ takes $\mathbf{D}_{\vec{c}}(\mathcal{X})$ to $\mathbf{D}_{\vec{c}}(\mathcal{Y})$, and this restricted functor has a right adjoint $f_{\vec{c}}^{\#}\colon \mathbf{D}_{\vec{c}}(\mathcal{Y}) \to \mathbf{D}_{\vec{c}}(\mathcal{X})$ given by*

$$f_{\vec{c}}^{\#}\mathcal{F} := \kappa_{\mathcal{X}}^*(\varphi_J^{\#}\mathbf{R}\Gamma(\mathcal{Y},\mathcal{F}))^{\sim} = \kappa_{\mathcal{X}}^*(\mathbf{R}\mathrm{Hom}^{\bullet}_{R}(\mathbf{R}\Gamma_J S, \mathbf{R}\Gamma(\mathcal{Y},\mathcal{F})))^{\sim} \qquad (\mathcal{F} \in \mathbf{D}_{\vec{c}}(\mathcal{Y})).$$

(c) *There are natural isomorphisms*

$$\begin{aligned} \mathbf{R}\Gamma(\mathcal{X}, f_{\mathrm{t}}^{\times}\mathcal{F}) &\xrightarrow{\sim} \varphi_J^{\times}\mathbf{R}\Gamma(\mathcal{Y},\mathcal{F}) && (\mathcal{F} \in \mathbf{D}_{\mathrm{qct}}(\mathcal{Y})),\\ \mathbf{R}\Gamma(\mathcal{X}, f_{\vec{c}}^{\#}\mathcal{F}) &\xrightarrow{\sim} \varphi_J^{\#}\mathbf{R}\Gamma(\mathcal{Y},\mathcal{F}) && (\mathcal{F} \in \mathbf{D}_{\vec{c}}(\mathcal{Y})). \end{aligned}$$

PROOF. The functor $^\sim$ induces an equivalence of categories $\mathbf{D}(S) \to \mathbf{D}_{\mathrm{qc}}(X)$, with quasi-inverse $\mathbf{R}\Gamma_X := \mathbf{R}\Gamma(X, -)$ ([**BN**, p. 225, Thm. 5.1], [**AJL**, p. 12, Proposition (1.3)]); and Proposition 3.3.1 below implies that $\kappa_{\mathcal{X}}^*\colon \mathbf{D}_{\mathrm{qc}}(X) \to \mathbf{D}_{\vec{c}}(\mathcal{X})$ is an equivalence, with quasi-inverse $(\mathbf{R}\Gamma_X\kappa_{\mathcal{X}*}-)^{\sim} = (\mathbf{R}\Gamma_{\mathcal{X}}-)^{\sim}$.[11]

It follows that *the functor taking $G \in \mathbf{D}(S)$ to $\kappa_{\mathcal{X}}^*\widetilde{G}$ is an equivalence, with quasi-inverse* $\mathbf{R}\Gamma_{\mathcal{X}}\colon \mathbf{D}_{\vec{c}}(\mathcal{X}) \to \mathbf{D}(S)$, and similarly for $\mathcal{Y}$ and R. Moreover, there is an induced equivalence between $\mathbf{D}_J(S)$ and $\mathbf{D}_{\mathrm{qct}}(\mathcal{X})$ (see Proposition 5.2.4). In particular, (c) follows from (a) and (b).

Corresponding to (2.1.5) and (2.1.1) there are then functorial isomorphisms

$$\begin{aligned} &\mathrm{Hom}_{\mathbf{D}(\mathcal{X})}(\mathcal{E}, f_{\mathrm{t}}^{\times}\mathcal{F}) \xrightarrow{\sim} \mathrm{Hom}_{\mathbf{D}(\mathcal{Y})}(\kappa_{\mathcal{Y}}^*(\mathbf{R}\Gamma_{\mathcal{X}}\mathcal{E})^{\sim_R}, \mathcal{F}) && (\mathcal{E} \in \mathbf{D}_{\mathrm{qct}}(\mathcal{X}),\ \mathcal{F} \in \mathbf{D}_{\mathrm{qct}}(\mathcal{Y})),\\ &\mathrm{Hom}_{\mathbf{D}(\mathcal{X})}(\mathcal{E}, f_{\vec{c}}^{\#}\mathcal{F}) \xrightarrow{\sim} \mathrm{Hom}_{\mathbf{D}(\mathcal{Y})}(\kappa_{\mathcal{Y}}^*(\mathbf{R}\Gamma_J\mathbf{R}\Gamma_{\mathcal{X}}\mathcal{E})^{\sim_R}, \mathcal{F}) && (\mathcal{E} \in \mathbf{D}_{\vec{c}}(\mathcal{X}),\ \mathcal{F} \in \mathbf{D}_{\vec{c}}(\mathcal{Y})); \end{aligned}$$

and it remains to demonstrate functorial isomorphisms

$$\begin{aligned} \kappa_{\mathcal{Y}}^*(\mathbf{R}\Gamma_{\mathcal{X}}\mathcal{E})^{\sim_R} &\xrightarrow{\sim} \mathbf{R}f_*\mathcal{E} && (\mathcal{E} \in \mathbf{D}_{\mathrm{qct}}(\mathcal{X})),\\ \kappa_{\mathcal{Y}}^*(\mathbf{R}\Gamma_J\mathbf{R}\Gamma_{\mathcal{X}}\mathcal{E})^{\sim_R} &\xrightarrow{\sim} \mathbf{R}f_*\mathbf{R}\mathit{\Gamma}'_{\mathcal{X}}\mathcal{E} && (\mathcal{E} \in \mathbf{D}_{\vec{c}}(\mathcal{X})), \end{aligned}$$

the first a special case of the second.

[11] In checking this note that $\kappa_{\mathcal{X}*}$ has an exact left adjoint, hence preserves K-injectivity.

To prove the second, let $E := \mathbf{R}\Gamma_{\mathfrak{X}}\mathcal{E}$, let $Z := \mathrm{Spec}(S/J) \subset X$, and let $f_0\colon X \to Y$ be the scheme-map corresponding to φ. The desired isomorphism comes from the sequence of natural isomorphisms

$$\begin{aligned}
\mathbf{R}f_*\mathbf{R}\varGamma'_{\mathfrak{X}}\mathcal{E} &\cong \mathbf{R}f_*\mathbf{R}\varGamma'_{\mathfrak{X}}\kappa^*_{\mathfrak{X}}\widetilde{E} \\
&\cong \mathbf{R}f_*\kappa^*_{\mathfrak{X}}\mathbf{R}\varGamma_Z\widetilde{E} && \text{(Proposition 5.2.4(c))} \\
&\cong \kappa^*_{\mathfrak{Y}}\mathbf{R}f_{0*}\mathbf{R}\varGamma_Z\widetilde{E} && \text{(Corollary 5.2.7)} \\
&\cong \kappa^*_{\mathfrak{Y}}\mathbf{R}f_{0*}(\mathbf{R}\Gamma_J E)^{\sim} && \text{([\textbf{AJL}, p. 9, (0.4.5)])} \\
&\cong \kappa^*_{\mathfrak{Y}}(\mathbf{R}\Gamma_J E)^{\sim_R}.
\end{aligned}$$

(The last isomorphism—well-known for bounded-below E—can be checked via the equivalences $\mathbf{R}\Gamma_X$ and $\mathbf{R}\Gamma_Y$, which satisfy $\mathbf{R}\Gamma_Y\mathbf{R}f_{0*} \cong \mathbf{R}\Gamma_X$ (see [**Sp**, pp. 142–143, 5.15(b) and 5.17]). □

Theorem 2.2.3 below globalizes Proposition 2.1.6. But first some preparatory remarks are needed. Recall from 1.2.2 that a map $f\colon \mathfrak{X} \to \mathfrak{Y}$ of noetherian formal schemes is *pseudo-finite* if it is pseudo-proper and has finite fibers, or equivalently, if f is pseudo-proper and *affine.* Such an f corresponds locally to a homomorphism $\varphi\colon (R, I) \to (S, J)$ of noetherian adic rings such that $\varphi(I) \subset J$ and S/J is a finite R-module. This φ can be extended to a homomorphism from a power series ring $R[[\mathbf{t}]] := R[[t_1, t_2, \dots, t_e]]$ such that the images of the variables t_i together with $\varphi(I)$ generate J, and thereby S becomes a finite $R[[\mathbf{t}]]$-module. Pseudo-finiteness is preserved under arbitrary (noetherian) base change.

We say that a pseudo-finite map $f\colon \mathfrak{X} \to \mathfrak{Y}$ of noetherian formal schemes has relative dimension $\le d$ if each $y \in \mathfrak{Y}$ has an affine neighborhood $\mathfrak{U}$ such that the map $\varphi_{\mathfrak{U}}\colon R \to S$ of adic rings corresponding to $f^{-1}\mathfrak{U} \to \mathfrak{U}$ has a continuous extension $R[[t_1, \dots, t_d]] \to S$ making S into a finite $R[[t_1, \dots, t_d]]$-module, or equivalently, there is a topologically nilpotent sequence $\mathbf{u} = (u_1, \dots, u_d)$ in S (i.e., $\lim_{n\to\infty} u_i^n = 0$ $(1 \le i \le d)$) such that $S/\mathbf{u}S$ is finitely generated as an R-module. The *relative dimension* $\dim f$ is defined to be the least among the integers d such that f has relative dimension $\le d$.

For any pseudo-proper map $f\colon \mathfrak{X} \to \mathfrak{Y}$ of noetherian formal schemes, we have the functor $f^{\#}\colon \mathbf{D}(\mathfrak{Y}) \to \mathbf{D}(\mathfrak{X})$ of Corollary 6.1.4, commuting with open base change on $\mathfrak{Y}$ (Theorem 4). The next Lemma is a special case of Proposition 8.3.2.

LEMMA 2.2.2. *For a pseudo-finite map $f\colon \mathfrak{X} \to \mathfrak{Y}$ of noetherian formal schemes and for any $\mathcal{F} \in \mathbf{D}^+_{\mathrm{c}}(\mathfrak{Y})$, it holds that $f^{\#}\mathcal{F} \in \mathbf{D}^+_{\mathrm{c}}(\mathfrak{X})$.*

PROOF. Since $f^{\#}$ commutes with open base change, the question is local, so we may assume that f corresponds to $\varphi\colon (R, I) \to (S, J)$ as above. Moreover, the isomorphism $(gf)^{\#} \cong f^{\#}g^{\#}$ in Corollary 6.1.4 allows us to assume that *either* $S = R[[t_1, \dots, t_d]]$ and φ is the natural map *or* S is a finite R-module and $J = IS$. In either case f is obtained by completing a proper map $f_0\colon X \to \mathrm{Spec}(R)$ along a closed subscheme $Z \subset f_0^{-1}\mathrm{Spec}(R/I)$. (In the first case, take X to be the projective space $\mathbb{P}^d_R \supset \mathrm{Spec}(R[t_1, \dots, t_d])$, and $Z := \mathrm{Spec}(R[t_1, \dots, t_d]/(I, t_1, \dots, t_d))$.) The conclusion is given then by Corollary 6.2.3. □

THEOREM 2.2.3 (Pseudo-finite Duality). *Let $f\colon \mathcal{X} \to \mathcal{Y}$ be a pseudo-finite map of noetherian formal schemes, and let $\mathcal{F}$ be a coherent $\mathcal{O}_{\mathcal{Y}}$-module. Then:*

(a) $H^n f^{\#}\mathcal{F} = 0$ *for all* $n < -\dim f$.

(b) *If* $\dim f \le d$ *and* $\mathcal{X}$ *is covered by affine open subsets with d-generated defining ideals, then with* $f'_{\mathcal{X}*} := f_*\Gamma'_{\mathcal{X}}$ *and, for* $i \in \mathbb{Z}$ *and* $\mathcal{J}$ *a defining ideal of* $\mathcal{X}$,

$$R^i f'_{\mathcal{X}*} := H^i\mathbf{R}f'_{\mathcal{X}*} = H^i\mathbf{R}f_*\mathbf{R}\Gamma'_{\mathcal{X}} = \varinjlim_n H^i\mathbf{R}f_*\mathbf{R}\mathcal{H}om^{\bullet}(\mathcal{O}_{\mathcal{X}}/\mathcal{J}^n, -),^{12}$$

there is, for quasi-coherent $\mathcal{O}_{\mathcal{X}}$*-modules* $\mathcal{E}$, *a functorial isomorphism*

$$f_*\mathcal{H}om_{\mathcal{X}}(\mathcal{E}, H^{-d}f^{\#}\mathcal{F}) \xrightarrow{\sim} \mathcal{H}om_{\mathcal{Y}}(R^d f'_{\mathcal{X}*}\mathcal{E}, \mathcal{F}).$$

(*Here* $H^{-d}f^{\#}\mathcal{F}$ *is coherent* (*Lemma* 2.2.2), *and by* (a), *vanishes unless* $d = \dim f$.)

PROOF. Since $f^{\#}$ commutes with open base change we may assume that $\mathcal{Y}$ is affine and that f corresponds to a map $\varphi\colon (R, I) \to (S, J)$ as in Proposition 2.1.6. Then there is an isomorphism of functors

$$\boldsymbol{j}\mathbf{R}Q_{\mathcal{X}}f^{\#} \cong \kappa^*_{\mathcal{X}}(\varphi^{\#}_J\mathbf{R}\Gamma(\mathcal{Y}, -))^{\sim},$$

both of these functors being right-adjoint to $\mathbf{R}f_*\mathbf{R}\Gamma'_{\mathcal{X}}\colon \mathbf{D}_{\vec{\mathrm{c}}}(\mathcal{X}) \to \mathbf{D}_{\vec{\mathrm{c}}}(\mathcal{Y})$ (Proposition 2.2.1(b) and remarks about right adjoints preceding it). Since $f^{\#}\mathcal{F} \in \mathbf{D}^+_{\mathrm{c}}(\mathcal{X})$ (Lemma 2.2.2), therefore, by Corollary 3.3.4, the natural map is an isomorphism $\boldsymbol{j}\mathbf{R}Q_{\mathcal{X}}f^{\#}\mathcal{F} \xrightarrow{\sim} f^{\#}\mathcal{F}$; and so, since $\kappa^*_{\mathcal{X}}$ is exact, Proposition 2.1.6 gives (a).

Next, consider the presheaf map associating to each open $\mathcal{U} \subset \mathcal{Y}$ the natural composition (with $\mathcal{V} := f^{-1}\mathcal{U}$):

$$\begin{aligned}\mathrm{Hom}_{\mathcal{V}}(\mathcal{E}, H^{-d}f^{\#}\mathcal{F}) \underset{\text{by (a)}}{\xrightarrow{\sim}} \mathrm{Hom}_{\mathbf{D}(\mathcal{V})}(\mathcal{E}[d], f^{\#}\mathcal{F}) &\underset{6.1.4}{\xrightarrow{\sim}} \mathrm{Hom}_{\mathbf{D}(\mathcal{U})}(\mathbf{R}f_*\mathbf{R}\Gamma'_{\mathcal{X}}\mathcal{E}[d], \mathcal{F})\\ &\longrightarrow \mathrm{Hom}_{\mathcal{U}}(R^d f'_{\mathcal{X}*}\mathcal{E}, \mathcal{F}).\end{aligned}$$

To prove (b) by showing that the resulting sheaf map

$$f_*\mathcal{H}om_{\mathcal{X}}(\mathcal{E}, H^{-d}f^{\#}\mathcal{F}) \to \mathcal{H}om_{\mathcal{Y}}(R^d f'_{\mathcal{X}*}\mathcal{E}, \mathcal{F})$$

is an isomorphism, it suffices to show that $R^i f'_{\mathcal{X}*}\mathcal{E} = 0$ for all $i > d$, a local problem for which we can (and do) assume that f corresponds to $\varphi\colon R \to S$ as above.

Now $\mathbf{R}\Gamma'_{\mathcal{X}}\mathcal{E} \in \mathbf{D}_{\mathrm{qct}}(\mathcal{X})$ (Proposition 5.2.1), so Proposition 5.2.4 for $X := \mathrm{Spec}(S)$ and $Z := \mathrm{Spec}(S/J)$ gives $\mathbf{R}\Gamma'_{\mathcal{X}}\mathcal{E} \cong \kappa^*_{\mathcal{X}}\mathcal{E}_0$ with $\mathcal{E}_0 := \kappa_{\mathcal{X}*}\mathbf{R}\Gamma'_{\mathcal{X}}\mathcal{E} \in \mathbf{D}^+_{\mathrm{qc}Z}(X)$. Since $\mathcal{X}$ has, locally, a d-generated defining ideal, we can represent $\mathbf{R}\Gamma'_{\mathcal{X}}\mathcal{E}$ locally by a $\varinjlim$ of Koszul complexes on d elements [**AJL**, p. 18, Lemma 3.1.1], whence $H^i\mathbf{R}\Gamma'_{\mathcal{X}}\mathcal{E} = 0$ for all $i > d$, and so, $\kappa_{\mathcal{X}*}$ being exact, $H^i\mathcal{E}_0 = 0$. Since the map $f_0 := \mathrm{Spec}(\varphi)$ is affine, it follows that $H^i\mathbf{R}f_{0*}\mathcal{E}_0 = 0$, whereupon, $\kappa_{\mathcal{Y}}$ being flat, Corollary 5.2.7 yields

$$R^i f'_{\mathcal{X}*}\mathcal{E} \cong H^i\mathbf{R}f_*\kappa^*_{\mathcal{X}}\mathcal{E}_0 \cong H^i\kappa^*_{\mathcal{Y}}\mathbf{R}f_{0*}\mathcal{E}_0 \cong \kappa^*_{\mathcal{Y}}H^i\mathbf{R}f_{0*}\mathcal{E}_0 = 0 \qquad (i > d),$$

as desired. (Alternatively, use Lemmas 3.4.2 and 5.1.4.) □

---

[12]The equalities hold because $\mathcal{X}$ being noetherian, any $\varinjlim$ of flasque sheaves (for example, $\varinjlim \mathcal{H}om(\mathcal{O}_{\mathcal{X}}/\mathcal{J}^n, \mathcal{E})$ with $\mathcal{E}$ an injective $\mathcal{O}_{\mathcal{X}}$-module) is f_*-acyclic, and $\varinjlim$ commutes with f_*. (For an additive functor $\phi\colon \mathcal{A}(\mathcal{X}) \to \mathcal{A}(\mathcal{Y})$, an $\mathcal{A}(\mathcal{X})$-complex $\mathcal{F}$ is *ϕ-acyclic* if the natural map $\phi\mathcal{F} \to \mathbf{R}\phi\mathcal{F}$ is a $\mathbf{D}(\mathcal{Y})$-isomorphism. Using a standard spectral sequence, or otherwise (cf. [**L4**, (2.7.2)]), one sees that any bounded-below complex of ϕ-acyclic $\mathcal{O}_{\mathcal{X}}$-modules is ϕ-acyclic.

2.3. Our results provide a framework for "Residue Theorems" such as those appearing in [**L1**, pp. 87–88] and [**HüS**, pp. 750-752] (central theorems in those papers): roughly speaking, Theorems 1 and 2 in section 1 include both local and global duality, and Corollary 6.1.6 expresses the compatibility between these dualities. But the dualizing objects we deal with are determined only up to isomorphism. The Residue Theorems run deeper in that they include a *canonical realization* of dualizing data, via differential forms. (See the above remarks on the Hübl-Kunz treatment of local duality.) This extra dimension belongs properly to a theory of the "Fundamental Class" of a morphism, a canonical map from relative differential forms to the relative dualizing complex, which will be pursued in a separate paper.

2.3.1. Let us be more explicit, starting with some remarks about "Grothendieck Duality with supports" for a map $f\colon X \to Y$ of noetherian separated schemes with respective closed subschemes $W \subset Y$ and $Z \subset f^{-1}W$. Via the natural equivalence of categories $\mathbf{D}(\mathcal{A}_{\mathrm{qc}}(X)) \to \mathbf{D}_{\mathrm{qc}}(X)$ (see §3.3), we regard the functor $f^{\times}\colon \mathbf{D}(Y) \to \mathbf{D}(\mathcal{A}_{\vec{c}}(X)) = \mathbf{D}(\mathcal{A}_{\mathrm{qc}}(X))$ of Theorem 1 as being right-adjoint to $\mathbf{R}f_*\colon \mathbf{D}_{\mathrm{qc}}(X) \to \mathbf{D}(Y)$.[13] The functor $\mathbf{R}\varGamma'_Z$ can be regarded as being right-adjoint to the inclusion $\mathbf{D}_Z(X) \hookrightarrow \mathbf{D}(X)$ (cf. Proposition 5.2.1(c)); and its restriction to $\mathbf{D}_{\mathrm{qc}}(X)$ agrees naturally with that of $\mathbf{R}\varGamma_Z$, both restrictions being right-adjoint to the inclusion $\mathbf{D}_{\mathrm{qc}Z}(X) \hookrightarrow \mathbf{D}_{\mathrm{qc}}(X)$. Similar statements hold for $W \subset Y$. Since $\mathbf{R}f_*(\mathbf{D}_{\mathrm{qc}Z}(X)) \subset \mathbf{D}_W(Y)$ (cf. proof of Proposition 5.2.6), we find that the functors $\mathbf{R}\varGamma_Z f^{\times}$ and $\mathbf{R}\varGamma_Z f^{\times}\mathbf{R}\varGamma'_W$ are both right-adjoint to $\mathbf{R}f_*\colon \mathbf{D}_{\mathrm{qc}Z}(X) \to \mathbf{D}(Y)$, so are isomorphic. We define the *local integral* (a generalized residue map, cf. [**HüK**, §4])

$$\rho(\mathcal{G})\colon \mathbf{R}f_*\mathbf{R}\varGamma_Z f^{\times}\mathcal{G} \to \mathbf{R}\varGamma'_W\mathcal{G} \qquad (\mathcal{G} \in \mathbf{D}(Y))$$

to be the natural composition

$$\mathbf{R}f_*\mathbf{R}\varGamma_Z f^{\times}\mathcal{G} \xrightarrow{\ \sim\ } \mathbf{R}f_*\mathbf{R}\varGamma_Z f^{\times}\mathbf{R}\varGamma'_W\mathcal{G} \to \mathbf{R}f_* f^{\times}\mathbf{R}\varGamma'_W\mathcal{G} \to \mathbf{R}\varGamma'_W\mathcal{G}.$$

Noting that for $\mathcal{F} \in \mathbf{D}_W(Y)$ there is a canonical isomorphism $\mathbf{R}\varGamma'_W\mathcal{F} \xrightarrow{\ \sim\ } \mathcal{F}$ (proof similar to that of Proposition 5.2.1(a)), we have then:

PROPOSITION 2.3.2 (Duality with supports). *For $\mathcal{E} \in \mathbf{D}_{\mathrm{qc}Z}(X)$, $\mathcal{F} \in \mathbf{D}_W(Y)$, the natural composition*

$$\begin{aligned}\mathrm{Hom}_{\mathbf{D}_{\mathrm{qc}Z}(X)}(\mathcal{E}, \mathbf{R}\varGamma_Z f^{\times}\mathcal{F}) &\longrightarrow \mathrm{Hom}_{\mathbf{D}_W(Y)}(\mathbf{R}f_*\mathcal{E}, \mathbf{R}f_*\mathbf{R}\varGamma_Z f^{\times}\mathcal{F}) \\ &\xrightarrow[\rho(\mathcal{F})]{} \mathrm{Hom}_{\mathbf{D}_W(Y)}(\mathbf{R}f_*\mathcal{E}, \mathcal{F})\end{aligned}$$

is an isomorphism.

This follows from adjointness of $\mathbf{R}f_*$ and $f^{\times}$, via the natural diagram

$$\begin{array}{ccc} \mathbf{R}f_*\mathbf{R}\varGamma_Z f^{\times}\mathcal{G} & \longrightarrow & \mathbf{R}f_* f^{\times}\mathcal{G} \\ {\scriptstyle\rho(\mathcal{G})}\big\downarrow & & \big\downarrow \\ \mathbf{R}\varGamma'_W\mathcal{G} & \longrightarrow & \mathcal{G} \end{array} \qquad (\mathcal{G} \in \mathbf{D}(Y)),$$

whose commutativity is a cheap version of the Residue Theorem [**HüS**, pp. 750-752].

Again, however, to be worthy of the name a Residue Theorem should involve *canonical realizations* of dualizing objects. For instance, when V is a proper d-dimensional variety over a field k and $v \in V$ is a closed point, taking $X = V$,

[13] For ordinary schemes, this functor $f^{\times}$ is well-known, and usually denoted $f^!$ when f is proper. When f is an open immersion, the functors $f^{\times}$ and $f^!(= f^*)$ need not agree.

$Z = \{v\}$, $W = Y = \mathrm{Spec}(k)$, $\mathcal{G} = k$, and setting $\omega_V := H^{-d}f^{\times}k$, we get an $\mathcal{O}_{V,v}$-module $\omega_{V,v}$ (commonly called "canonical", though defined only up to isomorphism) together with the k-linear map induced by $\rho(k)$:

$$H^d_v(\omega_{V,v}) \to k,$$

a map whose truly-canonical realization via differentials and residues is indicated in [**L1**, p. 86, (9.5)].

2.3.3. With preceding notation, consider the completion diagram

$$\begin{array}{ccc} X_{/Z} =: \mathfrak{X} & \xrightarrow{\kappa_{\mathfrak{X}}} & X \\ {\scriptstyle \hat{f}}\big\downarrow & & \big\downarrow{\scriptstyle f} \\ Y_{/W} =: \mathfrak{Y} & \xrightarrow[\kappa_{\mathfrak{Y}}]{} & Y \end{array}$$

Duality with supports can be regarded more intrinsically—via $\hat{f}$ rather than f—as a special case of the Torsion-Duality Theorem 6.1 ($\cong$ Theorem 2 of §1) for $\hat{f}$:

First of all, the local integral ρ is completely determined by $\kappa_{\mathfrak{Y}}^*(\rho)$: for $\mathcal{G} \in \mathbf{D}(\mathfrak{Y})$, the natural map $\mathbf{R}\varGamma'_W\mathcal{G} \to \kappa_{\mathfrak{Y}*}\kappa_{\mathfrak{Y}}^*\mathbf{R}\varGamma'_W\mathcal{G}$ is an isomorphism (Proposition 5.2.4); and the same holds for $\mathbf{R}f_*\mathbf{R}\varGamma_Z f^{\times}\mathcal{G} \to \kappa_{\mathfrak{Y}*}\kappa_{\mathfrak{Y}}^*\mathbf{R}f_*\mathbf{R}\varGamma_Z f^{\times}\mathcal{G}$ since as above,

$$\mathbf{R}f_*\mathbf{R}\varGamma_Z f^{\times}\mathcal{G} \in \mathbf{R}f_*(\mathbf{D}_{\mathrm{qc}Z}(X)) \subset \mathbf{D}_W(Y)$$

—and so $\rho = \kappa_{\mathfrak{Y}*}\kappa_{\mathfrak{Y}}^*(\rho)$. Furthermore, $\kappa_{\mathfrak{Y}}^*(\rho)$ is determined by the "trace" map $\tau_{\mathrm{t}} \colon \mathbf{R}\hat{f}_*\hat{f}_{\mathrm{t}}^{\times} \to \mathbf{1}$, as per the following natural commutative diagram, whose rows are isomorphisms:

$$\begin{array}{ccccccc} \kappa_{\mathfrak{Y}}^*\mathbf{R}f_*\mathbf{R}\varGamma_Z f^{\times}\mathcal{G} & \underset{5.2.7}{\xrightarrow{\ \sim\ }} & \mathbf{R}\hat{f}_*\kappa_{\mathfrak{X}}^*\mathbf{R}\varGamma_Z f^{\times}\kappa_{\mathfrak{Y}*}\kappa_{\mathfrak{Y}}^*\mathcal{G} & \underset{6.1.6}{\xrightarrow{\ \sim\ }} & \mathbf{R}\hat{f}_*\hat{f}_{\mathrm{t}}^{\times}\kappa_{\mathfrak{Y}}^*\mathcal{G} & \underset{6.1.5(\mathrm{b})}{\xrightarrow{\ \sim\ }} & \mathbf{R}\hat{f}_*\hat{f}_{\mathrm{t}}^{\times}\mathbf{R}\varGamma'_{\mathfrak{Y}}\kappa_{\mathfrak{Y}}^*\mathcal{G} \\ {\scriptstyle \kappa_{\mathfrak{Y}}^*(\rho)}\big\downarrow & & & & & & \big\downarrow{\scriptstyle \tau_{\mathrm{t}}} \\ \kappa_{\mathfrak{Y}}^*\mathbf{R}\varGamma'_W\mathcal{G} & & & \underset{5.2.4}{\xleftarrow{\qquad\qquad \sim \qquad\qquad}} & & & \mathbf{R}\varGamma'_{\mathfrak{Y}}\kappa_{\mathfrak{Y}}^*\mathcal{G} \end{array}$$

(To see that the natural map $\mathbf{R}\varGamma_Z f^{\times}\mathcal{G} \to \mathbf{R}\varGamma_Z f^{\times}\kappa_{\mathfrak{Y}*}\kappa_{\mathfrak{Y}}^*\mathcal{G}$ is an isomorphism, replace $\mathbf{R}\varGamma_Z f^{\times}$ by the isomorphic functor $\mathbf{R}\varGamma_Z f^{\times}\mathbf{R}\varGamma'_W$ and apply Proposition 5.2.4.)

Finally, we have isomorphisms (for $\mathcal{E} \in \mathbf{D}_{\mathrm{qc}Z}(X)$, $\mathcal{F} \in \mathbf{D}_W(Y)$),

$$\begin{aligned} \mathrm{Hom}_{\mathbf{D}(X)}(\mathcal{E}, \mathbf{R}\varGamma_Z f^{\times}\mathcal{F}) &\xrightarrow{\sim} \mathrm{Hom}_{\mathbf{D}(\mathfrak{X})}(\kappa_{\mathfrak{X}}^*\mathcal{E}, \kappa_{\mathfrak{X}}^*\mathbf{R}\varGamma_Z f^{\times}\kappa_{\mathfrak{Y}*}\kappa_{\mathfrak{Y}}^*\mathcal{F}) && (5.2.4)\\ &\xrightarrow{\sim} \mathrm{Hom}_{\mathbf{D}(\mathfrak{X})}(\kappa_{\mathfrak{X}}^*\mathcal{E}, \hat{f}_{\mathrm{t}}^{\times}\kappa_{\mathfrak{Y}}^*\mathcal{F}) && (6.1.6)\\ &\xrightarrow{\sim} \mathrm{Hom}_{\mathbf{D}(\mathfrak{Y})}(\mathbf{R}\hat{f}_*\kappa_{\mathfrak{X}}^*\mathcal{E}, \kappa_{\mathfrak{Y}}^*\mathcal{F}) && (6.1)\\ &\xrightarrow{\sim} \mathrm{Hom}_{\mathbf{D}(\mathfrak{Y})}(\kappa_{\mathfrak{Y}}^*\mathbf{R}f_*\mathcal{E}, \kappa_{\mathfrak{Y}}^*\mathcal{F}) && (5.2.7)\\ &\xrightarrow{\sim} \mathrm{Hom}_{\mathbf{D}(Y)}(\mathbf{R}f_*\mathcal{E}, \mathcal{F}) && (5.2.4), \end{aligned}$$

whose composition can be checked, via the preceding diagram, to be the same as the isomorphism of Proposition 2.3.2.

2.3.4. Proposition 2.3.6 expresses some homological consequences of the foregoing dualities, and furnishes a general context for [**L1**, pp. 87–88, Theorem (10.2)].

For any noetherian formal scheme $\mathfrak{X}$, $\mathcal{E} \in \mathbf{D}(\mathfrak{X})$, and $n \in \mathbb{Z}$, set

$$\mathrm{H}'^{\,n}_{\mathfrak{X}}(\mathcal{E}) := \mathrm{H}^n\mathbf{R}\Gamma(\mathfrak{X}, \mathbf{R}\varGamma'_{\mathfrak{X}}\mathcal{E}).$$

For instance, if $\mathfrak{X} = X_{/Z} \xrightarrow{\kappa} X$ is the completion of a noetherian scheme X along a closed $Z \subset X$, then for $\mathcal{F} \in \mathbf{D}(X)$, Proposition 5.2.4 yields natural isomorphisms

$$\begin{aligned}\mathbf{R}\Gamma(\mathfrak{X}, \mathbf{R}\varGamma'_{\mathfrak{X}}\kappa^*\mathcal{F}) &= \mathbf{R}\Gamma(X, \kappa_*\mathbf{R}\varGamma'_{\mathfrak{X}}\kappa^*\mathcal{F})\\ &\cong \mathbf{R}\Gamma(X, \kappa_*\kappa^*\mathbf{R}\varGamma'_Z\mathcal{F}) \cong \mathbf{R}\Gamma(X, \mathbf{R}\varGamma'_Z\mathcal{F}),\end{aligned}$$

and so if $\mathcal{F} \in \mathbf{D}_{\mathrm{qc}}(X)$, then with $\mathrm{H}^\bullet_Z$ the usual cohomology with supports in Z,

$$\mathrm{H}'^n_{\mathfrak{X}}(\kappa^*\mathcal{F}) \cong \mathrm{H}^n_Z(\mathcal{F}).$$

Let $\mathcal{J} \subset \mathcal{O}_{\mathfrak{X}}$ be an ideal of definition. Writing $\Gamma_{\mathfrak{X}}$ for the functor $\Gamma(\mathfrak{X}, -)$, we have a functorial map

$$\gamma(\mathcal{E})\colon \mathbf{R}(\Gamma_{\mathfrak{X}} \circ \varGamma'_{\mathfrak{X}})\mathcal{E} \to \mathbf{R}\Gamma_{\mathfrak{X}} \circ \mathbf{R}\varGamma'_{\mathfrak{X}}\mathcal{E} \qquad (\mathcal{E} \in \mathbf{D}(\mathfrak{X})),$$

which is an *isomorphism* when $\mathcal{E}$ is bounded-below, since for any injective $\mathcal{O}_{\mathfrak{X}}$-module $\mathcal{I}$, $\varinjlim_i$ of the flasque modules $\mathcal{H}om(\mathcal{O}_{\mathfrak{X}}/\mathcal{J}^i, \mathcal{I})$ is $\Gamma_{\mathfrak{X}}$-acyclic. Whenever $\gamma(\mathcal{E})$ is an isomorphism, the induced homology maps are isomorphisms

$$\varinjlim_i \mathrm{Ext}^n(\mathcal{O}_{\mathfrak{X}}/\mathcal{J}^i, \mathcal{E}) \xrightarrow{\ \sim\ } \mathrm{H}'^n_{\mathfrak{X}}(\mathcal{E}).$$

If $\mathcal{E} \in \mathbf{D}_{\mathrm{qc}}(\mathfrak{X})$, then $\mathbf{R}\varGamma'_{\mathfrak{X}}\mathcal{E} \in \mathbf{D}_{\mathrm{qct}}(\mathfrak{X})$ (Proposition 5.2.1). For any map $g\colon \mathfrak{X} \to \mathfrak{Y}$ satisfying the hypotheses of Theorem 6.1, for $\mathcal{G} \in \mathbf{D}(\mathfrak{Y})$, and with $R := \mathrm{H}^0(\mathfrak{Y}, \mathcal{O}_{\mathfrak{Y}})$, there are natural maps

$$\begin{aligned}\mathrm{Hom}_{\mathbf{D}(\mathfrak{X})}(\mathbf{R}\varGamma'_{\mathfrak{X}}\mathcal{E}, g_{\mathrm{t}}^{\times}\mathcal{G}) &\xrightarrow{\ \sim\ } \mathrm{Hom}_{\mathbf{D}(\mathfrak{X})}(\mathbf{R}\varGamma'_{\mathfrak{X}}\mathcal{E}, g_{\mathrm{t}}^{\times}\mathbf{R}\varGamma'_{\mathfrak{Y}}\mathcal{G}) && (6.1.5(\mathrm{b}))\\ &\xrightarrow{\ \sim\ } \mathrm{Hom}_{\mathbf{D}(\mathfrak{Y})}(\mathbf{R}g_*\mathbf{R}\varGamma'_{\mathfrak{X}}\mathcal{E}, \mathbf{R}\varGamma'_{\mathfrak{Y}}\mathcal{G}) && (2.3.4.1)\\ &\longrightarrow \mathrm{Hom}_R(\mathrm{H}'^n_{\mathfrak{X}}\mathcal{E}, \mathrm{H}'^n_{\mathfrak{Y}}\mathcal{G})\end{aligned}$$

where the last map arises via the functor $\mathrm{H}^n\mathbf{R}\Gamma(\mathfrak{Y}, -)$ $(n \in \mathbb{Z})$.

In particular, if $g = \hat{f}$ in the completion situation of §2.3.3, and if $\mathcal{E} := \kappa_{\mathfrak{X}}^*\mathcal{E}_0$, $\mathcal{G} = \kappa_{\mathfrak{Y}}^*\mathcal{G}_0$ ($\mathcal{E}_0 \in \mathbf{D}_{\mathrm{qc}}(X)$, $\mathcal{G}_0 \in \mathbf{D}_{\mathrm{qc}}(Y)$), then preceding considerations show that this composed map operates via Duality with Supports for f (Proposition 2.3.2), i.e., it can be identified with the natural composition

$$\begin{aligned}\mathrm{Hom}_{\mathbf{D}(X)}(\mathbf{R}\varGamma'_Z\mathcal{E}_0, \mathbf{R}\varGamma'_Z f^{\times}\mathcal{G}_0) &\underset{2.3.2}{\xrightarrow{\ \sim\ }} \mathrm{Hom}_{\mathbf{D}(Y)}(\mathbf{R}f_*\mathbf{R}\varGamma_Z\mathcal{E}_0, \mathbf{R}\varGamma_W\mathcal{G}_0)\\ &\longrightarrow \mathrm{Hom}_{\mathrm{H}^0(Y,\mathcal{O}_Y)}(\mathrm{H}^n_Z\mathcal{E}_0, \mathrm{H}^n_W\mathcal{G}_0).\end{aligned}$$

2.3.5. Next, let R be a complete noetherian local ring topologized as usual by its maximal ideal I, let (S, J) be a noetherian adic ring, let $\varphi\colon (R, I) \to (S, J)$ be a continuous homomorphism, and let

$$\mathfrak{Y} := \mathrm{Spf}(S) \xrightarrow{\ f\ } \mathrm{Spf}(R) =: \mathcal{V}$$

be the corresponding formal-scheme map. As before, $g\colon \mathfrak{X} \to \mathfrak{Y}$ is a map as in Theorem 6.1, and we set $h := fg$. Since the underlying space of $\mathcal{V}$ is a single point, at which the stalk of $\mathcal{O}_{\mathcal{V}}$ is just R, therefore the categories of $\mathcal{O}_{\mathcal{V}}$-modules and of R-modules are identical, and accordingly, for any $\mathcal{E} \in \mathbf{D}(\mathfrak{X})$ we can identify $\mathbf{R}h_*\mathcal{E}$ with $\mathbf{R}\Gamma(\mathfrak{X}, \mathcal{E}) \in \mathbf{D}(R)$.

Let K be an injective R-module, and $\mathcal{K}$ the corresponding injective $\mathcal{O}_{\mathcal{V}}$-module. There exist integers r, s such that $H^i(f^{\#}\mathcal{K}) = 0$ for all $i < -r$ (resp. $H^i(h^{\#}\mathcal{K}) = 0$ for all $i < -s$) (Corollary 6.1.4). Set $\omega_{\mathfrak{Y}} := H^{-r}(f^{\#}\mathcal{K})$ (resp. $\omega_{\mathfrak{X}} := H^{-s}(h^{\#}\mathcal{K})$).

PROPOSITION 2.3.6. *In the preceding situation $\omega_{\mathcal X}$ represents—via* (2.3.4.1)*—the functor* $\mathrm{Hom}_S(\mathrm{H}'^{s}_{\mathcal X}\mathcal E, \mathrm{H}'^{0}_{\mathcal Y}(f^{\#}\mathcal K))$ *of quasi-coherent $\mathcal O_{\mathcal X}$-modules $\mathcal E$. If $\omega_{\mathcal Y}$ is the only non-zero homology of $f^{\#}\mathcal K$, this functor is isomorphic to* $\mathrm{Hom}_S(\mathrm{H}'^{s}_{\mathcal X}\mathcal E, \mathrm{H}'^{r}_{\mathcal Y}\omega_{\mathcal Y})$.

Proof. There are natural maps

$$\mathrm{H}'^{r}_{\mathcal Y}(\omega_{\mathcal Y}) = \mathrm{H}'^{0}_{\mathcal Y}(\omega_{\mathcal Y}[r]) \xrightarrow{\ h\ } \mathrm{H}'^{0}_{\mathcal Y}(f^{\#}\mathcal K) \xrightarrow{\ \sim\ } \mathrm{Hom}_{R,J}(S,K)$$

where the last isomorphism results from Proposition 2.2.1(a), in view of the identity $\mathbf R\varGamma'_{\mathcal Y}f^{\#} = f_{\mathrm t}^{\times}$ (Corollary 6.1.5(a)) and the natural isomorphisms

$$\mathbf R\Gamma(\mathcal Y, \kappa_{\mathcal Y}^{*}\widetilde G) \xrightarrow{\ \sim\ } \mathbf R\Gamma(Y, \kappa_{\mathcal Y*}\kappa_{\mathcal Y}^{*}\widetilde G) \underset{5.2.4}{\xrightarrow{\ \sim\ }} \mathbf R\Gamma(Y, \widetilde G) \xrightarrow{\ \sim\ } G \qquad (G \in \mathbf D^{+}_{J}(S)),$$

for $G := \mathbf R\mathrm{Hom}^{\bullet}_{R,J}(S, \mathbf R\Gamma(\mathcal V, \mathcal K))$. (In fact $\mathbf R\Gamma(\mathcal Y, \kappa_{\mathcal Y}^{*}\widetilde G) \cong G$ for any $G \in \mathbf D(S)$, see Corollary 3.3.2 and the beginning of §3.3.) In case $\omega_{\mathcal Y}$ is the only non-vanishing homology of $f^{\#}\mathcal K$, then h is an isomorphism too.

The assertions follow from the (easily checked) commutativity, for any quasi-coherent $\mathcal O_{\mathcal X}$-module $\mathcal E$, of the diagram

$$\begin{array}{ccc}
\mathrm{Hom}_{\mathcal X}(\mathcal E, \omega_{\mathcal X}) = \mathrm{Hom}_{\mathbf D(\mathcal X)}(\mathcal E[s], g^{\#}f^{\#}\mathcal K) & \xrightarrow{\ 6.1.5(\mathrm a)\ } & \mathrm{Hom}_{\mathbf D(\mathcal X)}(\mathbf R\varGamma'_{\mathcal X}\mathcal E[s], g_{\mathrm t}^{\times}f^{\#}\mathcal K) \\
\simeq\big\downarrow & & \big\downarrow{\scriptstyle (2.3.4.1)} \\
\mathrm{Hom}_{\mathbf D(\mathcal V)}(\mathbf R h_{*}\mathbf R\varGamma'_{\mathcal X}\mathcal E[s], \mathcal K) & & \mathrm{Hom}_S(\mathrm{H}'^{0}_{\mathcal X}(\mathcal E[s]), \mathrm{H}'^{0}_{\mathcal Y}(f^{\#}\mathcal K)) \\
\Big\| & & \big\downarrow\simeq \\
\mathrm{Hom}_R(\mathrm{H}'^{s}_{\mathcal X}\mathcal E, K) & \xrightarrow{\ \ \sim\ \ } & \mathrm{Hom}_S(\mathrm{H}'^{s}_{\mathcal X}\mathcal E, \mathrm{Hom}_{R,J}(S,K))
\end{array}$$

2.3.7. Now let us fit [**L1**, pp. 87–88, Theorem (10.2)] into the preceding setup. The cited Theorem has both local and global components. The first deals with maps $\varphi\colon R \to S$ of local domains essentially of finite type over a perfect field k, with residue fields finite over k. To each such ring T one associates the canonical module ω_T of "regular" k-differentials of degree $\dim T$. Under mild restrictions on φ, the assertion is that the functor

$$\mathrm{Hom}_R(\mathrm{H}^{\dim S}_{m_{\hat S}}G,\ \mathrm{H}^{\dim R}_{m_R}\omega_R) \qquad (m := \text{maximal ideal})$$

of $\hat S$-modules G is represented by the completion $\widehat{\omega_S}$ together with a canonical map, the *relative residue*

$$\rho_\varphi\colon \mathrm{H}^{\dim S}_{m_{\hat S}}\widehat{\omega_S} = \mathrm{H}^{\dim S}_{m_S}\omega_S \to \mathrm{H}^{\dim R}_{m_R}\omega_R.$$

This may be viewed as a consequence of *concrete* local duality over k (§2.1.7).

The global aspect concerns a proper map of irreducible k-varieties $g\colon V \to W$ of respective dimensions s and r with all fibers over codimension 1 points of W having dimension $s - r$, a closed point $w \in W$, the fiber $E := g^{-1}(w)$, and the completion $\widehat V := V_{/E}$. The assertion is that the functor

$$\mathrm{Hom}_R(\mathrm{H}'^{s}_{\widehat V}\mathcal G,\ \mathrm{H}^{r}_{m_R}\omega_R) \qquad (R := \mathcal O_{W,w})$$

of coherent $\mathcal O_{\widehat V}$-modules $\mathcal G$ is represented by the completion $\widehat{\omega_V}$ along E of the canonical sheaf ω_V of regular differentials, together with a canonical map

$$\theta\colon \mathrm{H}'^{s}_{\widehat V}\widehat{\omega_V} = \mathrm{H}^{s}_{E}\omega_V \to \mathrm{H}^{r}_{m_R}\omega_R.$$

Moreover, the local and global representations are *compatible* in the sense that if $v \in E$ is any closed point and $\varphi_v\colon R \to S := \mathcal O_{V,v}$ is the canonical map, then the residue $\rho_v := \rho_{\varphi_v}$ factors as the natural map $\mathrm{H}^s_{m_S}\omega_S \to \mathrm{H}^s_E\omega_V$ followed by θ. This compatibility determines θ uniquely if the ρ_v $(v \in E)$ are given [**L1**, p. 95, (10.6)]; and of course conversely.

Basically, all this—*without the explicit description of the ω's and the maps ρ_v via differentials and residues*—is contained in Proposition 2.3.6, as follows.

In the completion situation of §2.3.3, take X and Y to be finite-type separated schemes over an artinian local ring R, of respective pure dimensions s and r, let $W = \{w\}$ with w a closed point of Y, write g in place of f, and assume that $Z \subset g^{-1}W$ is proper over R (which is so, e.g., if g is proper and Z is closed). Let K be an injective hull of the residue field of R, and let $\mathcal{K}$ be the corresponding injective sheaf on $\mathrm{Spec}(R) = \mathrm{Spf}(R)$. With $f\colon Y \to \mathrm{Spec}(R)$ the canonical map, and $h = fg$, define the *dualizing sheaves*

$$\omega_X := H^{-s}h^!\mathcal{K}, \qquad \omega_Y := H^{-r}f^!\mathcal{K},$$

where $h^!$ is the Grothendieck duality functor (compatible with open immersions, and equal to $h^\times$ when h is proper), and similarly for $f^!$. It is well-known (for example via a local factorization of h as smooth$\circ$finite) that $h^!\mathcal{K}$ has coherent homology, vanishing in all degrees $< -s$; and similarly $f^!\mathcal{K}$ has coherent homology, vanishing in all degrees $< -r$.

Let

$$\hat{f}\colon \mathcal{Y} := \mathrm{Spf}(\widehat{\mathcal{O}_{W,w}}) \to \mathrm{Spf}(R) =: \mathcal{V}$$

be the completion of f. We may assume, after compactifying f and g—which does not affect $\hat{f}$ or $\hat{g}$ (see [**Lü**]), that f and g are proper maps. Then Corollary 6.2.3 shows that $\hat{h}^{\#}\mathcal{K} = \kappa_{\mathcal{X}}^{*}h^!\mathcal{K}$, and so κ_Y being flat, we see that

$$\kappa_{\mathcal{X}}^{*}\omega_X = \omega_{\mathcal{X}} \tag{2.3.7.1}$$

where $\omega_{\mathcal{X}}$ is as in Proposition 2.3.6; and similarly $\kappa_{\mathcal{Y}}^{*}\omega_Y = \omega_{\mathcal{Y}}$.

Once again, some form of the theory of the Fundamental Class will enable us to represent ω_X by means of regular differential forms; and then both the local and global components of the cited Theorem (10.2) become special cases of Proposition 2.3.6 (modulo some technicalities [**L1**, p. 89, Lemma (10.3)] which allow a weakening of the condition that $\omega_{\mathcal{Y}}$ be the only non-vanishing homology of $\hat{f}^{\#}\mathcal{K}$).

As for the local-global compatibility, consider quite generally a pair of maps

$$\mathcal{X}_1 \xrightarrow{q} \mathcal{X} \xrightarrow{p} \mathcal{Y}$$

of noetherian formal schemes. In the above situation, for instance, we could take p to be $\hat{g}$, $\mathcal{X}_1$ to be the completion of X at a closed point $v \in Z$, and q to be the natural map. Theorem 2 gives us the adjunction

$$\mathbf{D}_{\mathrm{qct}}(\mathcal{X}) \underset{p_{\mathrm{t}}^{\times}}{\overset{\mathbf{R}p_*}{\rightleftarrows}} \mathbf{D}_{\mathrm{qct}}(\mathcal{Y}).$$

The natural isomorphism $\mathbf{R}(pq)_* \xrightarrow{\sim} \mathbf{R}p_*\mathbf{R}q_*$ gives rise then to an adjoint isomorphism $q_{\mathrm{t}}^{\times}p_{\mathrm{t}}^{\times} \xrightarrow{\sim} (pq)_{\mathrm{t}}^{\times}$; and for $\mathcal{E} \in \mathbf{D}_{\mathrm{qct}}(\mathcal{Y})$ the natural map $\mathbf{R}(pq)_*(pq)_{\mathrm{t}}^{\times}\mathcal{E} \to \mathcal{E}$ factors as

$$\mathbf{R}(pq)_*(pq)_{\mathrm{t}}^{\times}\mathcal{E} \xrightarrow{\sim} \mathbf{R}p_*\mathbf{R}q_*q_{\mathrm{t}}^{\times}p_{\mathrm{t}}^{\times}\mathcal{E} \to \mathbf{R}p_*p_{\mathrm{t}}^{\times}\mathcal{E} \to \mathcal{E}.$$

This factorization contains the compatibility between the above maps θ and ρ_v, as one sees by interpreting them as homological derivatives of maps of the type $\mathbf{R}p_*p_{\mathrm{t}}^{\times}\mathcal{E} \to \mathcal{E}$ (with $\mathcal{E} := \mathbf{R}\varGamma'_{\mathcal{Y}}\hat{f}^{\#}\mathcal{K}$). Details are left to the reader.

REMARK 2.3.8. In the preceding situation, suppose further that $Y = \operatorname{Spec}(R)$ (with R artinian) and $f =$ identity, so that $h = g\colon X \to Y$ is a finite-type separated map, X being of pure dimension s, and $\kappa_{\mathcal{X}}\colon \mathcal{X} \to X$ is the completion of X along a closed subset Z proper over Y. Again, K is an injective R-module, $\mathcal{K}$ is the corresponding $\mathcal{O}_Y$-module, and $\omega_X := H^{-s}g^!\mathcal{K}$ is a "dualizing sheaf" on X. Now Proposition 2.3.6 is just the instance $i = s$ of the canonical isomorphisms, for $\mathcal{E} \in \mathbf{D}_{\mathrm{qc}}(\mathcal{X})$, $i \in \mathbb{Z}$ (and with $\mathrm{H}'^{\bullet}_{\mathcal{X}} := \mathrm{H}^{\bullet}\mathbf{R}\Gamma(\mathcal{X}, \mathbf{R}\varGamma'_{\mathcal{X}})$, see §2.3.4, and $\hat{g} := g \circ \kappa_{\mathcal{X}}$):

$$\operatorname{Hom}_{\mathbf{D}(\mathcal{X})}(\mathcal{E}[i], \hat{g}^{\#}\mathcal{K}) \underset{\text{Thm. 2}}{\xrightarrow{\;\sim\;}} \operatorname{Hom}_{\mathbf{D}(\mathcal{Y})}(\mathbf{R}\hat{g}_*\mathbf{R}\varGamma'_{\mathcal{X}}\mathcal{E}[i], \mathcal{K}) \xrightarrow{\;\sim\;} \operatorname{Hom}_R(\mathrm{H}'^{i}_{\mathcal{X}}\mathcal{E}, K) =: (\mathrm{H}'^{i}_{\mathcal{X}}\mathcal{E})\check{}.$$

If X is Cohen-Macaulay then all the homology of $g^!\mathcal{K}$ other than ω_X vanishes, so all the homology of $\hat{g}^{\#}\mathcal{K} \cong \kappa^*_{\mathcal{X}}g^!\mathcal{K}$ other than $\omega_{\mathcal{X}} = \kappa^*_{\mathcal{X}}\omega_X$ vanishes (see (2.3.7.1)), and the preceding composed isomorphism becomes

$$\operatorname{Ext}^{s-i}_{\mathcal{X}}(\mathcal{E}, \omega_{\mathcal{X}}) \xrightarrow{\;\sim\;} (\mathrm{H}'^{i}_{\mathcal{X}}\mathcal{E})\check{}.$$

In particular, when $Z = X$ (so that $\mathcal{X} = X$) this is the usual duality isomorphism

$$\operatorname{Ext}^{s-i}_{X}(\mathcal{E}, \omega_{X}) \xrightarrow{\;\sim\;} \mathrm{H}^{i}(X, \mathcal{E})\check{}.$$

If X is Gorenstein and $\mathcal{F}$ is a locally free $\mathcal{O}_{\mathcal{X}}$-module of finite rank, then ω_X is invertible; and taking $\mathcal{E} := \mathcal{H}om_{\mathcal{X}}(\mathcal{F}, \omega_{\mathcal{X}}) = \check{\mathcal{F}} \otimes \omega_{\mathcal{X}}$ we get the isomorphism

$$\mathrm{H}^{s-i}(\mathcal{X}, \mathcal{F}) \xrightarrow{\;\sim\;} (\mathrm{H}'^{i}_{\mathcal{X}}(\check{\mathcal{F}} \otimes \omega_{\mathcal{X}}))\check{},$$

which generalizes the Formal Duality theorem [**H2**, p. 48, Proposition (5.2)].

2.4. Both [**H2**, p. 48; Proposition (5.2)] (Formal Duality) and the Theorem in [**L3**, p. 188] (Local-Global Duality) are contained in Proposition 2.4.1, see [**AJL**, §5.3].

Let R be a noetherian ring, discretely topologized, and set

$$Y := \operatorname{Spec}(R) = \operatorname{Spf}(R) =: \mathcal{Y}.$$

Let $g\colon X \to Y$ be a finite-type separated map, let $Z \subset X$ be *proper* over Y, let $\kappa\colon \mathcal{X} = X_{/Z} \to X$ be the completion of X along Z, and set $\hat{g} := g \circ \kappa\colon \mathcal{X} \to \mathcal{Y}$.

Assume that R has a *residual complex* $\mathcal{R}$ [**H1**, p. 304]. Then the corresponding quasi-coherent $\mathcal{O}_Y$-complex $\mathcal{R}_Y := \widetilde{\mathcal{R}}$ is a *dualizing complex,* and $\mathcal{R}_X := g^!\mathcal{R}_Y$ is a dualizing complex on X [**V**, p. 396, Corollary 3]. For any $\mathcal{F} \in \mathbf{D}_{\mathrm{c}}(X)$ set

$$\mathcal{F}' := \mathbf{R}\mathcal{H}om^{\bullet}_X(\mathcal{F}, \mathcal{R}_X) \in \mathbf{D}_{\mathrm{c}}(X),$$

so that $\mathcal{F} \cong \mathcal{F}'' = \mathbf{R}\mathcal{H}om^{\bullet}_X(\mathcal{F}', \mathcal{R}_X)$.

PROPOSITION 2.4.1. *In the preceding situation, with* $\Gamma_Z(-) := \Gamma(X, \varGamma_Z(-))$ *there is a functorial isomorphism*

$$\mathbf{R}\Gamma(\mathcal{X}, \kappa^*\mathcal{F}) \cong \mathbf{R}\mathrm{Hom}^{\bullet}_R(\mathbf{R}\Gamma_Z\mathcal{F}', \mathcal{R}) \qquad (\mathcal{F} \in \mathbf{D}_{\mathrm{c}}(X)).$$

PROOF. Replacing g by a compactification ([**Lü**]) doesn't affect $\mathcal{X}$ or $\mathbf{R}\Gamma_Z$, so assume that g is proper. Then Corollary 6.2.3 gives an isomorphism $\kappa^*\mathcal{R}_X \cong \hat{g}^{\#}\mathcal{R}_Y$.

Now just compose the chain of functorial isomorphisms

$$\begin{aligned}
\mathbf{R}\Gamma(\mathfrak{X},\kappa^*\mathcal{F}) &\cong \mathbf{R}\Gamma(\mathfrak{X},\kappa^*\mathbf{R}\mathcal{H}om^\bullet_X(\mathcal{F}',\mathcal{R}_X)) && \text{(see above)}\\
&\cong \mathbf{R}\Gamma(\mathfrak{X},\mathbf{R}\mathcal{H}om^\bullet_{\mathfrak{X}}(\kappa^*\mathcal{F}',\kappa^*\mathcal{R}_X)) && \text{(Lemma 2.4.2)}\\
&\cong \mathbf{R}\mathrm{Hom}^\bullet_{\mathfrak{X}}(\kappa^*\mathcal{F}',\hat{g}^\#\mathcal{R}_Y)) && \text{(see above)}\\
&\cong \mathbf{R}\mathrm{Hom}^\bullet_{\mathfrak{Y}}(\mathbf{R}\hat{g}_*\mathbf{R}\varGamma'_{\mathfrak{X}}\kappa^*\mathcal{F}',\mathcal{R}_Y) && \text{(Theorem 2)}\\
&\cong \mathbf{R}\mathrm{Hom}^\bullet_{Y}(\mathbf{R}g_*\mathbf{R}\varGamma_Z\mathcal{F}',\mathcal{R}_Y) && \text{(Proposition 5.2.4)}\\
&\cong \mathbf{R}\mathrm{Hom}^\bullet_{Y}(\widetilde{\mathbf{R}\Gamma_Z\mathcal{F}'},\mathcal{R}_Y) && \text{[**AJL**, footnote, §5.3]}\\
&\cong \mathbf{R}\mathrm{Hom}^\bullet_{R}(\mathbf{R}\Gamma_Z\mathcal{F}',\mathcal{R}) && \text{[**AJL**, p. 9, (0.4.4)].}
\end{aligned}$$
□

LEMMA 2.4.2. *Let X be a locally noetherian scheme, and let $\kappa\colon \mathfrak{X}\to X$ be its completion along some closed subset Z. Then for $\mathcal{G}\in\mathbf{D}_{\mathrm{qc}}(X)$ of finite injective dimension and for $\mathcal{F}\in\mathbf{D}_{\mathrm{c}}(X)$, the natural map is an isomorphism*

$$\kappa^*\mathbf{R}\mathcal{H}om^\bullet_X(\mathcal{F},\mathcal{G}) \xrightarrow{\sim} \mathbf{R}\mathcal{H}om^\bullet_{\mathfrak{X}}(\kappa^*\mathcal{F},\kappa^*\mathcal{G}).$$

PROOF. By [**H1**, p. 134, Proposition 7.20] we may assume that $\mathcal{G}$ is a bounded complex of quasi-coherent injective $\mathcal{O}_X$-modules, vanishing, say, in all degrees $> n$.

When $\mathcal{F}$ is bounded-above the (well-known) assertion is proved by localizing to the affine case and applying [**H1**, p. 68, Proposition 7.1] to reduce to the trivial case $\mathcal{F}=\mathcal{O}_X^m$ $(0<m\in\mathbb{Z})$. To do the same for unbounded $\mathcal{F}$ we must first show, for fixed $\mathcal{G}$, that the contravariant functor $\mathbf{R}\mathcal{H}om^\bullet_{\mathfrak{X}}(\kappa^*\mathcal{F},\kappa^*\mathcal{G})$ is bounded-above.

In fact we will show that if $H^i\mathcal{F}=0$ for all $i<i_0$ then for all $j>n-i_0$,

$$H^j\mathbf{R}\mathcal{H}om^\bullet_{\mathfrak{X}}(\kappa^*\mathcal{F},\kappa^*\mathcal{G}) = H^j\kappa_*\mathbf{R}\mathcal{H}om^\bullet_{\mathfrak{X}}(\kappa^*\mathcal{F},\kappa^*\mathcal{G}) = H^j\mathbf{R}\mathcal{H}om^\bullet_X(\mathcal{F},\kappa_*\kappa^*\mathcal{G}) = 0.$$

The homology in question is the sheaf associated to the presheaf which assigns

$$\mathrm{Hom}_{\mathbf{D}(U)}(\mathcal{F}|_U[-j],(\kappa_*\kappa^*\mathcal{G})|_U) = \mathrm{Hom}_{\mathbf{D}(U)}(\mathcal{F}|_U[-j],\mathbf{R}Q_U(\kappa_*\kappa^*\mathcal{G})|_U)$$

to each affine open subset $U=\mathrm{Spec}(A)$ in X. (Here we abuse notation by omitting $\boldsymbol{j}_U$ in front of $\mathbf{R}Q_U$, see beginning of §3.3).

Let $\mathfrak{U}:=\kappa^{-1}U$, and $\hat{A}:=\Gamma(\mathfrak{U},\mathcal{O}_{\mathfrak{X}})$, so that $\kappa|_{\mathfrak{U}}$ factors naturally as

$$\mathfrak{U}=\mathrm{Spf}(\hat{A}) \xrightarrow{\kappa_1} U_1 := \mathrm{Spec}(\hat{A}) \xrightarrow{k} \mathrm{Spec}(A)=U.$$

The functors $\mathbf{R}Q_U k_*$ and $k_*\mathbf{R}Q_{U_1}$, both right-adjoint to the natural composition $\mathbf{D}_{\mathrm{qc}}(U)\xrightarrow{k^*}\mathbf{D}_{\mathrm{qc}}(U_1)\hookrightarrow\mathbf{D}(U_1)$, are isomorphic; so there are natural isomorphisms

$$\mathbf{R}Q_U(\kappa_*\kappa^*\mathcal{G})|_U = \mathbf{R}Q_U k_*\kappa_{1*}\kappa_1^*k^*(\mathcal{G}|_U) \xrightarrow{\sim} k_*\mathbf{R}Q_{U_1}\kappa_{1*}\kappa_1^*k^*(\mathcal{G}|_U) \underset{3.3.1}{\xrightarrow{\sim}} k_*k^*(\mathcal{G}|_U)$$

and the presheaf becomes $U\mapsto \mathrm{Hom}_{\mathbf{D}(U)}(\mathcal{F}|_U[-j],k_*k^*(\mathcal{G}|_U))$.

The equivalence of categories $\mathbf{D}_{\mathrm{qc}}(U)\cong\mathbf{D}(\mathcal{A}_{\mathrm{qc}}(U))=\mathbf{D}(A)$ indicated at the beginning of §3.3 yields an isomorphism

$$\mathrm{Hom}_{\mathbf{D}(U)}(\mathcal{F}|_U[-j],k_*k^*(\mathcal{G}|_U)) \xrightarrow{\sim} \mathrm{Hom}_{\mathbf{D}(A)}(F[-j],G\otimes_A\hat{A})$$

where F is a complex of A-modules with $\mathrm{H}^iF=0$ for $i<i_0$, and both G and $G\otimes_A\hat{A}$ are complexes of injective A-modules vanishing in all degrees $>n$ (the latter since $\hat{A}$ is A-flat). Hence the presheaf vanishes, and the conclusion follows. □

2.5. (Dualizing complexes.) Let $\mathcal{X}$ be a noetherian formal scheme, and write $\mathbf{D}$ for $\mathbf{D}(\mathcal{X})$, etc. The derived functor $\boldsymbol{\varGamma} := \mathbf{R}\varGamma'_{\mathcal{X}}\colon \mathbf{D} \to \mathbf{D}$ (see Section 1.2.1) has a right adjoint $\boldsymbol{\Lambda} = \boldsymbol{\Lambda}_{\mathcal{X}} := \mathbf{R}\mathcal{H}om^{\bullet}(\mathbf{R}\varGamma'_{\mathcal{X}}\mathcal{O}_{\mathcal{X}}, -)$. This adjunction is given by (5.2.10.1), a natural isomorphism of which we'll need the sheafified form, proved similarly:

$$\mathbf{R}\mathcal{H}om^{\bullet}(\mathcal{M}, \boldsymbol{\Lambda}\mathcal{R}) \cong \mathbf{R}\mathcal{H}om^{\bullet}(\boldsymbol{\varGamma}\mathcal{M}, \mathcal{R}). \tag{2.5.0.1}$$

There are natural maps $\boldsymbol{\varGamma} \to \mathbf{1} \to \boldsymbol{\Lambda}$ inducing isomorphisms $\boldsymbol{\Lambda}\boldsymbol{\varGamma} \xrightarrow{\sim} \boldsymbol{\Lambda} \xrightarrow{\sim} \boldsymbol{\Lambda}\boldsymbol{\Lambda}$, $\boldsymbol{\varGamma}\boldsymbol{\varGamma} \xrightarrow{\sim} \boldsymbol{\varGamma} \xrightarrow{\sim} \boldsymbol{\varGamma}\boldsymbol{\Lambda}$ (Remark 6.3.1 (1)). Proposition 6.2.1, a form of Greenlees-May duality, shows that $\boldsymbol{\Lambda}(\mathbf{D}_{\mathrm{c}}) \subset \mathbf{D}_{\mathrm{c}}$. (Recall that the objects of the Δ-subcategory $\mathbf{D}_{\mathrm{c}} \subset \mathbf{D}$ are the complexes whose homology sheaves are all coherent.)

Let $\mathbf{D}_{\mathrm{c}}^{*}$ be the essential image of $\boldsymbol{\varGamma}|_{\mathbf{D}_{\mathrm{c}}}$, i.e., the full subcategory of $\mathbf{D}$ such that $\mathcal{E} \in \mathbf{D}_{\mathrm{c}}^{*} \Leftrightarrow \mathcal{E} \cong \boldsymbol{\varGamma}\mathcal{F}$ with $\mathcal{F} \in \mathbf{D}_{\mathrm{c}}$. Proposition 5.2.1 shows that $\mathbf{D}_{\mathrm{c}}^{*} \subset \mathbf{D}_{\mathrm{qct}}$. It follows from the preceding paragraph that

$$\mathcal{E} \in \mathbf{D}_{\mathrm{c}}^{*} \iff \boldsymbol{\varGamma}\mathcal{E} \xrightarrow{\sim} \mathcal{E} \quad \text{and } \boldsymbol{\Lambda}\mathcal{E} \in \mathbf{D}_{\mathrm{c}},$$
$$\mathcal{F} \in \mathbf{D}_{\mathrm{c}} \iff \mathcal{F} \xrightarrow{\sim} \boldsymbol{\Lambda}\mathcal{F} \quad \text{and } \boldsymbol{\varGamma}\mathcal{F} \in \mathbf{D}_{\mathrm{c}}^{*}.$$

(In particular, $\mathbf{D}_{\mathrm{c}}^{*}$ is a Δ-subcategory of $\mathbf{D}$.) Moreover $\boldsymbol{\varGamma}$ and $\boldsymbol{\Lambda}$ are quasi-inverse equivalences between the categories $\mathbf{D}_{\mathrm{c}}$ and $\mathbf{D}_{\mathrm{c}}^{*}$.

Definition 2.5.1. A complex $\mathcal{R}$ is a *c-dualizing complex on* $\mathcal{X}$ if

(i) $\mathcal{R} \in \mathbf{D}_{\mathrm{c}}^{+}(\mathcal{X})$.
(ii) The natural map is an isomorphism $\mathcal{O}_{\mathcal{X}} \xrightarrow{\sim} \mathbf{R}\mathcal{H}om^{\bullet}(\mathcal{R}, \mathcal{R})$.
(iii) There is an integer b such that for every coherent torsion sheaf $\mathcal{M}$ and for every $i > b$, $\mathcal{E}xt^{i}(\mathcal{M}, \mathcal{R}) := H^{i}\mathbf{R}\mathcal{H}om^{\bullet}(\mathcal{M}, \mathcal{R}) = 0$.

A complex $\mathcal{R}$ is a *t-dualizing complex on* $\mathcal{X}$ if

(i) $\mathcal{R} \in \mathbf{D}_{\mathrm{t}}^{+}(\mathcal{X})$.
(ii) The natural map is an isomorphism $\mathcal{O}_{\mathcal{X}} \xrightarrow{\sim} \mathbf{R}\mathcal{H}om^{\bullet}(\mathcal{R}, \mathcal{R})$.
(iii) There is an integer b such that for every coherent torsion sheaf $\mathcal{M}$ and for every $i > b$, $\mathcal{E}xt^{i}(\mathcal{M}, \mathcal{R}) := H^{i}\mathbf{R}\mathcal{H}om^{\bullet}(\mathcal{M}, \mathcal{R}) = 0$.
(iv) For some ideal of definition $\mathcal{J}$ of $\mathcal{X}$, $\mathbf{R}\mathcal{H}om^{\bullet}(\mathcal{O}_{\mathcal{X}}/\mathcal{J}, \mathcal{R}) \in \mathbf{D}_{\mathrm{c}}(\mathcal{X})$. (Equivalently—by simple arguments—$\mathbf{R}\mathcal{H}om^{\bullet}(\mathcal{M}, \mathcal{R}) \in \mathbf{D}_{\mathrm{c}}(\mathcal{X})$ for every coherent torsion sheaf $\mathcal{M}$.)

Remarks. (1) On an ordinary scheme, (iii) signifies *finite injective dimension* [**H1**, p. 83, Definition, and p. 134, $(\mathrm{iii})_{\mathrm{c}}$], so both c-dualizing and t-dualizing mean the same as what is called "dualizing" in [**H1**, p. 258, Definition]. (For the extension to arbitrary noetherian formal schemes, see (4) below.)

(2) By (i) and (iv), Proposition 5.2.1(a), and Corollary 5.1.3, any t-dualizing complex $\mathcal{R}$ is in $\mathbf{D}_{\mathrm{qct}}^{+}(\mathcal{X})$; and then (iii) implies that $\mathcal{R}$ *is isomorphic in* $\mathbf{D}$ *to a bounded complex of* $\mathcal{A}_{\mathrm{qct}}$*-injectives.*

To see this, begin by imitating the proof of [**H1**, p. 80, (iii)⇒(i)], using [**Y**, Theorem 4.8] and Lemma 2.5.6 below, to reduce to showing that *if* $\mathcal{N} \in \mathcal{A}_{\mathrm{qct}}(\mathcal{X})$ *is such that* $\mathcal{E}xt^{1}(\mathcal{M}, \mathcal{N}) = 0$ *for every coherent torsion sheaf* $\mathcal{M}$ *then* $\mathcal{N}$ *is* $\mathcal{A}_{\mathrm{qct}}$*-injective.*

For the last assertion, suppose first that $\mathcal{X}$ is affine. Lemma 5.1.4 implies that $\mathcal{H}om(\mathcal{M}, \mathcal{N}) \in \mathcal{A}_{\vec{\mathrm{c}}}(\mathcal{X})$; and then $\mathrm{Ext}^{1}(\mathcal{M}, \mathcal{N}) = 0$, by the natural exact sequence

$$0 \underset{(3.1.8)}{=} \mathrm{H}^{1}(\mathcal{X}, \mathcal{H}om(\mathcal{M}, \mathcal{N})) \to \mathrm{Ext}^{1}(\mathcal{M}, \mathcal{N}) \to \mathrm{H}^{0}(\mathcal{X}, \mathcal{E}xt^{1}(\mathcal{M}, \mathcal{N})).$$

Since coherent torsion sheaves generate $\mathcal{A}_{\mathrm{qct}}(\mathfrak{X})$ (Corollary 5.1.3, Lemma 5.1.4), a standard argument using Zorn's Lemma shows that $\mathcal{N}$ is indeed $\mathcal{A}_{\mathrm{qct}}$-injective.

In the general case, let $\mathfrak{U} \subset \mathfrak{X}$ be any affine open subset. For any coherent torsion $\mathcal{O}_{\mathfrak{U}}$-module $\mathcal{M}_0$, Proposition 5.1.1 and Lemma 5.1.4 imply there is a coherent torsion $\mathcal{O}_{\mathfrak{X}}$-module $\mathcal{M}$ restricting on $\mathfrak{U}$ to $\mathcal{M}_0$, whence $\mathcal{E}xt^1_{\mathfrak{U}}(\mathcal{M}_0, \mathcal{N}|_{\mathfrak{U}}) = 0$. By the affine case, then, $\mathcal{N}$ is $\mathcal{A}_{\mathrm{qct}}(\mathfrak{U})$-injective, hence $\mathcal{A}_{\mathrm{t}}(\mathfrak{U})$-injective [**Y**, Proposition 4.2]. Finally, as in [**H1**, p. 131, Lemma 7.16], using [**Y**, Lemma 4.1],[14] one concludes that $\mathcal{N}$ is $\mathcal{A}_{\mathrm{t}}(\mathfrak{X})$-injective, hence $\mathcal{A}_{\mathrm{qct}}(\mathfrak{X})$-injective.

(3) With (2) in mind, one finds that what is called here "t-dualizing complex" is what Yekutieli calls in [**Y**, §5] "dualizing complex."

(4) *A c-dualizing complex $\mathcal{R}$ has finite injective dimension*: there is an integer n_0 such that for any $i > n_0$ and any $\mathcal{O}_{\mathfrak{X}}$-module $\mathcal{E}$, $\mathrm{Hom}_{\mathbf{D}}(\mathcal{E}, \mathcal{R}[i]) = 0$. To see this, note first that

$$\mathrm{Hom}_{\mathbf{D}}(\mathcal{E}, \mathcal{R}[i]) \cong \mathrm{Hom}_{\mathbf{D}}(\mathcal{E}, \mathbf{\Lambda}\boldsymbol{\Gamma}\mathcal{R}[i]) \cong \mathrm{Hom}_{\mathbf{D}}(\boldsymbol{\Gamma}\mathcal{E}, \boldsymbol{\Gamma}\mathcal{R}[i]).$$

Lemma 2.5.3(b) below and (2) above show that $\boldsymbol{\Gamma}\mathcal{R}$ is isomorphic to a bounded complex of $\mathcal{A}_{\mathrm{qct}}$-injectives. The complex $\boldsymbol{\Gamma}\mathcal{E}$—obtained by applying the functor $\Gamma'_{\mathfrak{X}}$ to an injective resolution of $\mathcal{E}$—consists of torsion $\mathcal{O}_{\mathfrak{X}}$-modules, and so as in [**Y**, Corollary 4.3] (see also the proof of Lemma 2.5.6 below, with Proposition 5.1.2 in place of Proposition 3.1.1), the natural map

$$\mathrm{H}^i(\mathrm{Hom}^\bullet(\boldsymbol{\Gamma}\mathcal{E}, \boldsymbol{\Gamma}\mathcal{R})) \to \mathrm{H}^i(\mathbf{R}\mathrm{Hom}^\bullet(\boldsymbol{\Gamma}\mathcal{E}, \boldsymbol{\Gamma}\mathcal{R})) = \mathrm{Hom}_{\mathbf{D}}(\boldsymbol{\Gamma}\mathcal{E}, \boldsymbol{\Gamma}\mathcal{R}[i])$$

is an *isomorphism.* Since $\boldsymbol{\Gamma}\mathcal{E}$ vanishes in degrees < 0, the asserted result holds for any n_0 such that $H^i(\boldsymbol{\Gamma}\mathcal{R}) = 0$ for $i > n_0$.

(5) For a complex $\mathcal{R} \in \mathbf{D}_{\mathrm{c}}^+ \cap \mathbf{D}_{\mathrm{c}}^-$, conditions (ii) and (iii) in Definition 2.5.1 hold iff they hold stalkwise for $x \in \mathfrak{X}$, with an integer b *independent of* x. (The idea is that such an $\mathcal{R}$ is locally resolvable by a bounded-above complex $\mathcal{F}$ of finite-rank locally free $\mathcal{O}_{\mathfrak{X}}$-modules, as is $\mathcal{M}$ in (iii), and $\mathcal{H}om^\bullet(\mathcal{F}, \mathcal{R}) \simeq \mathbf{R}\mathcal{H}om^\bullet(\mathcal{F}, \mathcal{R}) \dots$.) Proceeding as in the proofs of [**H1**], Proposition 8.2, p. 288, and Corollary 7.2, p. 283, one concludes that $\mathcal{R}$ is c-dualizing iff $\mathfrak{X}$ has finite Krull dimension and $\mathcal{R}_x$ is a dualizing complex for the category of $\mathcal{O}_{\mathfrak{X},x}$-modules for every $x \in \mathfrak{X}$. (It is enough that the latter hold for all *closed* points $x \in \mathfrak{X}$.)

EXAMPLES 2.5.2. (1) If $\mathcal{R}$ is c-(or t-)dualizing then so is $\mathcal{R} \otimes \mathcal{L}[n]$ for any invertible $\mathcal{O}_{\mathfrak{X}}$-module and $n \in \mathbb{Z}$. The converse also holds, see Proposition 2.5.4.

(2) (Cf. [**Y**, Example 5.12].) If X is an ordinary scheme and $\kappa \colon \mathfrak{X} \to X$ is its completion along some closed subscheme Z, then for any dualizing $\mathcal{O}_X$-complex $\mathcal{R}$, $\kappa^*\mathcal{R}$ is c-dualizing on $\mathfrak{X}$, and $\boldsymbol{\Gamma}\kappa^*\mathcal{R} \cong \kappa^*\mathbf{R}\Gamma_Z\mathcal{R}$ (see Proposition 5.2.4(c)) is a t-dualizing complex lying in $\mathbf{D}_{\mathrm{c}}^*(\mathfrak{X})$.

Proof. For $\kappa^*\mathcal{R}$, conditions (i) and (ii) in the definition of c-dualizing follow easily from the same for $\mathcal{R}$ (because of Lemma 2.4.2). So does (iii), after we reduce to the case X affine, where Proposition 3.1.1 allows us to write $\mathcal{M} = \kappa^*\mathcal{M}_0$ with $\mathcal{M}_0 \in \mathcal{A}(X)$. (Recall from remark (1) above that $\mathcal{R}$ has finite injective dimension.) The last assertion is given by Lemma 2.5.3(b).

(3) If $\mathfrak{X} = \mathrm{Spf}(A)$ where A is a complete local noetherian ring topologized by its maximal ideal m—so that $\mathcal{A}(\mathfrak{X})$ is just the category of A-modules—then a c-dualizing $\mathcal{O}_{\mathfrak{X}}$-complex is an A-dualizing complex in the usual sense; and by (2)

[14] where one may assume that X and $\mathfrak{X}$ have the same underlying space

(via [**H1**, p. 276, 6.1]), or directly from Definition 2.5.1, the injective hull of A/m is a t-dualizing complex lying in $\mathbf{D}_{\mathrm{c}}^{*}(\mathfrak{X})$.

(4) It is clear from Definition 2.5.1 and remark (4) above that $\mathcal{O}_{\mathfrak{X}}$ is c-dualizing iff $\mathcal{O}_{\mathfrak{X}}$ has finite injective dimension over itself. By remark (5), $\mathcal{O}_{\mathfrak{X}}$ is c-dualizing iff $\mathfrak{X}$ is finite dimensional and $\mathcal{O}_{\mathfrak{X},x}$ is *Gorenstein* for all $x \in \mathfrak{X}$ [**H1**, p. 295, Definition].

(5) For instance, if the finite-dimensional noetherian formal scheme $\mathfrak{Y}$ is *regular* (i.e., the local rings $\mathcal{O}_{\mathfrak{Y},y}$ $(y \in \mathfrak{Y})$ are all regular), and $\mathcal{I}$ is a coherent $\mathcal{O}_{\mathfrak{Y}}$-ideal, defining a closed formal subscheme $i\colon \mathfrak{X} \hookrightarrow \mathfrak{Y}$ [**GD**, p. 441,(10.14.2)], then by remark (3), $\mathbf{R}\mathcal{H}om^{\bullet}(i_*\mathcal{O}_{\mathfrak{X}}, \mathcal{O}_{\mathfrak{Y}})$ is c-dualizing on $\mathfrak{X}$. So Lemma 2.5.3 gives that

$$\mathbf{R}\mathcal{H}om^{\bullet}(\mathcal{O}_{\mathfrak{Y}}/\mathcal{I}, \mathbf{R}\mathit{\Gamma}'_{\mathfrak{Y}}\mathcal{O}_{\mathfrak{Y}}) \underset{5.2.10(4)}{\cong} \mathbf{R}\mathit{\Gamma}'_{\mathfrak{X}}\mathbf{R}\mathcal{H}om^{\bullet}(i_*\mathcal{O}_{\mathfrak{X}}, \mathcal{O}_{\mathfrak{Y}}) \in \mathbf{D}_{\mathrm{c}}^{*}(\mathfrak{X})$$

is t-dualizing on $\mathfrak{X}$. (This is also shown in [**Y**, Proposition 5.11, Theorem 5.14].)

LEMMA 2.5.3. (a) *If $\mathcal{R} \in \mathbf{D}_{\mathrm{c}}^{*}$ is t-dualizing then $\mathbf{\Lambda}\mathcal{R}$ is c-dualizing.*

(b) *If $\mathcal{R}$ is c-dualizing then $\boldsymbol{\mathit{\Gamma}}\mathcal{R}$ is t-dualizing, and lies in $\mathbf{D}_{\mathrm{c}}^{*}$.*

PROOF. (a) If $\mathcal{R} \in \mathbf{D}_{\mathrm{c}}^{*}$ then of course $\mathbf{\Lambda}\mathcal{R} \in \mathbf{D}_{\mathrm{c}}$. Also, $\mathbf{\Lambda}(\mathbf{D}^{+}) \subset \mathbf{D}^{+}$ because $\mathbf{R}\mathit{\Gamma}'_{\mathfrak{X}}\mathcal{O}_{\mathfrak{X}}$ is given locally by a finite complex $\mathcal{K}_{\infty}^{\bullet}$, see proof of Proposition 5.2.1(a).

For condition (ii), note that if $\mathcal{R} \in \mathbf{D}_{\mathrm{qct}}^{+}(\mathfrak{X})$ then $\boldsymbol{\mathit{\Gamma}}\mathcal{R} \cong \mathcal{R}$ (Proposition 5.2.1), then use the natural isomorphisms (see (2.5.0.1):

$$\mathbf{R}\mathcal{H}om^{\bullet}(\mathbf{\Lambda}\mathcal{R}, \mathbf{\Lambda}\mathcal{R}) \cong \mathbf{R}\mathcal{H}om^{\bullet}(\boldsymbol{\mathit{\Gamma}}\mathbf{\Lambda}\mathcal{R}, \mathcal{R}) \cong \mathbf{R}\mathcal{H}om^{\bullet}(\boldsymbol{\mathit{\Gamma}}\mathcal{R}, \mathcal{R}) \cong \mathbf{R}\mathcal{H}om^{\bullet}(\mathcal{R}, \mathcal{R}).$$

For (iii) note that $\boldsymbol{\mathit{\Gamma}}\mathcal{M} \cong \mathcal{M}$ (Proposition 5.2.1), then use (2.5.0.1).

(b) Proposition 5.2.1 makes clear that if $\mathcal{R} \in \mathbf{D}_{\mathrm{c}}^{+}$ then $\boldsymbol{\mathit{\Gamma}}\mathcal{R} \in \mathbf{D}_{\mathrm{qct}}^{+} \cap \mathbf{D}_{\mathrm{c}}^{*}$.

For (ii) use the isomorphisms (the second holding because $\mathcal{R} \in \mathbf{D}_{\mathrm{c}}$):

$$\mathbf{R}\mathcal{H}om^{\bullet}(\boldsymbol{\mathit{\Gamma}}\mathcal{R}, \boldsymbol{\mathit{\Gamma}}\mathcal{R}) \underset{5.2.3}{\cong} \mathbf{R}\mathcal{H}om^{\bullet}(\boldsymbol{\mathit{\Gamma}}\mathcal{R}, \mathcal{R}) \underset{6.2.1}{\cong} \mathbf{R}\mathcal{H}om^{\bullet}(\mathcal{R}, \mathcal{R}).$$

For (iii) use the isomorphism $\mathbf{R}\mathcal{H}om^{\bullet}(\mathcal{M}, \boldsymbol{\mathit{\Gamma}}\mathcal{R}) \underset{5.2.3}{\cong} \mathbf{R}\mathcal{H}om^{\bullet}(\mathcal{M}, \mathcal{R})$. For (iv), note that when $\mathcal{M} = \mathcal{O}_{\mathfrak{X}}/\mathfrak{J}$ ($\mathfrak{J}$ any ideal of definition) this isomorphism gives

$$\mathbf{R}\mathcal{H}om^{\bullet}(\mathcal{O}_{\mathfrak{X}}/\mathfrak{J}, \boldsymbol{\mathit{\Gamma}}\mathcal{R}) \cong \mathbf{R}\mathcal{H}om^{\bullet}(\mathcal{O}_{\mathfrak{X}}/\mathfrak{J}, \mathcal{R}) \underset{3.2.4}{\in} \mathbf{D}_{\mathrm{c}}, \qquad \square$$

The essential *uniqueness* of t-(resp. c-)dualizing complexes is expressed by:

PROPOSITION 2.5.4. (a) (Yekutieli) *If $\mathcal{R}$ is t-dualizing then a complex $\mathcal{R}'$ is t-dualizing iff there is an invertible sheaf $\mathcal{L}$ and an integer n such that $\mathcal{R}' \cong \mathcal{R} \otimes \mathcal{L}[n]$.*

(b) *If $\mathcal{R}$ is c-dualizing then a complex $\mathcal{R}'$ is c-dualizing iff there is an invertible sheaf $\mathcal{L}$ and an integer n such that $\mathcal{R}' \cong \mathcal{R} \otimes \mathcal{L}[n]$.*

PROOF. Part (a) is proved in [**Y**, Theorem 5.6].

Now for a fixed invertible sheaf $\mathcal{L}$ there is a natural isomorphism of functors

$$\mathbf{\Lambda}(\mathcal{F} \otimes \mathcal{L}) \xrightarrow{\sim} \mathbf{\Lambda}\mathcal{F} \otimes \mathcal{L} \qquad (\mathcal{F} \in \mathbf{D}), \tag{2.5.4.1}$$

as one deduces, e.g., from a readily-established natural isomorphism between the respective right adjoints

$$\boldsymbol{\mathit{\Gamma}}\mathcal{E} \otimes \mathcal{L}^{-1} \xleftarrow{\sim} \boldsymbol{\mathit{\Gamma}}(\mathcal{E} \otimes \mathcal{L}^{-1}) \qquad (\mathcal{E} \in \mathbf{D}).$$

Part (b) results, because $\boldsymbol{\Gamma}\mathcal{R}'$ and $\boldsymbol{\Gamma}\mathcal{R}$ are t-dualizing (Lemma 2.5.3), so that by (a) (and taking $\mathcal{F} := \boldsymbol{\Gamma}\mathcal{R}[n]$ in (2.5.4.1)) we have isomorphisms

$$\begin{aligned}\mathcal{R}' \cong \boldsymbol{\Lambda}(\boldsymbol{\Gamma}\mathcal{R}') &\cong \boldsymbol{\Lambda}(\boldsymbol{\Gamma}\mathcal{R} \otimes \mathcal{L}[n]) \qquad (\mathcal{L} \text{ invertible}, n \in \mathbb{Z})\\ &\cong (\boldsymbol{\Lambda}\boldsymbol{\Gamma}\mathcal{R}) \otimes \mathcal{L}[n] \cong \mathcal{R} \otimes \mathcal{L}[n].\end{aligned}$$

□

COROLLARY 2.5.5. *If $\mathfrak{X}$ is locally embeddable in a regular finite-dimensional formal scheme then any t-dualizing complex on $\mathfrak{X}$ lies in $\mathbf{D}_{\mathrm{c}}^{*}$.*

PROOF. Whether a t-dualizing complex $\mathcal{R}$ satisfies $\boldsymbol{\Lambda}\mathcal{R} \in \mathbf{D}_{\mathrm{c}}$ is a local question, so we may assume that $\mathfrak{X}$ is a closed subscheme of a finite-dimensional regular formal scheme, and then Example 2.5.2(5) shows that *some*—hence by Proposition 2.5.4, *any*—t-dualizing complex lies in $\mathbf{D}_{\mathrm{c}}^{*}$. □

LEMMA 2.5.6 (cf. [**Y**, Corollary 4.3]). *Let $\mathcal{F} \in \mathbf{D}_{\vec{\mathrm{c}}}$ and let $\mathcal{I}$ be a bounded-below complex of $\mathcal{A}_{\mathrm{qct}}$-injectives. Then the canonical map is a $\mathbf{D}$-isomorphism*

$$\gamma\colon \mathcal{H}om^{\bullet}(\mathcal{F},\mathcal{I}) \xrightarrow{\sim} \mathbf{R}\mathcal{H}om^{\bullet}(\mathcal{F},\mathcal{I}).$$

PROOF. By the proof of [**Y**, Proposition 4.2], $\mathcal{A}_{\mathrm{qct}}$-injectives are just direct sums of sheaves of the form $\mathcal{J}(x)$ $(x \in \mathfrak{X})$, where for any open $\mathfrak{U} \subset \mathfrak{X}$, $\Gamma(\mathfrak{U},\mathcal{J}(x))$ is a fixed injective hull of the residue field of $\mathcal{O}_{\mathfrak{X},x}$ if $x \in \mathfrak{U}$, and vanishes otherwise. Hence the question is local, and we may assume that $\mathfrak{X} = \mathrm{Spf}(A)$ for some noetherian adic ring A. Let $\kappa\colon \mathfrak{X} \to X := \mathrm{Spec}(A)$ be the natural map. Then by Proposition 3.1.1, $\mathcal{F} \cong \kappa^{*}F$ for some $F \in \mathcal{A}_{\mathrm{qc}}(X)$. Looking at homology sheaves we see it's enough that $\kappa_{*}(\gamma)$ be an isomorphism—which it is, because it factors naturally as

$$\begin{aligned}\kappa_{*}\mathcal{H}om^{\bullet}_{\mathfrak{X}}(\kappa^{*}F,\mathcal{I}) \cong \mathcal{H}om^{\bullet}_{X}(F,\kappa_{*}\mathcal{I}) &\to \mathbf{R}\mathcal{H}om^{\bullet}_{X}(F,\kappa_{*}\mathcal{I})\\ &\cong \kappa_{*}\mathbf{R}\mathcal{H}om^{\bullet}_{\mathfrak{X}}(\kappa^{*}F,\mathcal{I}) \qquad ([\mathbf{Sp}, \text{p.147, 6.7(2)}]),\end{aligned}$$

where $\kappa_{*}\mathcal{I}$ (a direct sum of injective hulls) is injective on X. □

Proposition 2.5.8 below brings out the basic property of the *dualizing functors* associated with dualizing complexes. (For illustration, one might keep in mind the special case of Example 2.5.2(3).)

LEMMA 2.5.7. *Let $\mathcal{R}$ be a c-dualizing complex on $\mathfrak{X}$, let $\mathcal{R}_{\mathrm{t}}$ be the t-dualizing complex $\mathcal{R}_{\mathrm{t}} := \boldsymbol{\Gamma}\mathcal{R}$, and for any $\mathcal{E} \in \mathbf{D}$ set*

$$\mathcal{D}\mathcal{E} := \mathbf{R}\mathcal{H}om^{\bullet}(\mathcal{E},\mathcal{R}), \qquad \mathcal{D}_{\mathrm{t}}\mathcal{E} := \mathbf{R}\mathcal{H}om^{\bullet}(\mathcal{E},\mathcal{R}_{\mathrm{t}}).$$

(a) *There are functorial isomorphisms*

$$\boldsymbol{\Lambda}\mathcal{D}_{\mathrm{t}} \cong \boldsymbol{\Lambda}\mathcal{D} \cong \mathcal{D}\boldsymbol{\Lambda} \cong \mathcal{D} \cong \mathcal{D}\boldsymbol{\Gamma} \cong \mathcal{D}_{\mathrm{t}}\boldsymbol{\Gamma}.$$

(b) *For all $\mathcal{F} \in \mathbf{D}_{\mathrm{c}}$, $\mathcal{D}\mathcal{F} \in \mathbf{D}_{\mathrm{c}}$ and there is a natural isomorphism $\mathcal{D}_{\mathrm{t}}\mathcal{F} \cong \boldsymbol{\Gamma}\mathcal{D}\mathcal{F}$.*

PROOF. (a) For any $\mathcal{E} \in \mathbf{D}$, Proposition 6.2.1 gives the isomorphism

$$\mathcal{D}\mathcal{E} = \mathbf{R}\mathcal{H}om^{\bullet}(\mathcal{E},\mathcal{R}) \cong \mathbf{R}\mathcal{H}om^{\bullet}(\boldsymbol{\Gamma}\mathcal{E},\mathcal{R}) = \mathcal{D}\boldsymbol{\Gamma}\mathcal{E}.$$

In particular, $\mathcal{D}\boldsymbol{\Lambda}\mathcal{E} \cong \mathcal{D}\boldsymbol{\Gamma}\boldsymbol{\Lambda}\mathcal{E} \cong \mathcal{D}\boldsymbol{\Gamma}\mathcal{E}$. Thus $\mathcal{D} \cong \mathcal{D}\boldsymbol{\Gamma} \cong \mathcal{D}\boldsymbol{\Lambda}$.

Furthermore, using that the natural map $\boldsymbol{\Gamma}\mathcal{O}_{\mathfrak{X}} \underline{\underline{\otimes}} \mathcal{E} \to \boldsymbol{\Gamma}\mathcal{E}$ is an *isomorphism* (localize, and see [**AJL**, p. 20, Corollary (3.1.2)]) we get natural isomorphisms

$$\begin{aligned}\mathbf{R}\mathcal{H}om^{\bullet}(\boldsymbol{\Gamma}\mathcal{O}_{\mathfrak{X}},\mathcal{H}om^{\bullet}(\mathcal{E},\mathcal{R})) \xrightarrow{\sim} \mathcal{H}om^{\bullet}(\boldsymbol{\Gamma}\mathcal{E},\mathcal{R}) &\underset{5.2.3}{\cong} \mathbf{R}\mathcal{H}om^{\bullet}(\boldsymbol{\Gamma}\mathcal{E},\boldsymbol{\Gamma}\mathcal{R})\\ &\cong \mathbf{R}\mathcal{H}om^{\bullet}(\boldsymbol{\Gamma}\mathcal{O}_{\mathfrak{X}},\mathcal{H}om^{\bullet}(\mathcal{E},\boldsymbol{\Gamma}\mathcal{R})),\end{aligned}$$

giving $\boldsymbol{\Lambda}\mathcal{D} \cong \mathcal{D}\boldsymbol{\Gamma} \cong \mathcal{D}_{\mathrm{t}}\boldsymbol{\Gamma} \cong \boldsymbol{\Lambda}\mathcal{D}_{\mathrm{t}}$.

(b) Given remark (2) following Definition 2.5.1, Lemma 2.5.6 implies that the functor $\mathcal{D}_{\mathrm{t}} := \mathbf{R}\mathcal{H}om^\bullet(-,\mathcal{R}_{\mathrm{t}})$ is bounded on $\mathbf{D}_{\vec{\mathrm{c}}}$. The same holds for $\mathcal{D} = \mathcal{D}_{\mathrm{t}}\boldsymbol{\varGamma}$ (see (a)), because $\boldsymbol{\varGamma}(\mathbf{D}_{\vec{\mathrm{c}}}) \subset \mathbf{D}_{\mathrm{qct}} \subset \mathbf{D}_{\vec{\mathrm{c}}}$ (Lemma 5.1.4), and $\boldsymbol{\varGamma}$ is bounded. ($\boldsymbol{\varGamma}$ is given locally by tensoring with a bounded flat complex $\mathcal{K}^\bullet_\infty$, see proof of Proposition 5.2.1(a)).

Arguing as in Proposition 3.2.4, we see that $\mathcal{D}_{\mathrm{t}}\mathcal{F} := \mathbf{R}\mathcal{H}om^\bullet(\mathcal{F},\mathcal{R}_{\mathrm{t}}) \in \mathbf{D}_{\mathrm{qct}}$, so that $\boldsymbol{\varGamma}\mathcal{D}_{\mathrm{t}}\mathcal{F} \xrightarrow{\sim} \mathcal{D}_{\mathrm{t}}\mathcal{F}$ (Proposition 5.2.1(a)); and similarly, $\mathcal{D}\mathcal{F} \in \mathbf{D}_{\mathrm{c}}$. Furthermore, the argument in Remark 5.2.10(4) gives an isomorphism $\boldsymbol{\varGamma}\mathcal{D}_{\mathrm{t}}\mathcal{F} \cong \boldsymbol{\varGamma}\mathcal{D}\mathcal{F}$. □

PROPOSITION 2.5.8. *With notation as in Lemma 2.5.7 we have, for $\mathcal{E},\mathcal{F} \in \mathbf{D}$:*

(a) $\mathcal{E} \in \mathbf{D}_{\mathrm{c}}^* \iff \mathcal{D}_{\mathrm{t}}\mathcal{E} \in \mathbf{D}_{\mathrm{c}}$ *and the natural map is an isomorphism* $\mathcal{E} \xrightarrow{\sim} \mathcal{D}_{\mathrm{t}}\mathcal{D}_{\mathrm{t}}\mathcal{E}$.

(b) $\mathcal{F} \in \mathbf{D}_{\mathrm{c}} \iff \mathcal{D}\mathcal{F} \in \mathbf{D}_{\mathrm{c}}$ *and the natural map is an isomorphism* $\mathcal{F} \xrightarrow{\sim} \mathcal{D}\mathcal{D}\mathcal{F}$.

(c) $\mathcal{F} \in \mathbf{D}_{\mathrm{c}} \iff \mathcal{D}_{\mathrm{t}}\mathcal{F} \in \mathbf{D}_{\mathrm{c}}^*$ *and the natural map is an isomorphism* $\mathcal{F} \xrightarrow{\sim} \mathcal{D}_{\mathrm{t}}\mathcal{D}_{\mathrm{t}}\mathcal{F}$.

Remark. The isomorphism $\mathcal{F} \xrightarrow{\sim} \mathcal{D}_{\mathrm{t}}\mathcal{D}_{\mathrm{t}}\mathcal{F}$ is a formal version of "Affine Duality, " see [**AJL**, §5.2].

PROOF. For $\mathcal{F} \in \mathbf{D}_{\mathrm{c}}$, Lemma 2.5.7(b) gives $\mathcal{D}\mathcal{F} \in \mathbf{D}_{\mathrm{c}}$, so $\mathcal{D}_{\mathrm{t}}\mathcal{F} \cong \boldsymbol{\varGamma}\mathcal{D}\mathcal{F} \in \mathbf{D}_{\mathrm{c}}^*$. Moreover, from the isomorphism $\mathcal{D}_{\mathrm{t}}\boldsymbol{\varGamma}\mathcal{F} \cong \mathcal{D}\mathcal{F}$ of Lemma 2.5.7(a) it follows that $\mathcal{D}_{\mathrm{t}}(\mathbf{D}_{\mathrm{c}}^*) \subset \mathbf{D}_{\mathrm{c}}$. The $\Longleftarrow$ implications in (a), (b) and (c) result, as do the first parts of the $\Longrightarrow$ implications.

Establishing the isomorphisms $\mathcal{D}\mathcal{D}\mathcal{F} \xleftarrow{\sim} \mathcal{F} \xrightarrow{\sim} \mathcal{D}_{\mathrm{t}}\mathcal{D}_{\mathrm{t}}\mathcal{F}$ is a local problem, so we may assume $\mathfrak{X}$ affine. Since the functors $\mathcal{D}$ and $\mathcal{D}_{\mathrm{t}}$ are bounded on $\mathbf{D}_{\vec{\mathrm{c}}}$ (see proof of Lemma 2.5.7(b)), and both take $\mathbf{D}_{\mathrm{c}}$ into $\mathbf{D}_{\vec{\mathrm{c}}}$, therefore the functors $\mathcal{D}\mathcal{D}$ and $\mathcal{D}_{\mathrm{t}}\mathcal{D}_{\mathrm{t}}$ are bounded on $\mathbf{D}_{\mathrm{c}}$, and so [**H1**, p. 68, 7.1] (dualized) reduces the problem to the tautological case $\mathcal{F} = \mathcal{O}_{\mathfrak{X}}$ (cf. [**H1**, p. 258, Proposition 2.1].)

For assertion (a) one may assume that $\mathcal{E} = \boldsymbol{\varGamma}\mathcal{F}$ ($\mathcal{F} \in \mathbf{D}_{\mathrm{c}}$), so that there is a composed isomorphism (which one checks to be the natural map):

$$\mathcal{E} = \boldsymbol{\varGamma}\mathcal{F} \cong \boldsymbol{\varGamma}\mathcal{D}\mathcal{D}\mathcal{F} \underset{2.5.7(\mathrm{b})}{\cong} \mathcal{D}_{\mathrm{t}}\mathcal{D}\mathcal{F} \underset{2.5.7(\mathrm{a})}{\cong} \mathcal{D}_{\mathrm{t}}\mathcal{D}_{\mathrm{t}}\boldsymbol{\varGamma}\mathcal{F} = \mathcal{D}_{\mathrm{t}}\mathcal{D}_{\mathrm{t}}\mathcal{E}. \qquad \square$$

COROLLARY 2.5.9. *With the preceding notation,*

(a) *The functor $\mathcal{D}$ induces an involutive anti-equivalence of $\mathbf{D}_{\mathrm{c}}$ with itself.*

(b) *The functor $\mathcal{D}_{\mathrm{t}}$ induces quasi-inverse anti-equivalences between $\mathbf{D}_{\mathrm{c}}$ and $\mathbf{D}_{\mathrm{c}}^*$.*

LEMMA 2.5.10. *Let $\mathcal{J}$ be an ideal of definition of $\mathfrak{X}$. Then a complex $\mathcal{R} \in \mathbf{D}_{\mathrm{c}}$ (resp. $\mathcal{R} \in \mathbf{D}_{\mathrm{qct}}$) is c-dualizing (resp. t-dualizing) iff for every $n > 0$ the complex $\mathbf{R}\mathcal{H}om^\bullet(\mathcal{O}_{\mathfrak{X}}/\mathcal{J}^n,\mathcal{R})$ is dualizing on the scheme $X_n := (\mathfrak{X},\mathcal{O}_{\mathfrak{X}}/\mathcal{J}^n)$.*

PROOF. Remark (1) after Definition 2.5.1 makes it straightforward to see that if $\mathcal{R}$ is either c- or t-dualizing on $\mathfrak{X}$ then $\mathbf{R}\mathcal{H}om^\bullet(\mathcal{O}_{\mathfrak{X}}/\mathcal{J}^n,\mathcal{R})$ is dualizing on X_n.

For the converse, to begin with, Corollary 5.2.3 gives

$$\mathbf{R}\mathcal{H}om^\bullet(\mathcal{O}_{\mathfrak{X}}/\mathcal{J}^n,\mathcal{R}) = \mathbf{R}\mathcal{H}om^\bullet(\mathcal{O}_{\mathfrak{X}}/\mathcal{J}^n,\boldsymbol{\varGamma}\mathcal{R}),$$

and it follows from Lemma 2.5.3 that it suffices to consider the t-dualizing case. So suppose that $\mathcal{R} \in \mathbf{D}_{\mathrm{qct}}$ and that for all n, $\mathbf{R}\mathcal{H}om^\bullet(\mathcal{O}_{\mathfrak{X}}/\mathcal{J}^n,\mathcal{R})$ is dualizing on X_n. Taking $\tilde{\mathcal{R}} = \mathcal{R}$ in the proof of [**Y**, Theorem 5.6], one gets $\mathcal{O}_{\mathfrak{X}} \xrightarrow{\sim} \mathbf{R}\mathcal{H}om^\bullet(\mathcal{R},\mathcal{R})$.

It remains to check condition (iii) in Definition 2.5.1. We may assume $\mathcal{R}$ to be K-injective, so that $\mathcal{R}_n := \mathcal{H}om^\bullet(\mathcal{O}_\mathfrak{X}/\mathcal{J}^n, \mathcal{R})$ is K-injective on X_n for all n. Then, since $\varGamma'_\mathfrak{X}\mathcal{R} \cong \mathbf{R}\varGamma'_\mathfrak{X}\mathcal{R} \cong \mathcal{R}$ (Proposition 5.2.1(a)),

$$H^i\mathcal{R} \cong H^i\varGamma'_\mathfrak{X}\mathcal{R} \cong H^i \varinjlim_n \mathcal{R}_n \cong \varinjlim_n H^i\mathcal{R}_n \qquad (i \in \mathbb{Z}).$$

For each n, $\mathcal{R}_n$ is quasi-isomorphic to a *residual complex,* which is an injective $\mathcal{O}_{X_n}$-complex vanishing in degrees outside a certain finite interval $I := [a, b]$ ([**H1**, pp. 304–306]). If $m \le n$, the same holds—with the same I—for the complex $\mathcal{R}_m \cong \mathcal{H}om_{X_n}(\mathcal{O}_\mathfrak{X}/\mathcal{J}^m, \mathcal{R}_n)$. It follows that $H^i\mathcal{R} = 0$ for $i \notin I$. In particular, $\mathcal{R} \in \mathbf{D}^+_{\mathrm{qct}}$.

So now we we may assume that $\mathcal{R}$ is a bounded-below complex of $\mathcal{A}_{\mathrm{qct}}$-injectives [**Y**, Theorem 4.8]. Then for any coherent torsion sheaf $\mathcal{M}$, the homology of

$$\mathbf{R}\mathcal{H}om^\bullet_\mathfrak{X}(\mathcal{M}, \mathcal{R}) \underset{2.5.6}{\cong} \mathcal{H}om^\bullet_\mathfrak{X}(\mathcal{M}, \mathcal{R}) \cong \mathcal{H}om^\bullet_\mathfrak{X}(\mathcal{M}, \varinjlim_n \mathcal{R}_n) \cong \varinjlim_n \mathcal{H}om^\bullet_\mathfrak{X}(\mathcal{M}, \mathcal{R}_n)$$

vanishes in all degrees $> b$, as required by (iii). □

PROPOSITION 2.5.11. *Let $f\colon \mathfrak{X} \to \mathfrak{Y}$ be a pseudo-proper map of noetherian formal schemes.*

(a) *If $\mathcal{R}$ is a t-dualizing complex on $\mathfrak{Y}$, then $f_\mathrm{t}^\times\mathcal{R}$ is t-dualizing on $\mathfrak{X}$.*

(b) *If $\mathcal{R}$ is a c-dualizing complex on $\mathfrak{Y}$, then $f^\#\mathcal{R}$ is c-dualizing on $\mathfrak{X}$.*

PROOF. (a) Let $\mathcal{J}$ be a defining ideal of $\mathfrak{X}$, and let $\mathcal{I}$ be a defining ideal of $\mathfrak{Y}$ such that $\mathcal{I}\mathcal{O}_\mathfrak{X} \subset \mathcal{J}$. Let $X_\mathcal{J} := (\mathfrak{X}, \mathcal{O}_\mathfrak{X}/\mathcal{J}) \overset{j}{\hookrightarrow} \mathfrak{X}$ and $Y_\mathcal{I} := (\mathfrak{Y}, \mathcal{O}_\mathfrak{Y}/\mathcal{I}) \overset{i}{\hookrightarrow} \mathfrak{Y}$ be the resulting closed immersions. Example 6.1.3(4) shows that $i_\mathrm{t}^\times\mathcal{R} \cong \mathbf{R}\mathcal{H}om^\bullet(\mathcal{O}_\mathfrak{Y}/\mathcal{I}, \mathcal{R})$, which is a dualizing complex on $Y_\mathcal{I}$. Pseudo-properness of f means the map $f_{\mathcal{I}\mathcal{J}}\colon X_\mathcal{J} \to Y_\mathcal{I}$ induced by f is proper, so as in [**V**, p. 396, Corollary 3] (hypotheses about finite Krull dimension being unnecessary here for the existence of $f_\mathrm{t}^\times$, etc.),

$$\mathbf{R}\mathcal{H}om^\bullet(\mathcal{O}_\mathfrak{X}/\mathcal{J}, f_\mathrm{t}^\times\mathcal{R}) \cong j_\mathrm{t}^\times f_\mathrm{t}^\times\mathcal{R} \cong (f_{\mathcal{I}\mathcal{J}})_\mathrm{t}^\times i_\mathrm{t}^\times\mathcal{R}$$

is a dualizing complex on $X_\mathcal{J}$. The assertion is given then by Lemma 2.5.10.

(b) By Proposition 8.3.2, $f^\#\mathcal{R} \in \mathbf{D}_\mathrm{c}(\mathfrak{X})$. By Corollary 6.1.5, Lemma 2.5.3(b), and the just-proved assertion (a),

$$\mathbf{R}\varGamma'_\mathfrak{X} f^\#\mathcal{R} \cong f_\mathrm{t}^\times\mathcal{R} \cong f_\mathrm{t}^\times\mathbf{R}\varGamma'_\mathfrak{Y}\mathcal{R},$$

is t-dualizing on $\mathfrak{X}$. So by Lemma 2.5.3(a), $f^\#\mathcal{R} \cong \mathbf{\Lambda}_\mathfrak{X}\mathbf{R}\varGamma'_\mathfrak{X} f^\#\mathcal{R}$ is c-dualizing. □

The following proposition generalizes [**H1**, p. 291, 8.5] (see also [**H1**, middle of p. 384] and [**V**, p. 396, Corollary 3]).

PROPOSITION 2.5.12. *Let $f\colon \mathfrak{X} \to \mathfrak{Y}$ be a pseudo-proper map of noetherian formal schemes. Suppose that $\mathfrak{Y}$ has a c-dualizing complex $\mathcal{R}_\mathrm{c}$, or equivalently* (*by Lemma* 2.5.3), *a t-dualizing complex $\mathcal{R}_\mathrm{t} \in \mathbf{D}^*_\mathrm{c}(\mathfrak{Y})$, so that $f^\#\mathcal{R}_\mathrm{c}$ is c-dualizing* (*resp. $f_\mathrm{t}^\times\mathcal{R}_\mathrm{t}$ is t-dualizing*) *on $\mathfrak{X}$* (*Proposition* 2.5.11). *Define dualizing functors*

$$\mathcal{D}_\mathrm{t}^\mathfrak{Y}(-) := \mathbf{R}\mathcal{H}om^\bullet_\mathfrak{Y}(-, \mathcal{R}_\mathrm{t}), \qquad \mathcal{D}_\mathrm{c}^\mathfrak{Y}(-) := \mathbf{R}\mathcal{H}om^\bullet_\mathfrak{Y}(-, \mathcal{R}_\mathrm{c}),$$
$$\mathcal{D}_\mathrm{t}^\mathfrak{X}(-) := \mathbf{R}\mathcal{H}om^\bullet_\mathfrak{X}(-, f_\mathrm{t}^\times\mathcal{R}_\mathrm{t}), \qquad \mathcal{D}_\mathrm{c}^\mathfrak{X}(-) := \mathbf{R}\mathcal{H}om^\bullet_\mathfrak{X}(-, f^\#\mathcal{R}_\mathrm{c}).$$

Then there are natural isomorphisms

$$f_\mathrm{t}^\times\mathcal{E} \cong \mathcal{D}_\mathrm{t}^\mathfrak{X}\mathbf{L}f^*\mathcal{D}_\mathrm{t}^\mathfrak{Y}\mathcal{E}, \qquad (\mathcal{E} \in \mathbf{D}^*_\mathrm{c}(\mathfrak{Y}) \cap \mathbf{D}^+(\mathfrak{Y})),$$
$$f^\#\mathcal{E} \cong \mathcal{D}_\mathrm{c}^\mathfrak{X}\mathbf{L}f^*\mathcal{D}_\mathrm{c}^\mathfrak{Y}\mathcal{E} \qquad (\mathcal{E} \in \mathbf{D}^+_\mathrm{c}(\mathfrak{Y})).$$

PROOF. When $\mathcal{E} \in \mathbf{D}_{\mathrm{c}}^{*}(\mathcal{Y}) \cap \mathbf{D}^{+}(\mathcal{Y})$ (resp. $\mathcal{E} \in \mathbf{D}_{\mathrm{c}}^{+}(\mathcal{Y})$) set $\mathcal{F} := \mathcal{D}_{\mathrm{t}}^{\mathcal{Y}}\mathcal{E}$ (resp. $\mathcal{F} := \mathcal{D}_{\mathrm{c}}^{\mathcal{Y}}\mathcal{E}$). In either case, $\mathcal{F} \in \mathbf{D}_{\mathrm{c}}(\mathcal{Y})$ (Proposition 2.5.8), and also $\mathcal{F} \in \mathbf{D}^{-}(\mathcal{Y})$—in the first case by remark (2) following Definition 2.5.1 and Lemma 2.5.6, in the second case by remark (1) following Definition 2.5.1. So, by Proposition 2.5.8, we need to find natural isomorphisms

$$f_{\mathrm{t}}^{\times}\mathcal{D}_{\mathrm{t}}^{\mathcal{Y}}\mathcal{F} \cong \mathcal{D}_{\mathrm{t}}^{\mathcal{X}}\mathbf{L}f^{*}\mathcal{F},$$
$$f^{\#}\mathcal{D}_{\mathrm{c}}^{\mathcal{Y}}\mathcal{F} \cong \mathcal{D}_{\mathrm{c}}^{\mathcal{X}}\mathbf{L}f^{*}\mathcal{F}.$$

Such isomorphisms are given by the next result—a generalization of [**H1**, p. 194, 8.8(7)]—for $\mathcal{G} := \mathcal{R}_{\mathrm{t}}$ (resp. $\mathcal{R}_{\mathrm{c}}$). □

PROPOSITION 2.5.13. *Let $f\colon \mathcal{X} \to \mathcal{Y}$ be a map of noetherian formal schemes. Then for $\mathcal{F} \in \mathbf{D}_{\mathrm{c}}^{-}(\mathcal{Y})$ and $\mathcal{G} \in \mathbf{D}^{+}(\mathcal{Y})$ there are natural isomorphisms*

$$\mathbf{R}\mathcal{H}om_{\mathcal{X}}^{\bullet}(\mathbf{L}f^{*}\mathcal{F}, f_{\mathrm{t}}^{\times}\mathcal{G}) \xrightarrow{\ \sim\ } f_{\mathrm{t}}^{\times}\mathbf{R}\mathcal{H}om_{\mathcal{Y}}^{\bullet}(\mathcal{F},\mathcal{G}),$$
$$\mathbf{R}\mathcal{H}om_{\mathcal{X}}^{\bullet}(\mathbf{L}f^{*}\mathcal{F}, f^{\#}\mathcal{G}) \xrightarrow{\ \sim\ } f^{\#}\mathbf{R}\mathcal{H}om_{\mathcal{Y}}^{\bullet}(\mathcal{F},\mathcal{G}).$$

PROOF. The second isomorphism follows from the first, since $f^{\#} = \mathbf{\Lambda} f_{\mathrm{t}}^{\times}$ and since there are natural isomorphisms

$$\begin{aligned}\mathbf{\Lambda}\mathbf{R}\mathcal{H}om_{\mathcal{X}}^{\bullet}(\mathbf{L}f^{*}\mathcal{F}, f_{\mathrm{t}}^{\times}\mathcal{G}) &= \mathbf{R}\mathcal{H}om_{\mathcal{X}}^{\bullet}(\mathbf{R}\varGamma_{\mathcal{X}}'\mathcal{O}_{\mathcal{X}}, \mathbf{R}\mathcal{H}om_{\mathcal{X}}^{\bullet}(\mathbf{L}f^{*}\mathcal{F}, f_{\mathrm{t}}^{\times}\mathcal{G}))\\ &\cong \mathbf{R}\mathcal{H}om_{\mathcal{X}}^{\bullet}(\mathbf{R}\varGamma_{\mathcal{X}}'\mathcal{O}_{\mathcal{X}} \underline{\underline{\otimes}}\, \mathbf{L}f^{*}\mathcal{F}, f_{\mathrm{t}}^{\times}\mathcal{G}))\\ &\cong \mathbf{R}\mathcal{H}om_{\mathcal{X}}^{\bullet}(\mathbf{L}f^{*}\mathcal{F}, \mathbf{R}\mathcal{H}om_{\mathcal{X}}^{\bullet}(\mathbf{R}\varGamma_{\mathcal{X}}'\mathcal{O}_{\mathcal{X}}, f_{\mathrm{t}}^{\times}\mathcal{G}))\\ &= \mathbf{R}\mathcal{H}om_{\mathcal{X}}^{\bullet}(\mathbf{L}f^{*}\mathcal{F}, f^{\#}\mathcal{G}).\end{aligned}$$

For fixed $\mathcal{F}$ the source and target of the first isomorphism in Proposition 2.5.13 are functors from $\mathbf{D}^{+}(\mathcal{Y})$ to $\mathbf{D}_{\mathrm{qct}}(\mathcal{X})$ (see Proposition 3.2.4), right adjoint, respectively, to the functors $\mathbf{R}f_{*}(\mathcal{E} \underline{\underline{\otimes}}\, \mathbf{L}f^{*}\mathcal{F})$ and $\mathbf{R}f_{*}\mathcal{E} \underline{\underline{\otimes}}\, \mathcal{F}$ ($\mathcal{E} \in \mathbf{D}_{\mathrm{qct}}(\mathcal{X})$). The functorial "projection" map

$$\mathbf{R}f_{*}\mathcal{E} \underline{\underline{\otimes}}\, \mathcal{F} \to \mathbf{R}f_{*}(\mathcal{E} \underline{\underline{\otimes}}\, \mathbf{L}f^{*}\mathcal{F}),$$

is, by definition, adjoint to the natural composition

$$\mathbf{L}f^{*}(\mathbf{R}f_{*}\mathcal{E} \underline{\underline{\otimes}}\, \mathcal{F}) \to \mathbf{L}f^{*}\mathbf{R}f_{*}\mathcal{E} \underline{\underline{\otimes}}\, \mathbf{L}f^{*}\mathcal{F} \to \mathcal{E} \underline{\underline{\otimes}}\, \mathbf{L}f^{*}\mathcal{F};$$

and it will suffice to show that this projection map is an isomorphism.

For this, the standard strategy is to localize to where $\mathcal{Y}$ is affine, then use boundedness of some functors, and compatibilities with direct sums, to reduce to the trivial case $\mathcal{F} = \mathcal{O}_{\mathcal{Y}}$. Details appear, e.g., in [**L4**, pp. 123–125, Proposition 3.9.4], modulo the following substitutions: use $\mathbf{D}_{\vec{\mathrm{c}}}$ in place of $\mathbf{D}_{\mathrm{qc}}$, and for boundedness and direct sums use Lemma 5.1.4 and Propositions 3.4.3(b) and 3.5.2 below. □

3. Direct limits of coherent sheaves on formal schemes.

In this section we establish, for a locally noetherian formal scheme $\mathcal{X}$, properties of $\mathcal{A}_{\vec{\mathrm{c}}}(\mathcal{X})$ needed in §4 to adapt Deligne's proof of global Grothendieck Duality to the formal context. The basic result, Proposition 3.2.2, is that $\mathcal{A}_{\vec{\mathrm{c}}}(\mathcal{X})$ is *plump* (see opening remarks in §1), hence abelian, and so (being closed under $\varinjlim$) cocomplete, i.e., it has arbitrary small colimits. This enables us to speak about $\mathbf{D}(\mathcal{A}_{\vec{\mathrm{c}}}(\mathcal{X}))$, and to apply standard adjoint functor theorems to colimit-preserving functors on $\mathcal{A}_{\vec{\mathrm{c}}}(\mathcal{X})$. (See e.g., Proposition 3.2.3, Grothendieck Duality for the identity map of $\mathcal{X}$).

The preliminary paragraph 3.1 sets up an equivalence of categories which allows us to reduce local questions about the (globally defined) category $\mathcal{A}_{\vec{c}}(\mathfrak{X})$ to corresponding questions about quasi-coherent sheaves on ordinary noetherian schemes. Paragraph 3.3 extends this equivalence to derived categories. As one immediate application, Corollary 3.3.4 asserts that the natural functor $\mathbf{D}(\mathcal{A}_{\vec{c}}(\mathfrak{X})) \to \mathbf{D}_{\vec{c}}(\mathfrak{X})$ is an equivalence of categories when $\mathfrak{X}$ is *properly algebraic*, i.e., the J-adic completion of a proper B-scheme with B a noetherian ring and J a B-ideal. This will yield a stronger version of Grothendieck Duality on such formal schemes—for $\mathbf{D}_{\vec{c}}(\mathfrak{X})$ rather than $\mathbf{D}(\mathcal{A}_{\vec{c}}(\mathfrak{X}))$, see Corollary 4.1.1. We do not know whether such global results hold over arbitrary noetherian formal schemes.

Paragraph 3.4 establishes boundedness for some derived functors, a condition which allows us to apply them freely to unbounded complexes, as illustrated, e.g., in Paragraph 3.5.

3.1. For X a noetherian ordinary scheme, $\mathcal{A}_{\vec{c}}(X) = \mathcal{A}_{\mathrm{qc}}(X)$ [**GD**, p. 319, 6.9.9]. The inclusion $j_X\colon \mathcal{A}_{\mathrm{qc}}(X) \to \mathcal{A}(X)$ has a right adjoint $Q_X\colon \mathcal{A}(X) \to \mathcal{A}_{\mathrm{qc}}(X)$, the "quasi-coherator," necessarily left exact [**I**, p. 187, Lemme 3.2]. (See Proposition 3.2.3 and Corollary 5.1.5 for generalizations to formal schemes.)

PROPOSITION 3.1.1. *Let A be a noetherian adic ring with ideal of definition I, let $f_0\colon X \to \mathrm{Spec}(A)$ be a proper map, set $Z := f_0^{-1}\mathrm{Spec}(A/I)$, and let*

$$\kappa\colon \mathfrak{X} = X_{/Z} \to X$$

be the formal completion of X along Z. Let $Q := Q_X$ be as above. Then κ^* induces equivalences of categories from $\mathcal{A}_{\mathrm{qc}}(X)$ to $\mathcal{A}_{\vec{c}}(\mathfrak{X})$ and from $\mathcal{A}_{\mathrm{c}}(X)$ to $\mathcal{A}_{\mathrm{c}}(\mathfrak{X})$, both with quasi-inverse $Q\kappa_*$.

PROOF. For any quasi-coherent $\mathcal{O}_X$-module $\mathcal{G}$ the canonical maps are *isomorphisms*

$$\mathrm{H}^i(X,\mathcal{G}) \xrightarrow{\sim} \mathrm{H}^i(X,\kappa_*\kappa^*\mathcal{G}) = \mathrm{H}^i(\mathfrak{X},\kappa^*\mathcal{G}) \qquad (i \ge 0). \tag{3.1.2}$$

(The equality holds because κ_* transforms any flasque resolution of $\kappa^*\mathcal{G}$ into one of $\kappa_*\kappa^*\mathcal{G}$.)

For, if $(\mathcal{G}_\lambda)$ is the family of coherent submodules of $\mathcal{G}$, ordered by inclusion, then X and $\mathfrak{X}$ being noetherian, one checks that (3.1.2) is the composition of the sequence of natural isomorphisms

$$\begin{aligned}
\mathrm{H}^i(X,\mathcal{G}) &\xrightarrow{\sim} \mathrm{H}^i(X,\varinjlim_\lambda \mathcal{G}_\lambda) && [\textbf{GD}, \text{p. 319, (6.9.9)}]\\
&\xrightarrow{\sim} \varinjlim_\lambda \mathrm{H}^i(X,\mathcal{G}_\lambda)\\
&\xrightarrow{\sim} \varinjlim_\lambda \mathrm{H}^i(\mathfrak{X},\kappa^*\mathcal{G}_\lambda) && [\textbf{EGA}, \text{p. 125, (4.1.7)}]\\
&\xrightarrow{\sim} \mathrm{H}^i(\mathfrak{X},\varinjlim_\lambda \kappa^*\mathcal{G}_\lambda)\\
&\xrightarrow{\sim} \mathrm{H}^i(\mathfrak{X},\kappa^*\varinjlim_\lambda \mathcal{G}_\lambda) \xrightarrow{\sim} \mathrm{H}^i(\mathfrak{X},\kappa^*\mathcal{G}).
\end{aligned}$$

Next, for any $\mathcal{G}$ and $\mathcal{H}$ in $\mathcal{A}_{\mathrm{qc}}(X)$ the natural map is an *isomorphism*

$$\mathrm{Hom}_X(\mathcal{G},\mathcal{H}) \xrightarrow{\sim} \mathrm{Hom}_{\mathfrak{X}}(\kappa^*\mathcal{G},\kappa^*\mathcal{H}) \tag{3.1.3}$$

For, with $\mathcal{G}_\lambda$ as above, (3.1.3) factors as the sequence of natural isomorphisms

$$\begin{aligned}\mathrm{Hom}_X(\mathcal{G},\mathcal{H}) &\xrightarrow{\sim} \varprojlim_\lambda \mathrm{Hom}_X(\mathcal{G}_\lambda,\mathcal{H})\\ &\xrightarrow{\sim} \varprojlim_\lambda \mathrm{H}^0(X,\mathcal{H}om_X(\mathcal{G}_\lambda,\mathcal{H}))\\ &\xrightarrow{\sim} \varprojlim_\lambda \mathrm{H}^0(\mathfrak{X},\kappa^*\mathcal{H}om_X(\mathcal{G}_\lambda,\mathcal{H})) \qquad \text{(see (3.1.2))}\\ &\xrightarrow{\sim} \varprojlim_\lambda \mathrm{H}^0(\mathfrak{X},\mathcal{H}om_\mathfrak{X}(\kappa^*\mathcal{G}_\lambda,\kappa^*\mathcal{H}))\\ &\xrightarrow{\sim} \varprojlim_\lambda \mathrm{Hom}_\mathfrak{X}(\kappa^*\mathcal{G}_\lambda,\kappa^*\mathcal{H})\\ &\xrightarrow{\sim} \mathrm{Hom}_\mathfrak{X}(\varinjlim_\lambda \kappa^*\mathcal{G}_\lambda,\kappa^*\mathcal{H}) \xrightarrow{\sim} \mathrm{Hom}_\mathfrak{X}(\kappa^*\mathcal{G},\kappa^*\mathcal{H}).\end{aligned}$$

Finally, we show the equivalence of the following conditions, for $\mathcal{F}\in\mathcal{A}(\mathfrak{X})$:

(1) The functorial map $\alpha(\mathcal{F})\colon \kappa^*Q\kappa_*\mathcal{F}\to\mathcal{F}$ (adjoint to the canonical map $Q\kappa_*\mathcal{F}\to\kappa_*\mathcal{F}$) is an isomorphism.
(2) There exists an isomorphism $\kappa^*\mathcal{G}\xrightarrow{\sim}\mathcal{F}$ with $\mathcal{G}\in\mathcal{A}_{\mathrm{qc}}(X)$.
(3) $\mathcal{F}\in\mathcal{A}_{\vec{\mathrm{c}}}(\mathfrak{X})$.

Clearly (1) $\Rightarrow$ (2); and (2) $\Rightarrow$ (3) because $\varinjlim_\lambda \kappa^*\mathcal{G}_\lambda \xrightarrow{\sim} \kappa^*\mathcal{G}$ ($\mathcal{G}_\lambda$ as before).

Since κ^* commutes with $\varinjlim$ and induces an equivalence of categories from $\mathcal{A}_\mathrm{c}(X)$ to $\mathcal{A}_\mathrm{c}(\mathfrak{X})$ [**EGA**, p. 150, (5.1.6)], we see that (3) $\Rightarrow$ (2).

For $\mathcal{G}\in\mathcal{A}_{\mathrm{qc}}(X)$, let $\beta(\mathcal{G})\colon\mathcal{G}\to Q\kappa_*\kappa^*\mathcal{G}$ be the canonical map (the unique one whose composition with $Q\kappa_*\kappa^*\mathcal{G}\to\kappa_*\kappa^*\mathcal{G}$ is the canonical map $\mathcal{G}\to\kappa_*\kappa^*\mathcal{G}$). Then for any $\mathcal{H}\in\mathcal{A}_{\mathrm{qc}}(X)$ we have the natural commutative diagram

$$\begin{array}{ccc}\mathrm{Hom}(\mathcal{H},\mathcal{G}) & \xrightarrow{\text{via }\beta} & \mathrm{Hom}(\mathcal{H},Q\kappa_*\kappa^*\mathcal{G})\\ \simeq\big\downarrow & & \big\downarrow\simeq\\ \mathrm{Hom}(\kappa^*\mathcal{H},\kappa^*\mathcal{G}) & \xrightarrow{\ \sim\ } & \mathrm{Hom}(\mathcal{H},\kappa_*\kappa^*\mathcal{G})\end{array}$$

where the left vertical arrow is an isomorphism by (3.1.3), the right one is an isomorphism because Q is right-adjoint to $\mathcal{A}_{\mathrm{qc}}(X)\hookrightarrow\mathcal{A}(X)$, and the bottom arrow is an isomorphism because κ_* is right-adjoint to κ^*; so "via β" is an isomorphism for all $\mathcal{H}$, whence $\beta(\mathcal{G})$ *is an isomorphism.* The implication (2) $\Rightarrow$ (1) follows now from the easily checked fact that $\alpha(\kappa^*\mathcal{G})\circ\kappa^*\beta(\mathcal{G})$ is the identity map of $\kappa^*\mathcal{G}$.

We see also that $Q\kappa_*(\mathcal{A}_\mathrm{c}(\mathfrak{X}))\subset\mathcal{A}_\mathrm{c}(X)$, since by [**EGA**, p. 150, (5.1.6)] every $\mathcal{F}\in\mathcal{A}_\mathrm{c}(\mathfrak{X})$ is isomorphic to $\kappa^*\mathcal{G}$ for some $\mathcal{G}\in\mathcal{A}_\mathrm{c}(X)$, and $\beta(\mathcal{G})$ is an isomorphism.

Thus we have the functors $\kappa^*\colon\mathcal{A}_{\mathrm{qc}}(X)\to\mathcal{A}_{\vec{\mathrm{c}}}(\mathfrak{X})$ and $Q\kappa_*\colon\mathcal{A}_{\vec{\mathrm{c}}}(\mathfrak{X})\to\mathcal{A}_{\mathrm{qc}}(X)$, both of which preserve coherence, and the functorial isomorphisms

$$\alpha(\mathcal{F})\colon\kappa^*Q\kappa_*\mathcal{F}\xrightarrow{\sim}\mathcal{F}\ \ (\mathcal{F}\in\mathcal{A}_{\vec{\mathrm{c}}}(\mathfrak{X}));\qquad \beta(\mathcal{G})\colon\mathcal{G}\xrightarrow{\sim}Q\kappa_*\kappa^*\mathcal{G}\ \ (\mathcal{G}\in\mathcal{A}_{\mathrm{qc}}(X)).$$

Proposition 3.1.1 results. □

Since κ^* is right-exact, we deduce:

COROLLARY 3.1.4. *For any affine noetherian formal scheme $\mathfrak{X}$, $\mathcal{F}\in\mathcal{A}_{\vec{\mathrm{c}}}(\mathfrak{X})$ iff $\mathcal{F}$ is a cokernel of a map of free $\mathcal{O}_\mathfrak{X}$-modules (i.e., direct sums of copies of $\mathcal{O}_\mathfrak{X}$).*

COROLLARY 3.1.5. *For a locally noetherian formal scheme $\mathfrak{X}$, $\mathcal{A}_{\vec{\mathrm{c}}}(\mathfrak{X})\subset\mathcal{A}_{\mathrm{qc}}(\mathfrak{X})$, i.e., any $\varinjlim$ of coherent $\mathcal{O}_\mathfrak{X}$-modules is quasi-coherent.*

PROOF. Being local, the assertion follows from Corollary 3.1.4. □

COROLLARY 3.1.6 (cf. [**Y**, 3.4, 3.5]). *For a locally noetherian formal scheme* $\mathfrak{X}$ *let* $\mathcal{F}$ *and* $\mathcal{G}$ *be quasi-coherent* $\mathcal{O}_{\mathfrak{X}}$*-modules. Then:*

(a) *The kernel, cokernel, and image of any* $\mathcal{O}_{\mathfrak{X}}$*-homomorphism* $\mathcal{F} \to \mathcal{G}$ *are quasi-coherent.*

(b) $\mathcal{F}$ *is coherent iff* $\mathcal{F}$ *is locally finitely generated.*

(c) *If* $\mathcal{F}$ *is coherent and* $\mathcal{G}$ *is a sub- or quotient module of* $\mathcal{F}$ *then* $\mathcal{G}$ *is coherent.*

(d) *If* $\mathcal{F}$ *is coherent then* $\mathcal{H}om(\mathcal{F}, \mathcal{G})$ *is quasi-coherent; and if also* $\mathcal{G}$ *is coherent then* $\mathcal{H}om(\mathcal{F}, \mathcal{G})$ *is coherent. (For a generalization, see Proposition* 3.2.4.)

PROOF. The questions being local, we may assume $\mathfrak{X} = \mathrm{Spf}(A)$ (A noetherian adic), and, by Corollary 3.1.4, that $\mathcal{F}$ and $\mathcal{G}$ are in $\mathcal{A}_{\vec{c}}(\mathfrak{X})$. Then, κ^* being exact, Proposition 3.1.1 with $X := \mathrm{Spec}(A)$ and $f_0 :=$ identity reduces the problem to noting that the corresponding statements about coherent and quasi-coherent sheaves on X are true. (These statements are in [**GD**, p. 217, Cor. (2.2.2) and p. 228, §(2.7.1)]. Observe also that if F and G are $\mathcal{O}_X$-modules with F coherent then $\mathcal{H}om_{\mathfrak{X}}(\kappa^*F, \kappa^*G) \cong \kappa^*\mathcal{H}om_X(F, G)$.) $\square$

COROLLARY 3.1.7. *For a locally noetherian formal scheme* $\mathfrak{X}$, *any* $\mathcal{F} \in \mathcal{A}_{\vec{c}}(\mathfrak{X})$ *is the* $\varinjlim$ *of its coherent* $\mathcal{O}_{\mathfrak{X}}$*-submodules.*

PROOF. Note that by Corollary 3.1.6(a) and (b) the sum of any two coherent submodules of $\mathcal{F}$ is again coherent. By definition, $\mathcal{F} = \varinjlim_{\mu} \mathcal{F}_{\mu}$ with $\mathcal{F}_{\mu}$ coherent, and from Corollary 3.1.6(a) and (b) it follows that the canonical image of $\mathcal{F}_{\mu}$ is a coherent submodule of $\mathcal{F}$, whence the conclusion. $\square$

COROLLARY 3.1.8. *For any affine noetherian formal scheme* $\mathfrak{X}$, *any* $\mathcal{F} \in \mathcal{A}_{\vec{c}}(\mathfrak{X})$ *and any* $i > 0$,

$$\mathrm{H}^i(\mathfrak{X}, \mathcal{F}) = 0.$$

PROOF. Taking f_0 in Proposition 3.1.1 to be the identity map, we have $\mathcal{F} \cong \kappa^*\mathcal{G}$ with $\mathcal{G}$ quasi-coherent; and so by (3.1.2), $\mathrm{H}^i(\mathfrak{X}, \mathcal{F}) \cong \mathrm{H}^i(\mathrm{Spec}(A), \mathcal{G}) = 0$. $\square$

3.2. Proposition 3.1.1 will now be used to show, for locally noetherian formal schemes $\mathfrak{X}$, that $\mathcal{A}_{\vec{c}}(\mathfrak{X}) \subset \mathcal{A}(\mathfrak{X})$ is plump, and that this inclusion has a right adjoint, extending to derived categories.

LEMMA 3.2.1. *Let* $\mathfrak{X}$ *be a noetherian formal scheme, let* $\mathcal{F} \in \mathcal{A}_{\mathrm{c}}(\mathfrak{X})$, *and let* $(\mathcal{G}_{\alpha}, \gamma_{\alpha\beta} \colon \mathcal{G}_{\beta} \to \mathcal{G}_{\alpha})_{\alpha,\beta \in \Omega}$ *be a directed system in* $\mathcal{A}_{\mathrm{c}}(\mathfrak{X})$. *Then for every* $q \geq 0$ *the natural map is an isomorphism*

$$\varinjlim_{\alpha} \mathrm{Ext}^q(\mathcal{F}, \mathcal{G}_{\alpha}) \xrightarrow{\sim} \mathrm{Ext}^q(\mathcal{F}, \varinjlim_{\alpha} \mathcal{G}_{\alpha}).$$

PROOF. For an $\mathcal{O}_{\mathfrak{X}}$-module $\mathcal{M}$, let $\mathrm{E}(\mathcal{M})$ denote the usual spectral sequence

$$\mathrm{E}_2^{pq}(\mathcal{M}) := \mathrm{H}^p(\mathfrak{X}, \mathcal{E}xt^q(\mathcal{F}, \mathcal{M})) \Rightarrow \mathrm{Ext}^{p+q}(\mathcal{F}, \mathcal{M}).$$

It suffices that the natural map of spectral sequences be an isomorphism

$$\varinjlim \mathrm{E}(\mathcal{G}_{\alpha}) \xrightarrow{\sim} \mathrm{E}(\varinjlim \mathcal{G}_{\alpha}) \qquad (\varinjlim := \varinjlim_{\alpha}),$$

and for that we need only check out the E_2^{pq} terms, i.e., show that the natural maps

$$\varinjlim \mathrm{H}^p(\mathfrak{X}, \mathcal{E}xt^q(\mathcal{F}, \mathcal{G}_{\alpha})) \to \mathrm{H}^p(\mathfrak{X}, \varinjlim \mathcal{E}xt^q(\mathcal{F}, \mathcal{G}_{\alpha})) \to \mathrm{H}^p(\mathfrak{X}, \mathcal{E}xt^q(\mathcal{F}, \varinjlim \mathcal{G}_{\alpha}))$$

are isomorphisms. The first one is, because $\mathfrak{X}$ is noetherian. So we need only show that the natural map is an isomorphism

$$\varinjlim \mathcal{E}xt^q(\mathcal{F}, \mathcal{G}_{\alpha}) \xrightarrow{\sim} \mathcal{E}xt^q(\mathcal{F}, \varinjlim \mathcal{G}_{\alpha}).$$

For this localized question we may assume that $\mathcal{X} = \mathrm{Spf}(A)$ with A a noetherian adic ring. By Proposition 3.1.1 (with f_0 the identity map of $X := \mathrm{Spec}(A)$) there is a coherent $\mathcal{O}_X$-module F and a directed system $(G_\alpha, g_{\alpha\beta} \colon G_\beta \to G_\alpha)_{\alpha,\beta\in\Omega}$ of coherent $\mathcal{O}_X$-modules such that $\mathcal{F} = \kappa^*F$, $\mathcal{G}_\alpha = \kappa^*G_\alpha$, and $\gamma_{\alpha,\beta} = \kappa^*g_{\alpha,\beta}$. Then the well-known natural isomorphisms (see [**EGA**, (Chapter 0), p. 61, Prop. (12.3.5)]—or the proof of Corollary 3.3.2 below)

$$\begin{aligned}\varinjlim \mathcal{E}xt^q_{\mathcal{X}}(\mathcal{F},\mathcal{G}_\alpha) &\xrightarrow{\sim} \varinjlim \kappa^*\mathcal{E}xt^q_X(F,G_\alpha) \xrightarrow{\sim} \kappa^*\varinjlim \mathcal{E}xt^q_X(F,G_\alpha)\\ &\xrightarrow{\sim} \kappa^*\mathcal{E}xt^q_X(F,\varinjlim G_\alpha) \xrightarrow{\sim} \mathcal{E}xt^q_{\mathcal{X}}(\kappa^*F,\kappa^*\varinjlim G_\alpha) \xrightarrow{\sim} \mathcal{E}xt^q_{\mathcal{X}}(\mathcal{F},\varinjlim \mathcal{G}_\alpha)\end{aligned}$$

give the desired conclusion. □

Proposition 3.2.2. *Let $\mathcal{X}$ be a locally noetherian formal scheme. If*

$$\mathcal{F}_1 \to \mathcal{F}_2 \to \mathcal{F} \to \mathcal{F}_3 \to \mathcal{F}_4$$

is an exact sequence of $\mathcal{O}_{\mathcal{X}}$-modules and if $\mathcal{F}_1$, $\mathcal{F}_2$, $\mathcal{F}_3$ and $\mathcal{F}_4$ are all in $\mathcal{A}_{\mathrm{qc}}(\mathcal{X})$ (resp. $\mathcal{A}_{\vec{\mathrm{c}}}(\mathcal{X})$) then $\mathcal{F} \in \mathcal{A}_{\mathrm{qc}}(\mathcal{X})$ (resp. $\mathcal{A}_{\vec{\mathrm{c}}}(\mathcal{X})$). Thus $\mathcal{A}_{\mathrm{qc}}(\mathcal{X})$ and $\mathcal{A}_{\vec{\mathrm{c}}}(\mathcal{X})$ are plump—hence abelian—subcategories of $\mathcal{A}(\mathcal{X})$, and both $\mathbf{D}_{\mathrm{qc}}(\mathcal{X})$ and its subcategory $\mathbf{D}_{\vec{\mathrm{c}}}(\mathcal{X})$ are triangulated subcategories of $\mathbf{D}(\mathcal{X})$. Furthermore, $\mathcal{A}_{\vec{\mathrm{c}}}(\mathcal{X})$ is closed under arbitrary small $\mathcal{A}(\mathcal{X})$-colimits.

Proof. Part of the $\mathcal{A}_{\mathrm{qc}}$ case is covered by Corollary 3.1.6(a), and all of it by [**Y**, Proposition 3.5]. At any rate, since every quasi-coherent $\mathcal{O}_{\mathcal{X}}$-module is locally in $\mathcal{A}_{\vec{\mathrm{c}}} \subset \mathcal{A}_{\mathrm{qc}}$ (see Corollaries 3.1.4 and 3.1.5), it suffices to treat the $\mathcal{A}_{\vec{\mathrm{c}}}$ case.

Let us first show that the kernel $\mathcal{K}$ of an $\mathcal{A}_{\vec{\mathrm{c}}}$ map

$$\psi\colon \varinjlim{}_\beta \mathcal{H}_\beta = \mathcal{H} \to \mathcal{G} = \varinjlim{}_\alpha \mathcal{G}_\alpha \qquad (\mathcal{G}_\alpha, \mathcal{H}_\beta \in \mathcal{A}_{\mathrm{c}}(\mathcal{X}))$$

is itself in $\mathcal{A}_{\vec{\mathrm{c}}}(\mathcal{X})$. It will suffice to do so for the kernel $\mathcal{K}_\beta$ of the composition

$$\psi_\beta\colon \mathcal{H}_\beta \xrightarrow{\text{natural}} \mathcal{H} \xrightarrow{\psi} \mathcal{G},$$

since $\mathcal{K} = \varinjlim{}_\beta \mathcal{K}_\beta$.

By the case $q = 0$ of Corollary 3.2.1, there is an α such that ψ_β factors as

$$\mathcal{H}_\beta \xrightarrow{\psi_{\beta\alpha}} \mathcal{G}_\alpha \xrightarrow{\text{natural}} \mathcal{G};$$

and then with $\mathcal{K}_{\beta\alpha'}$ ($\alpha' > \alpha$) the (coherent) kernel of the composed map

$$\mathcal{H}_\beta \xrightarrow{\psi_{\beta\alpha}} \mathcal{G}_\alpha \xrightarrow{\text{natural}} \mathcal{G}_{\alpha'}$$

we have $\mathcal{K}_\beta = \varinjlim{}_{\alpha'} \mathcal{K}_{\beta\alpha'} \in \mathcal{A}_{\vec{\mathrm{c}}}(\mathcal{X})$.

Similarly, we find that $\mathrm{coker}(\psi) \in \mathcal{A}_{\vec{\mathrm{c}}}(\mathcal{X})$. Being closed under small direct sums, then, $\mathcal{A}_{\vec{\mathrm{c}}}(\mathcal{X})$ is closed under arbitrary small $\mathcal{A}(\mathcal{X})$-colimits [**M1**, Corollary 2, p. 109].

Consideration of the exact sequence

$$0 \longrightarrow \mathrm{coker}(\mathcal{F}_1 \to \mathcal{F}_2) \longrightarrow \mathcal{F} \longrightarrow \ker(\mathcal{F}_3 \to \mathcal{F}_4) \longrightarrow 0$$

now reduces the original question to where $\mathcal{F}_1 = \mathcal{F}_4 = 0$. Since $\mathcal{F}_3$ is the $\varinjlim$ of its coherent submodules (Corollary 3.1.7) and $\mathcal{F}$ is the $\varinjlim$ of the inverse images of those submodules, we need only show that each such inverse image is in $\mathcal{A}_{\vec{\mathrm{c}}}(\mathcal{X})$. Thus we may assume $\mathcal{F}_3$ coherent (and $\mathcal{F}_2 = \varinjlim{}_\alpha \mathcal{G}_\alpha$ with $\mathcal{G}_\alpha$ coherent).

The exact sequence $0 \to \mathcal{F}_2 \to \mathcal{F} \to \mathcal{F}_3 \to 0$ represents an element

$$\eta \in \operatorname{Ext}^1(\mathcal{F}_3, \mathcal{F}_2) = \operatorname{Ext}^1(\mathcal{F}_3, \varinjlim_\alpha \mathcal{G}_\alpha);$$

and by Corollary 3.2.1, there is an α such that η is the natural image of an element $\eta_\alpha \in \operatorname{Ext}^1(\mathcal{F}_3, \mathcal{G}_\alpha)$, represented by an exact sequence $0 \to \mathcal{G}_\alpha \to \mathcal{F}_\alpha \to \mathcal{F}_3 \to 0$. Then $\mathcal{F}_\alpha$ is coherent, and by [**M2**, p. 66, Lemma 1.4], we have an isomorphism

$$\mathcal{F} \xrightarrow{\sim} \mathcal{F}_2 \oplus_{\mathcal{G}_\alpha} \mathcal{F}_\alpha .$$

Thus $\mathcal{F}$ is the cokernel of a map in $\mathcal{A}_{\vec{c}}(\mathfrak{X})$, and so as above, $\mathcal{F} \in \mathcal{A}_{\vec{c}}(\mathfrak{X})$. □

PROPOSITION 3.2.3. *On a locally noetherian formal scheme $\mathfrak{X}$, the inclusion functor $j_{\mathfrak{X}} \colon \mathcal{A}_{\vec{c}}(\mathfrak{X}) \to \mathcal{A}(\mathfrak{X})$ has a right adjoint $Q_{\mathfrak{X}} \colon \mathcal{A}(\mathfrak{X}) \to \mathcal{A}_{\vec{c}}(\mathfrak{X})$; and $\mathbf{R}Q_{\mathfrak{X}}$ is right-adjoint to the natural functor $\mathbf{D}(\mathcal{A}_{\vec{c}}(\mathfrak{X})) \to \mathbf{D}(\mathfrak{X})$. In particular, if $\kappa \colon \mathfrak{X} \to X$ is as in Proposition* 3.1.1 *then $Q_{\mathfrak{X}} \cong \kappa^* Q_X \kappa_*$ and $\mathbf{R}Q_{\mathfrak{X}} \cong \kappa^* \mathbf{R}Q_X \kappa_*$.*

PROOF. Since $\mathcal{A}_{\vec{c}}(\mathfrak{X})$ has a small family of (coherent) generators, and is closed under arbitrary small $\mathcal{A}(\mathfrak{X})$-colimits, the existence of $Q_{\mathfrak{X}}$ follows from the Special Adjoint Functor Theorem ([**F**, p. 90] or [**M1**, p. 126, Corollary]).[15]

In an abelian category $\mathcal{A}$, a complex J is, by definition, K-injective if for each exact $\mathcal{A}$-complex G, the complex $\operatorname{Hom}^\bullet_{\mathcal{A}}(G, J)$ is exact too. Since $j_{\mathfrak{X}}$ is exact, it follows that its right adjoint $Q_{\mathfrak{X}}$ transforms K-injective $\mathcal{A}(\mathfrak{X})$-complexes into K-injective $\mathcal{A}_{\vec{c}}(\mathfrak{X})$-complexes, whence the derived functor $\mathbf{R}Q_{\mathfrak{X}}$ is right-adjoint to the natural functor $\mathbf{D}(\mathcal{A}_{\vec{c}}(\mathfrak{X})) \to \mathbf{D}(\mathfrak{X})$ (see [**Sp**, p. 129, Proposition 1.5(b)]).

The next assertion is a corollary of Proposition 3.1.1: any $\mathcal{M} \in \mathcal{A}_{\vec{c}}(\mathfrak{X})$ is isomorphic to $\kappa^*\mathcal{G}$ for some $\mathcal{G} \in \mathcal{A}_{\mathrm{qc}}(X)$, and then for any $\mathcal{N} \in \mathcal{A}(\mathfrak{X})$ there are natural isomorphisms

$$\begin{aligned}
\operatorname{Hom}_{\mathfrak{X}}(j_{\mathfrak{X}}\mathcal{M}, \mathcal{N}) &\cong \operatorname{Hom}_{\mathfrak{X}}(j_{\mathfrak{X}}\kappa^*\mathcal{G}, \mathcal{N}) \\
&\cong \operatorname{Hom}_X(j_X\mathcal{G}, \kappa_*\mathcal{N}) \cong \operatorname{Hom}_{\mathcal{A}_{\mathrm{qc}}(X)}(\mathcal{G}, Q_X\kappa_*\mathcal{N}) \\
&\cong \operatorname{Hom}_{\mathcal{A}_{\vec{c}}(\mathfrak{X})}(\kappa^*\mathcal{G}, \kappa^*Q_X\kappa_*\mathcal{N}) \cong \operatorname{Hom}_{\mathcal{A}_{\vec{c}}(\mathfrak{X})}(\mathcal{M}, \kappa^*Q_X\kappa_*\mathcal{N}).
\end{aligned}$$

Moreover, since κ_* has an exact left adjoint (viz. κ^*), therefore, as above, κ_* transforms K-injective $\mathcal{A}(\mathfrak{X})$-complexes into K-injective $\mathcal{A}(X)$-complexes, and it follows at once that $\mathbf{R}Q_{\mathfrak{X}} \cong \kappa^*\mathbf{R}Q_X\kappa_*$. □

Let $\mathfrak{X}$ be a locally noetherian formal scheme. A property $\mathbf{P}$ of sheaves of modules is *local* if it is defined on $\mathcal{A}(\mathfrak{U})$ for arbitrary open subsets $\mathfrak{U}$ of $\mathfrak{X}$, and is such that for any $\mathcal{E} \in \mathcal{A}(\mathfrak{U})$ and any open covering $(\mathfrak{U}_\alpha)$ of $\mathfrak{U}$, $\mathbf{P}(\mathcal{E})$ holds iff $\mathbf{P}(\mathcal{E}|_{\mathfrak{U}_\alpha})$ holds for all α.

For example, coherence and quasi-coherence are both local properties—to which by Proposition 3.2.2, the following Proposition applies.

PROPOSITION 3.2.4. *Let $\mathfrak{X}$ be a locally noetherian formal scheme, and let $\mathbf{P}$ be a local property of sheaves of modules. Suppose further that for all open $\mathfrak{U} \subset \mathfrak{X}$ the full subcategory $\mathcal{A}_{\mathbf{P}}(\mathfrak{U})$ of $\mathcal{A}(\mathfrak{U})$ whose objects are all the $\mathcal{E} \in \mathcal{A}(\mathfrak{U})$ for which $\mathbf{P}(\mathcal{E})$ holds is a* plump *subcategory of $\mathcal{A}(\mathfrak{U})$. Then for all $\mathcal{F} \in \mathbf{D}_{\mathrm{c}}^-(\mathfrak{X})$ and $\mathcal{G} \in \mathbf{D}_{\mathbf{P}}^+(\mathfrak{X})$, it holds that $\mathbf{R}\mathcal{H}om^\bullet(\mathcal{F}, \mathcal{G}) \in \mathbf{D}_{\mathbf{P}}^+(\mathfrak{X})$.*

---

[15] It follows that $\mathcal{A}_{\vec{c}}(\mathfrak{X})$ is closed under *all* $\mathcal{A}(\mathfrak{X})$-colimits (not necessarily small): if F is any functor into $\mathcal{A}_{\vec{c}}(X)$ and $\mathcal{F} \in \mathcal{A}(X)$ is a colimit of $j_{\mathfrak{X}} \circ F$, then $Q_{\mathfrak{X}}\mathcal{F}$ is a colimit of F, and the natural map is an isomorphism $\mathcal{F} \xrightarrow{\sim} j_{\mathfrak{X}}Q_{\mathfrak{X}}\mathcal{F}$. (Proof: exercise, given in dual form in [**F**, p. 80].)

PROOF. Plumpness implies that $\mathbf{D_P}(\mathcal{X})$ is a triangulated subcategory of $\mathbf{D}(\mathcal{X})$, as is $\mathbf{D}_c(\mathcal{X})$, so [**H1**, p. 68, Prop. 7.1] gives a "way-out" reduction to where $\mathcal{F}$ and $\mathcal{G}$ are $\mathcal{O}_\mathcal{X}$-modules. The question being local on $\mathcal{X}$, we may assume $\mathcal{X}$ affine and replace $\mathcal{F}$ by a quasi-isomorphic bounded-above complex $\mathcal{F}^\bullet$ of finite-rank free $\mathcal{O}_\mathcal{X}$-modules, see [**GD**, p. 427, (10.10.2)]. Then $\mathbf{R}\mathcal{H}om^\bullet(\mathcal{F}^\bullet, \mathcal{G}) = \mathcal{H}om^\bullet(\mathcal{F}^\bullet, \mathcal{G})$, and the conclusion follows easily. □

3.3. Proposition 3.2.3 applies in particular to any noetherian scheme X. When X is separated, j_X induces an *equivalence of categories* $\mathbf{j}_X\colon \mathbf{D}(\mathcal{A}_{\mathrm{qc}}(X)) \cong \mathbf{D}_{\mathrm{qc}}(X)$, with quasi-inverse $\mathbf{R}Q_X|_{\mathbf{D}_{\mathrm{qc}}(X)}$. (See [**H1**, p. 133, Corollary 7.19] for bounded-below complexes, and [**BN**, p. 230, Corollary 5.5] or [**AJL**, p. 12, Proposition (1.3)] for the general case.) We do not know if such an equivalence, with "$\vec{\mathrm{c}}$" in place of "qc," always holds for separated noetherian formal schemes. The next result will at least take care of the "properly algebraic" case, see Corollary 3.3.4.

PROPOSITION 3.3.1. *In Proposition* 3.1.1, *the functor* $\kappa^*\colon \mathbf{D}(X) \to \mathbf{D}(\mathcal{X})$ *induces equivalences from* $\mathbf{D}_{\mathrm{qc}}(X)$ *to* $\mathbf{D}_{\vec{\mathrm{c}}}(\mathcal{X})$ *and from* $\mathbf{D}_c(X)$ *to* $\mathbf{D}_c(\mathcal{X})$, *both with quasi-inverse* $\mathbf{R}Q\kappa_*$ (*where* $\mathbf{R}Q$ *stands for* $\mathbf{j}_X \circ \mathbf{R}Q_X$).

PROOF. Since κ^* is exact, Proposition 3.1.1 implies that $\kappa^*(\mathbf{D}_{\mathrm{qc}}(X)) \subset \mathbf{D}_{\vec{\mathrm{c}}}(\mathcal{X})$ and $\kappa^*(\mathbf{D}_c(X)) \subset \mathbf{D}_c(\mathcal{X})$. So it will be enough to show that:

(1) If $\mathcal{F} \in \mathbf{D}_{\vec{\mathrm{c}}}(\mathcal{X})$ then the functorial $\mathbf{D}(\mathcal{X})$-map $\kappa^*\mathbf{R}Q\kappa_*\mathcal{F} \to \mathcal{F}$ adjoint to the natural map $\mathbf{R}Q\kappa_*\mathcal{F} \to \kappa_*\mathcal{F}$ is an isomorphism.
(2) If $\mathcal{G} \in \mathbf{D}_{\mathrm{qc}}(X)$ then the natural map $\mathcal{G} \xrightarrow{\sim} \mathbf{R}Q\kappa_*\kappa^*\mathcal{G}$ is an isomorphism.
(3) If $\mathcal{F} \in \mathbf{D}_c(\mathcal{X})$ then $\mathbf{R}Q\kappa_*\mathcal{F} \in \mathbf{D}_c(X)$.

Since $\mathbf{D}_{\vec{\mathrm{c}}}(\mathcal{X})$ is triangulated (Proposition 3.2.2), we can use way-out reasoning [**H1**, p. 68, Proposition 7.1 and p. 73, Proposition 7.3] to reduce to where $\mathcal{F}$ or $\mathcal{G}$ is a single sheaf. (For bounded-below complexes we just need the obvious facts that κ^* and the restriction of $\mathbf{R}Q\kappa_*$ to $\mathbf{D}_{\vec{\mathrm{c}}}(\mathcal{X})$ are both bounded-below (= way-out right) functors. For unbounded complexes, we need those functors to be bounded-above as well, which is clear for the exact functor κ^*, and will be shown for $\mathbf{R}Q\kappa_*|_{\mathbf{D}_{\vec{\mathrm{c}}}(\mathcal{X})}$ in Proposition 3.4.4 below.)

Any $\mathcal{F} \in \mathcal{A}_{\vec{\mathrm{c}}}(\mathcal{X})$ is isomorphic to $\kappa^*\mathcal{G}$ for some $\mathcal{G} \in \mathcal{A}_{\mathrm{qc}}(X)$; and one checks that the natural composed map $\kappa^*\mathcal{G} \to \kappa^*\mathbf{R}Q\kappa_*\kappa^*\mathcal{G} \to \kappa^*\mathcal{G}$ is the identity, whence (2) ⇒ (1). Moreover, if $\mathcal{F} \in \mathcal{A}_c(\mathcal{X})$ then $\mathcal{G} \cong Q\kappa_*\mathcal{F} \in \mathcal{A}_c(X)$, whence (2) ⇒ (3).

Now a map $\varphi\colon \mathcal{G}_1 \to \mathcal{G}_2$ in $\mathbf{D}^+_{\mathrm{qc}}(X)$ is an isomorphism iff

($*$): the induced map $\mathrm{Hom}_{\mathbf{D}(X)}(\mathcal{E}[-n], \mathcal{G}_1) \to \mathrm{Hom}_{\mathbf{D}(X)}(\mathcal{E}[-n], \mathcal{G}_2)$ is an isomorphism for every $\mathcal{E} \in \mathcal{A}_c(X)$ and every $n \in \mathbb{Z}$.

(For, if $\mathcal{V}$ is the vertex of a triangle with base φ, then ($*$) says that for all $\mathcal{E}$, n, $\mathrm{Hom}_{\mathbf{D}(X)}(\mathcal{E}[-n], \mathcal{V}) = 0$; but if φ is not an isomorphism, i.e., $\mathcal{V}$ has non-vanishing homology, say $H^n(\mathcal{V}) \neq 0$ and $H^i(\mathcal{V}) = 0$ for all $i < n$, then the inclusion into $H^n(\mathcal{V})$ of any coherent non-zero submodule $\mathcal{E}$ gives a non-zero map $\mathcal{E}[-n] \to \mathcal{V}$.) So for (2) it's enough to check that the natural composition

$$\begin{aligned}\mathrm{Hom}_{\mathbf{D}(X)}(\mathcal{E}[-n], \mathcal{G}) &\longrightarrow \mathrm{Hom}_{\mathbf{D}(X)}(\mathcal{E}[-n], \mathbf{R}Q\kappa_*\kappa^*\mathcal{G})\\ &\xrightarrow{\sim} \mathrm{Hom}_{\mathbf{D}(X)}(\mathcal{E}[-n], \kappa_*\kappa^*\mathcal{G}) \xrightarrow{\sim} \mathrm{Hom}_{\mathbf{D}(\mathcal{X})}(\kappa^*\mathcal{E}[-n], \kappa^*\mathcal{G})\end{aligned}$$

is the *isomorphism* $\mathrm{Ext}^n_X(\mathcal{E}, \mathcal{G}) \xrightarrow{\sim} \mathrm{Ext}^n_\mathcal{X}(\kappa^*\mathcal{E}, \kappa^*\mathcal{G})$ in the following consequence of (3.1.2):

COROLLARY 3.3.2. *With* $\kappa\colon \mathcal{X} \to X$ *as in Proposition* 3.1.1 *and* $\mathcal{L} \in \mathbf{D}_{\mathrm{qc}}(X)$, *the natural map* $\mathbf{R}\Gamma(X,\mathcal{L}) \to \mathbf{R}\Gamma(\mathcal{X},\kappa^*\mathcal{L})$ *is an isomorphism. In particular, for* $\mathcal{E} \in \mathbf{D}_{\mathrm{c}}^-(X)$ *and* $\mathcal{G} \in \mathbf{D}_{\mathrm{qc}}^+(X)$ *the natural map* $\mathrm{Ext}^n_X(\mathcal{E},\mathcal{G}) \to \mathrm{Ext}^n_{\mathcal{X}}(\kappa^*\mathcal{E},\kappa^*\mathcal{G})$ *is an isomorphism.*

Proof. After "way-out" reduction to the case where $\mathcal{L} \in \mathcal{A}_{\mathrm{qc}}(X)$ (the $\mathbf{R}\Gamma$'s are bounded, by Corollary 3.4.3(a) below), the first assertion is given by (3.1.2). To get the second assertion, take $\mathcal{L} := \mathbf{R}\mathcal{H}om^\bullet_X(\mathcal{E},\mathcal{G})$ (which is in $\mathbf{D}^+_{\mathrm{qc}}(X)$, [**H1**, p. 92, Proposition 3.3]), so that $\kappa^*\mathcal{L} \cong \mathbf{R}\mathcal{H}om^\bullet_{\mathcal{X}}(\kappa^*\mathcal{E},\kappa^*\mathcal{G})$ (as one sees easily after way-out reduction to where $\mathcal{E}$ and $\mathcal{G}$ are $\mathcal{O}_X$-modules, and further reduction to where X is affine, so that $\mathcal{E}$ has a resolution by finite-rank free modules ...). □

DEFINITION 3.3.3. A formal scheme $\mathcal{X}$ is said to be *properly algebraic* if there exist a noetherian ring B, a B-ideal J, a proper B-scheme X, and an isomorphism from $\mathcal{X}$ to the J-adic completion of X.

COROLLARY 3.3.4. *On a properly algebraic formal scheme* $\mathcal{X}$ *the natural functor* $j_{\mathcal{X}}\colon \mathbf{D}(\mathcal{A}_{\vec{\mathrm{c}}}(\mathcal{X})) \to \mathbf{D}_{\vec{\mathrm{c}}}(\mathcal{X})$ *is an equivalence of categories with quasi-inverse* $\mathbf{R}Q_{\mathcal{X}}$; *and therefore* $j_{\mathcal{X}}\circ\mathbf{R}Q_{\mathcal{X}}$ *is right-adjoint to the inclusion* $\mathbf{D}_{\vec{\mathrm{c}}}(\mathcal{X}) \hookrightarrow \mathbf{D}(\mathcal{X})$.

PROOF. If $\mathcal{X}$ is properly algebraic, then with $A :=$ J-adic completion of B and $I := JA$, it holds that $\mathcal{X}$ is the I-adic completion of $X \otimes_B A$, and so we may assume the hypotheses and conclusions of Proposition 3.1.1. We have also, as above, the equivalence of categories $j_X : \mathbf{D}(\mathcal{A}_{\mathrm{qc}}(X)) \to \mathbf{D}_{\mathrm{qc}}(X)$; and so the assertion follows from Propositions 3.3.1 and 3.2.3. □

PROPOSITION 3.3.5. *For a map* $g\colon \mathcal{Z} \to \mathcal{X}$ *of locally noetherian formal schemes,*

$$\mathbf{L}g^*(\mathbf{D}_{\vec{\mathrm{c}}}(\mathcal{X})) \subset \mathbf{D}_{\mathrm{qc}}(\mathcal{Z}).$$

If $\mathcal{X}$ *is properly algebraic, then*

$$\mathbf{L}g^*(\mathbf{D}_{\vec{\mathrm{c}}}(\mathcal{X})) \subset \mathbf{D}_{\vec{\mathrm{c}}}(\mathcal{Z}).$$

PROOF. The first assertion, being local on $\mathcal{X}$, follows from the second. Assuming $\mathcal{X}$ properly algebraic we may, as in the proof of Corollary 3.3.4, place ourselves in the situation of Proposition 3.1.1, so that any $\mathcal{G} \in \mathbf{D}_{\vec{\mathrm{c}}}(\mathcal{X})$ is, by Corollary 3.3.4 and Proposition 3.1.1, isomorphic to $\kappa^*\mathcal{E}$ for some $\mathcal{E} \in \mathbf{D}_{\mathrm{qc}}(X)$. By [**AJL**, p. 10, Proposition (1.1)]), $\mathcal{E}$ is isomorphic to a $\varinjlim$ of bounded-above quasi-coherent flat complexes (see the very end of the proof of *ibid.*); and therefore $\mathcal{G} \cong \kappa^*\mathcal{E}$ is isomorphic to a K-flat complex of $\mathcal{A}_{\vec{\mathrm{c}}}(\mathcal{X})$-objects. Since $\mathbf{L}g^*$ agrees with g^* on K-flat complexes, and $g^*(\mathcal{A}_{\vec{\mathrm{c}}}(\mathcal{X})) \subset \mathcal{A}_{\vec{\mathrm{c}}}(\mathcal{Z})$, we are done. □

REMARKS 3.3.6. (1) Let $\mathcal{X}$ be a properly algebraic formal scheme (necessarily noetherian) with ideal of definition $\mathcal{I}$, and set $I := \mathrm{H}^0(\mathcal{X},\mathcal{I}) \subset A := \mathrm{H}^0(\mathcal{X},\mathcal{O}_{\mathcal{X}})$. Then *$A$ is a noetherian I-adic ring, and* $\mathcal{X}$ *is* $\mathrm{Spf}(A)$*-isomorphic to the I-adic completion of a proper A-scheme.* Hence $\mathcal{X}$ is proper over $\mathrm{Spf}(A)$, via the canonical map given by [**GD**, p. 407, (10.4.6)].

Indeed, with B, J and X as in Definition 3.3.3, [**EGA**, p. 125, Theorem (4.1.7)] implies that the topological ring

$$A = \varprojlim_{n>0} \mathrm{H}^0(\mathfrak{X}, \mathcal{O}_{\mathfrak{X}}/\mathcal{J}^n\mathcal{O}_{\mathfrak{X}}) = \varprojlim_{n>0} \mathrm{H}^0(X, \mathcal{O}_X/I^n\mathcal{O}_X)$$

is the J-adic completion of the noetherian B-algebra $A_0 := H^0(X, \mathcal{O}_X)$, and that the J-adic and I-adic topologies on A are the same; and then $\mathfrak{X}$ is the I-adic completion of $X \otimes_{A_0} A$.

(2) It follows that a quasi-compact formal scheme $\mathfrak{X}$ is properly algebraic iff so is each of its connected components.

(3) While (1) provides a less relaxed characterization of properly algebraic formal schemes than Definition 3.3.3, Corollary 3.3.8 below provides a more relaxed one.

LEMMA 3.3.7. *Let X be a locally noetherian scheme, $\mathcal{I}_1 \subset \mathcal{I}_2$ quasi-coherent $\mathcal{O}_X$-ideals, Z_i the support of $\mathcal{O}_X/\mathcal{I}_i$, and $\mathfrak{X}_i$ the completion $X_{/Z_i}$ $(i = 1, 2)$. Suppose that $\mathcal{I}_1\mathcal{O}_{\mathfrak{X}_2}$ is an ideal of definition of $\mathfrak{X}_2$. Then $\mathfrak{X}_2$ is a union of connected components of $\mathfrak{X}_1$ (with the induced formal-subscheme structure).*

PROOF. We need only show that Z_2 is open in Z_1. Locally we have a noetherian ring A and A-ideals $I \subset J$ equal to their own radicals such that with $\hat{A}$ the J-adic completion, $J^n\hat{A} \subset I\hat{A}$ for some $n > 0$; and we want the natural map $A/I \twoheadrightarrow A/J$ to be *flat.* (For then with $L := J/I$, $L/L^2 = \mathrm{Tor}_1^{A/I}(A/J, A/J) = 0$, whence $(1-\ell)L = (0)$ for some $\ell \in L$, whence $\ell = \ell^2$ and $L = \ell(A/I)$, so that $A/I \cong L \times (A/J)$ and $\mathrm{Spec}(A/J) \hookrightarrow \mathrm{Spec}(A/I)$ is open.)

So it suffices that the localization $(A/I)_{1+J} \to (A/J)_{1+J} = A/J$ by the multiplicatively closed set $1 + J$ be an isomorphism, i.e., that its kernel $J(A/I)_{1+J}$ be nilpotent (hence (0), since A/I is reduced.) But this is so because the natural map $A_{1+J} \to \hat{A}$ is faithfully flat, and therefore $J^nA_{1+J} \subset IA_{1+J}$. □

COROLLARY 3.3.8. *Let A be a noetherian ring, let I be an A-ideal, and let $\hat{A}$ be the I-adic completion of A. Let $f_0 \colon X \to \mathrm{Spec}(A)$ be a separated finite-type scheme-map, let Z be a closed subscheme of $f_0^{-1}(\mathrm{Spec}(A/I))$, let $\mathfrak{X} = X_{/Z}$ be the completion of X along Z, and let $f \colon \mathfrak{X} \to \mathrm{Spf}(\hat{A})$ be the formal-scheme map induced by f_0:*

$$\begin{array}{ccc} \mathfrak{X} := X_{/Z} & \longrightarrow & X \\ {\scriptstyle f}\big\downarrow & & \big\downarrow{\scriptstyle f_0} \\ \mathrm{Spf}(\hat{A}) & \longrightarrow & \mathrm{Spec}(A) \end{array}$$

If f is proper (see §1.2.2) then $\mathfrak{X}$ is properly algebraic.

PROOF. Consider a compactification of f_0 (see [**Lü**, Theorem 3.2]):

$$X \underset{\text{open}}{\hookrightarrow} \overline{X} \xrightarrow[\text{proper}]{\bar{f}_0} \mathrm{Spec}(A).$$

Since f is proper, therefore Z is proper over $\mathrm{Spec}(A)$, hence closed in $\overline{X}$. Thus we may replace f_0 by $\bar{f}_0$, i.e., we may assume f_0 proper. Since f, being proper, is adic, Lemma 3.3.7, with $Z_2 := Z$ and $Z_1 := f_0^{-1}(\mathrm{Spec}(A/I))$, shows that $\mathfrak{X}$ is a union of connected components of the properly algebraic formal scheme $X_{/Z_1}$. Conclude by Remark 3.3.6(2). □

3.4. To deal with unbounded complexes we need the following boundedness results on certain derived functors. (See, e.g., Propositions 3.5.1 and 3.5.3 below.)

3.4.1. Refer to §1.2.2 for the definitions of separated, resp. affine, maps.

A formal scheme $\mathcal{X}$ is *separated* if the natural map $f_{\mathcal{X}}\colon \mathcal{X} \to \mathrm{Spec}(\mathbb{Z})$ is separated, i.e., for some—hence any—ideal of definition $\mathcal{J}$, the scheme $(\mathcal{X}, \mathcal{O}_{\mathcal{X}}/\mathcal{J})$ is separated. For example, any locally noetherian affine formal scheme is separated.

A locally noetherian formal scheme $\mathcal{X}$ is affine if and only if the map $f_{\mathcal{X}}$ is affine, i.e., for some—hence any—ideal of definition $\mathcal{J}$, the scheme $(\mathcal{X}, \mathcal{O}_{\mathcal{X}}/\mathcal{J})$ is affine. Hence the intersection $\mathcal{V} \cap \mathcal{V}'$ of any two affine open subsets of a separated locally noetherian formal scheme $\mathcal{Y}$ is again affine. In other words, the inclusion $\mathcal{V} \hookrightarrow \mathcal{Y}$ is an affine map. More generally, if $f\colon \mathcal{X} \to \mathcal{Y}$ is a map of locally noetherian formal schemes, if $\mathcal{Y}$ is separated, and if $\mathcal{V}$ and $\mathcal{V}'$ are affine open subsets of $\mathcal{Y}$ and $\mathcal{X}$ respectively, then $f^{-1}\mathcal{V} \cap \mathcal{V}'$ is affine [**GD**, p. 282, (5.8.10)].

LEMMA 3.4.2. *If $g\colon \mathcal{X} \to \mathcal{Y}$ is an affine map of locally noetherian formal schemes, then every $\mathcal{M} \in \mathcal{A}_{\vec{c}}(\mathcal{X})$ is g_*-acyclic, i.e., $R^i g_* \mathcal{M} = 0$ for all $i > 0$. More generally, if $\mathcal{G} \in \mathbf{D}_{\vec{c}}(\mathcal{X})$ and $e \in \mathbb{Z}$ are such that $H^i(\mathcal{G}) = 0$ for all $i \geq e$, then $H^i(\mathbf{R}g_*\mathcal{G}) = 0$ for all $i \geq e$.*

PROOF. $R^i g_* \mathcal{M}$ is the sheaf associated to the presheaf $\mathcal{U} \mapsto \mathrm{H}^i(g^{-1}(\mathcal{U}), \mathcal{M})$, ($\mathcal{U}$ open in $\mathcal{Y}$) [**EGA**, Chap. 0, (12.2.1)]. If $\mathcal{U}$ is affine then so is $g^{-1}(\mathcal{U}) \subset \mathcal{X}$, and Corollary 3.1.8 gives $\mathrm{H}^i(g^{-1}(\mathcal{U}), \mathcal{M}) = 0$ for all $i > 0$.

Now consider in $\mathbf{K}(\mathcal{X})$ a quasi-isomorphism $\mathcal{G} \to I$ where I is a "special" inverse limit of injective resolutions I_{-e} of the truncations $\mathcal{G}^{\geq e}$ (see (4.1.3.2)), so that $H^i(\mathbf{R}g_*\mathcal{G})$ is the sheaf associated to the presheaf $\mathcal{U} \mapsto \mathrm{H}^i(\Gamma(g^{-1}\mathcal{U}, I))$, see [**Sp**, p. 134, 3.13]. If C_{-e} is the kernel of the split surjection $I_{-e} \to I_{1-e}$ then $C_{-e}[e]$ is an injective resolution of $H^e(\mathcal{G}) \in \mathcal{A}_{\vec{c}}(\mathcal{X})$, and so for any affine open $\mathcal{U} \subset \mathcal{Y}$ and $i > e$, $\mathrm{H}^i(\Gamma(g^{-1}\mathcal{U}, C_{-e})) = 0$. Applying [**Sp**, p. 126, Lemma], one finds then that for $i \geq e$ the natural map $\mathrm{H}^i(\Gamma(g^{-1}\mathcal{U}, I)) \to \mathrm{H}^i(\Gamma(g^{-1}\mathcal{U}, I_{-e}))$ is an isomorphism. Consequently if $H^i(\mathcal{G}) = 0$ for all $i \geq e$ (whence $I_{-e} \cong \mathcal{G}^{\geq e} = 0$ in $\mathbf{D}(\mathcal{X})$) then $\mathrm{H}^i(\Gamma(g^{-1}\mathcal{U}, I)) = 0$. □

PROPOSITION 3.4.3. *Let $\mathcal{X}$ be a noetherian formal scheme. Then:*

(a) *The functor $\mathbf{R}\Gamma(\mathcal{X}, -)$ is bounded above on $\mathbf{D}_{\vec{c}}(\mathcal{X})$. In other words, there is an integer $e \geq 0$ such that if $\mathcal{G} \in \mathbf{D}_{\vec{c}}(\mathcal{X})$ and $H^i(\mathcal{G}) = 0$ for all $i \geq i_0$ then $\mathrm{H}^i(\mathbf{R}\Gamma(\mathcal{X}, -)) = 0$ for all $i \geq i_0 + e$.*

(b) *For any formal-scheme map $f\colon \mathcal{X} \to \mathcal{Y}$ with $\mathcal{Y}$ quasi-compact, the functor $\mathbf{R}f_*$ is bounded-above on $\mathbf{D}_{\vec{c}}(\mathcal{X})$, i.e., there is an integer $e \geq 0$ such that if $\mathcal{G} \in \mathbf{D}_{\vec{c}}(\mathcal{X})$ and $H^i(\mathcal{G}) = 0$ for all $i \geq i_0$ then $H^i(\mathbf{R}f_*\mathcal{G}) = 0$ for all $i \geq i_0 + e$.*

PROOF. Let us prove (b). (The proof of (a) is the same, *mutatis mutandis.*) Suppose first that $\mathcal{X}$ is separated, see §3.4.1. Since $\mathcal{Y}$ has a finite affine open cover and $\mathbf{R}f_*$ commutes with open base change, we may assume that $\mathcal{Y}$ itself is affine. Let $n(\mathcal{X})$ be the least positive integer n such that there exists a finite affine open cover $\mathcal{X} = \cup_{i=1}^n \mathcal{X}_i$, and let us show by induction on $n(\mathcal{X})$ that $e := n(\mathcal{X}) - 1$ *will do.*

The case $n(\mathcal{X}) = 1$ is covered by Lemma 3.4.2. So assume that $n := n(\mathcal{X}) \geq 2$, let $\mathcal{X} = \cup_{i=1}^n \mathcal{X}_i$ be an affine open cover, and let $u_1\colon \mathcal{X}_1 \hookrightarrow \mathcal{X}$, $u_2\colon \cup_{i=2}^n \mathcal{X}_i \hookrightarrow \mathcal{X}$, $u_3\colon \cup_{i=2}^n (\mathcal{X}_1 \cap \mathcal{X}_i) \hookrightarrow \mathcal{X}$ be the respective inclusion maps. Note that $\mathcal{X}_1 \cap \mathcal{X}_i$ is affine because $\mathcal{X}$ is separated. So by the inductive hypothesis, the assertion holds for the maps $f_i := f \circ u_i$ $(i = 1, 2, 3)$.

Now apply the Δ-functor $\mathbf{R}f_*$ to the "Mayer-Vietoris" triangle

$$\mathcal{G} \longrightarrow \mathbf{R}u_{1*}u_1^*\mathcal{G} \oplus \mathbf{R}u_{2*}u_2^*\mathcal{G} \longrightarrow \mathbf{R}u_{3*}u_3^*\mathcal{G} \xrightarrow{+1}$$

(derived from the standard exact sequence

$$0 \to \mathcal{E} \to u_{1*}u_1^*\mathcal{E} \oplus u_{2*}u_2^*\mathcal{E} \to u_{3*}u_3^*\mathcal{E} \to 0$$

where $\mathcal{G} \to \mathcal{E}$ is a K-injective resolution) to get the $\mathbf{D}(\mathcal{Y})$-triangle

$$\mathbf{R}f_*\mathcal{G} \longrightarrow \mathbf{R}f_{1*}u_1^*\mathcal{G} \oplus \mathbf{R}f_{2*}u_2^*\mathcal{G} \longrightarrow \mathbf{R}f_{3*}u_3^*\mathcal{G} \xrightarrow{+1}$$

whose associated long exact homology sequence yields the assertion for f.

The general case can now be disposed of with a similar Mayer-Vietoris induction on the least number of *separated* open subsets needed to cover $\mathcal{X}$. □

PROPOSITION 3.4.4. *Let X be a separated noetherian scheme, let $Z \subset X$ be a closed subscheme, and let $\kappa_{\mathcal{X}}\colon \mathcal{X} = X_{/Z} \to X$ be the completion map. Then the functor $\mathbf{R}Q_X\kappa_*$ is bounded-above on $\mathbf{D}_{\vec{c}}(\mathcal{X})$.*

PROOF. Set $\kappa := \kappa_{\mathcal{X}}$. Let $n(X)$ be the least number of affine open subschemes needed to cover X. When X is affine, Q_X is the sheafification of the global section functor, and since κ_* is exact and, being right adjoint to the *exact* functor κ^*, preserves K-injectivity, we find that for any $\mathcal{F} \in \mathbf{D}(\mathcal{X})$, $\mathbf{R}Q_X\kappa_*\mathcal{F}$ is the sheafification of the complex $\mathbf{R}\Gamma(X, \kappa_*\mathcal{F}) = \mathbf{R}\Gamma(\mathcal{X}, \mathcal{F})$. Thus Proposition 3.4.3(a) yields the desired result for $n(X) = 1$.

Proceed by induction when $n(X) > 1$, using a "Mayer-Vietoris" argument as in the proof of Proposition 3.4.3. The enabling points are that if $v\colon V \hookrightarrow X$ is an open immersion with $n(V) < n(X)$, giving rise to the natural commutative diagram

$$\begin{array}{ccc} V_{/Z\cap V} =: \mathcal{V} & \xrightarrow{\kappa_{\mathcal{V}}} & V \\ {\scriptstyle \hat{v}}\downarrow & & \downarrow{\scriptstyle v} \\ \mathcal{X} & \xrightarrow[\kappa_{\mathcal{X}}]{} & X \end{array}$$

then there are natural isomorphisms, for $\mathcal{F} \in \mathbf{D}_{\vec{c}}(\mathcal{X})$ and $v_*^{\mathrm{qc}}\colon \mathcal{A}_{\mathrm{qc}}(V) \to \mathcal{A}_{\mathrm{qc}}(X)$ the restriction of v_*:

$$\mathbf{R}Q_X\kappa_{\mathcal{X}*}\mathbf{R}\hat{v}_*\hat{v}^*\mathcal{F} \cong \mathbf{R}Q_X\mathbf{R}v_*\kappa_{\mathcal{V}*}\hat{v}^*\mathcal{F} \cong \mathbf{R}v_*^{\mathrm{qc}}\mathbf{R}Q_V\kappa_{\mathcal{V}*}\hat{v}^*\mathcal{F},$$

and the functor $\mathbf{R}Q_V\kappa_{\mathcal{V}*}\hat{v}^*$ is bounded-above, by the inductive hypothesis on $n(V) < n(X)$, as is $\mathbf{R}v_*^{\mathrm{qc}}$, by the proof of [**AJL**, p. 12, Proposition (1.3)]. □

3.5. Here are some examples of how boundedness is used.

PROPOSITION 3.5.1. *Let $f\colon \mathcal{X} \to \mathcal{Y}$ be a proper map of noetherian formal schemes. Then*

$$\mathbf{R}f_*\mathbf{D}_{\mathrm{c}}(\mathcal{X}) \subset \mathbf{D}_{\mathrm{c}}(\mathcal{Y}) \quad \text{and} \quad \mathbf{R}f_*\mathbf{D}_{\vec{c}}(\mathcal{X}) \subset \mathbf{D}_{\vec{c}}(\mathcal{Y}).$$

PROOF. For a coherent $\mathcal{O}_{\mathcal{X}}$-module $\mathcal{M}$, $\mathbf{R}f_*\mathcal{M} \in \mathbf{D}_{\mathrm{c}}(\mathcal{Y})$ [**EGA**, p. 119, (3.4.2)]. Since $\mathcal{X}$ is noetherian, the homology functors $H^i\mathbf{R}f_*$ commute with $\varinjlim$ on $\mathcal{O}_{\mathcal{X}}$-modules, whence $\mathbf{R}f_*\mathcal{N} \in \mathbf{D}_{\vec{c}}(\mathcal{Y})$ for all $\mathcal{N} \in \mathcal{A}_{\vec{c}}(\mathcal{X})$. $\mathbf{R}f_*$ being bounded on $\mathbf{D}_{\vec{c}}(\mathcal{X})$ (Proposition 3.4.3(b)), way-out reasoning [**H1**, p. 74, (iii)] completes the proof. □

PROPOSITION 3.5.2. *Let $f\colon \mathfrak{X} \to \mathfrak{Y}$ be a map of quasi-compact formal schemes, with $\mathfrak{X}$ noetherian. Then the functor $\mathbf{R}f_*|_{\mathbf{D}_{\vec{c}}(\mathfrak{X})}$ commutes with small direct sums, i.e., for any small family $(\mathcal{E}_\alpha)$ in $\mathbf{D}_{\vec{c}}(\mathfrak{X})$ the natural map*

$$\oplus_\alpha(\mathbf{R}f_*\mathcal{E}_\alpha) \to \mathbf{R}f_*(\oplus_\alpha \mathcal{E}_\alpha)$$

is a $\mathbf{D}(\mathfrak{Y})$-isomorphism.

PROOF. It suffices to look at the induced homology maps in each degree, i.e., setting $R^if_* := H^i\mathbf{R}f_*$ $(i \in \mathbb{Z})$, we need to show that *the natural map*

$$\oplus_\alpha(R^if_*\mathcal{E}_\alpha) \xrightarrow{\sim} R^if_*(\oplus_\alpha \mathcal{E}_\alpha).$$

is an isomorphism.

For any $\mathcal{F} \in \mathbf{D}_{\vec{c}}(\mathfrak{X})$ and any integer $e \geq 0$, the vertex $\mathcal{G}$ of a triangle based on the natural map t_{i-e} from $\mathcal{F}$ to the truncation $\mathcal{F}^{\geq i-e}$ (see (4.1.3.2)) satisfies $H^j(\mathcal{G}) = 0$ for all $j \geq i-e-1$; so if e is the integer in Proposition 3.4.3(b), then $R^{i-1}f_*\mathcal{G} = R^if_*\mathcal{G} = 0$, and the map induced by t_{i-e} is an *isomorphism*

$$R^if_*\mathcal{F} \xrightarrow{\sim} R^if_*\mathcal{F}^{\geq i-e}.$$

We can therefore replace each $\mathcal{E}_\alpha$ by $\mathcal{E}_\alpha^{\geq i-e}$, i.e., we may assume that the $\mathcal{E}_\alpha$ are uniformly bounded below.

We may assume further that each complex $\mathcal{E}_\alpha$ is injective, hence f_*-acyclic (i.e., the canonical map is an *isomorphism* $f_*\mathcal{E}_\alpha \xrightarrow{\sim} \mathbf{R}f_*\mathcal{E}_\alpha$). Since $\mathfrak{X}$ is noetherian, R^if_* commutes with direct sums; and so each component of $\oplus_\alpha\mathcal{E}_\alpha$ is an f_*-acyclic $\mathcal{O}_\mathfrak{X}$-module. This implies that the bounded-below complex $\oplus_\alpha\mathcal{E}_\alpha$ is itself f_*-acyclic. Thus in the natural commutative diagram

$$\begin{array}{ccc} \oplus_\alpha(f_*\mathcal{E}_\alpha) & \xrightarrow{\ \sim\ } & f_*(\oplus_\alpha\mathcal{E}_\alpha) \\ \simeq\downarrow & & \downarrow\simeq \\ \oplus_\alpha(\mathbf{R}f_*\mathcal{E}_\alpha) & \longrightarrow & \mathbf{R}f_*(\oplus_\alpha\mathcal{E}_\alpha) \end{array}$$

the top and both sides are isomorphisms, whence so is the bottom. □

The following Proposition generalizes [**EGA**, p. 92, Theorem (4.1.5)].

PROPOSITION 3.5.3. *Let $f_0\colon X \to Y$ be a proper map of locally noetherian schemes, let $W \subset Y$ be a closed subset, let $Z := f_0^{-1}W$, let $\kappa_\mathfrak{Y}\colon \mathfrak{Y} = Y_{/W} \to Y$ and $\kappa_\mathfrak{X}\colon \mathfrak{X} = X_{/Z} \to X$ be the respective (flat) completion maps, and let $f\colon \mathfrak{X} \to \mathfrak{Y}$ be the map induced by f_0. Then for $\mathcal{E} \in \mathbf{D}_{\mathrm{qc}}(X)$ the map $\theta_\mathcal{E}$ adjoint to the natural composition*

$$\mathbf{R}f_{0*}\mathcal{E} \longrightarrow \mathbf{R}f_{0*}\kappa_{\mathfrak{X}*}\kappa_\mathfrak{X}^*\mathcal{E} \xrightarrow{\sim} \kappa_{\mathfrak{Y}*}\mathbf{R}f_*\kappa_\mathfrak{X}^*\mathcal{E}$$

is an isomorphism

$$\theta_\mathcal{E}\colon \kappa_\mathfrak{Y}^*\mathbf{R}f_{0*}\mathcal{E} \xrightarrow{\sim} \mathbf{R}f_*\kappa_\mathfrak{X}^*\mathcal{E}.$$

PROOF. We may assume Y affine, say $Y = \mathrm{Spec}(A)$, and then $W = \mathrm{Spec}(A/I)$ for some A-ideal I. Let $\hat{A}$ be the I-adic completion of A, so that there is a natural cartesian diagram

$$\begin{array}{ccc} X \otimes_A \hat{A} =: X_1 & \xrightarrow{k_X} & X \\ f_1\downarrow & & \downarrow f_0 \\ \mathrm{Spec}(\hat{A}) =: Y_1 & \xrightarrow[k_Y]{} & Y \end{array}$$

Here k_Y is flat, and the natural map is an isomorphism $k_Y^*\mathbf{R}f_{0*}\mathcal{E} \xrightarrow{\sim} \mathbf{R}f_{1*}k_X^*\mathcal{E}$: since $\mathbf{R}f_{0*}$ (resp. $\mathbf{R}f_{1*}$) is bounded-above on $\mathbf{D}_{\mathrm{qc}}(X)$ (resp. $\mathbf{D}_{\mathrm{qc}}(X_1)$), see Proposition 3.4.3(b), way-out reasoning reduces this assertion to the well-known case where $\mathcal{E}$ is a single quasi-coherent $\mathcal{O}_X$-module. Simple considerations show then that we can replace f_0 by f_1 and $\mathcal{E}$ by $k_X^*\mathcal{E}$; in other words, we can assume $A = \hat{A}$.

From Proposition 3.5.1 it follows that $\mathbf{R}f_{0*}\mathcal{E} \in \mathbf{D}_{\mathrm{qc}}(Y)$ and $\mathbf{R}f_*\kappa_{\mathcal{X}}^*\mathcal{E} \in \mathbf{D}_{\vec{c}}(\mathcal{Y})$. Recalling the equivalences in Proposition 3.3.1, we see that any $\mathcal{F} \in \mathbf{D}_{\vec{c}}(\mathcal{Y})$ is isomorphic to $\kappa_{\mathcal{Y}}^*\mathcal{F}_0$ for some $\mathcal{F}_0 \in \mathbf{D}_{\mathrm{qc}}(Y)$ (so that $\mathbf{L}f_0^*\mathcal{F}_0 \in \mathbf{D}_{\mathrm{qc}}(X)$), and that there is a sequence of natural isomorphisms

$$
\begin{aligned}
\mathrm{Hom}_{\mathcal{Y}}(\mathcal{F}, \kappa_{\mathcal{Y}}^*\mathbf{R}f_{0*}\mathcal{E}) &\xrightarrow{\sim} \mathrm{Hom}_Y(\mathcal{F}_0, \mathbf{R}f_{0*}\mathcal{E}) \\
&\xrightarrow{\sim} \mathrm{Hom}_X(\mathbf{L}f_0^*\mathcal{F}_0, \mathcal{E}) \\
&\xrightarrow{\sim} \mathrm{Hom}_{\mathcal{X}}(\kappa_{\mathcal{X}}^*\mathbf{L}f_0^*\mathcal{F}_0, \kappa_{\mathcal{X}}^*\mathcal{E}) \\
&\xrightarrow{\sim} \mathrm{Hom}_{\mathcal{X}}(\mathbf{L}f^*\kappa_{\mathcal{Y}}^*\mathcal{F}_0, \kappa_{\mathcal{X}}^*\mathcal{E}) \xrightarrow{\sim} \mathrm{Hom}_{\mathcal{Y}}(\mathcal{F}, \mathbf{R}f_*\kappa_{\mathcal{X}}^*\mathcal{E}).
\end{aligned}
$$

The conclusion follows. □

4. Global Grothendieck Duality.

THEOREM 4.1. *Let $f\colon \mathcal{X} \to \mathcal{Y}$ be a map of quasi-compact formal schemes, with $\mathcal{X}$ noetherian, and let $\boldsymbol{j}\colon \mathbf{D}(\mathcal{A}_{\vec{c}}(\mathcal{X})) \to \mathbf{D}(\mathcal{X})$ be the natural functor. Then the Δ-functor $\mathbf{R}f_* \circ \boldsymbol{j}$ has a right Δ-adjoint. In fact there is a bounded-below Δ-functor $f^\times\colon \mathbf{D}(\mathcal{Y}) \to \mathbf{D}(\mathcal{A}_{\vec{c}}(\mathcal{X}))$ and a map of Δ-functors $\tau\colon \mathbf{R}f_*\boldsymbol{j}f^\times \to \mathbf{1}$ such that for all $\mathcal{G} \in \mathbf{D}(\mathcal{A}_{\vec{c}}(\mathcal{X}))$ and $\mathcal{F} \in \mathbf{D}(\mathcal{Y})$, the composed map (in the derived category of abelian groups)*

$$
\begin{aligned}
\mathbf{R}\mathrm{Hom}^\bullet_{\mathcal{A}_{\vec{c}}(\mathcal{X})}(\mathcal{G}, f^\times\mathcal{F}) &\xrightarrow{\text{natural}} \mathbf{R}\mathrm{Hom}^\bullet_{\mathcal{A}(\mathcal{Y})}(\mathbf{R}f_*\boldsymbol{j}\mathcal{G}, \mathbf{R}f_*\boldsymbol{j}f^\times\mathcal{F}) \\
&\xrightarrow{\text{via } \tau} \mathbf{R}\mathrm{Hom}^\bullet_{\mathcal{A}(\mathcal{Y})}(\mathbf{R}f_*\boldsymbol{j}\mathcal{G}, \mathcal{F})
\end{aligned}
$$

is an isomorphism.

With Corollary 3.3.4 this gives:

COROLLARY 4.1.1. *If $\mathcal{X}$ is properly algebraic, the restriction of $\mathbf{R}f_*$ to $\mathbf{D}_{\vec{c}}(\mathcal{X})$ has a right Δ-adjoint (also to be denoted $f^\times$ when no confusion results).*

Remarks. 1. Recall that over any abelian category $\mathcal{A}$ in which each complex $\mathcal{F}$ has a K-injective resolution $\rho(\mathcal{F})$, we can set

$$
\mathbf{R}\mathrm{Hom}^\bullet_{\mathcal{A}}(\mathcal{G}, \mathcal{F}) := \mathrm{Hom}^\bullet_{\mathcal{A}}(\mathcal{G}, \rho(\mathcal{F})) \qquad (\mathcal{G}, \mathcal{F} \in \mathbf{D}(\mathcal{A}));
$$

and there are natural isomorphisms

$$
\mathrm{H}^i\mathbf{R}\mathrm{Hom}^\bullet_{\mathcal{A}}(\mathcal{G}, \mathcal{F}) \cong \mathrm{Hom}_{\mathbf{D}(\mathcal{A})}(\mathcal{G}, \mathcal{F}[i]) \qquad (i \in \mathbb{Z}).
$$

2. Application of homology to the second assertion in the Theorem reveals that it is equivalent to the first one.

3. We do not know in general (when $\mathcal{X}$ is not properly algebraic) has a right adjoint $(\text{identity})^\times \cong \mathbf{R}Q_{\mathcal{X}}$ (see Proposition 3.2.3), but it may be that for some $\mathcal{E} \in \mathcal{A}_{\vec{c}}(\mathcal{X})$ the natural map $\mathcal{E} \to \mathbf{R}Q_{\mathcal{X}}\boldsymbol{j}\mathcal{E}$ is not an isomorphism.

4. For a *proper* map $f_0\colon X \to Y$ of *ordinary* schemes it is customary to write $f_0^!$ instead of $f_0^\times$. (Our extension of this notation to maps of formal schemes—introduced immediately after Definition 7.3—is not what would be expected here.)

5. Theorem 4.1 includes the case when $\mathcal{X}$ and $\mathcal{Y}$ are ordinary noetherian schemes. (In fact the proof below applies with minor changes to arbitrary maps of quasi-compact, quasi-separated schemes, cf. [**L4**, Chapter 4].) The next Corollary relates the formal situation to the ordinary one.

COROLLARY 4.1.2. *Let A be a noetherian adic ring with ideal of definition I, set $Y := \mathrm{Spec}(A)$ and $W := \mathrm{Spec}(A/I) \subset Y$. Let $f_0\colon X \to Y$ be a proper map and set $Z := f_0^{-1}W$, so that there is a commutative diagram*

$$\begin{array}{ccc} \mathcal{X} := X_{/Z} & \xrightarrow{\kappa_{\mathcal{X}}} & X \\ {\scriptstyle f}\big\downarrow & & \big\downarrow{\scriptstyle f_0} \\ \mathcal{Y} := \mathrm{Spf}(A) & \xrightarrow[\kappa_{\mathcal{Y}}]{} & Y \end{array}$$

with $\kappa_{\mathcal{X}}$ and $\kappa_{\mathcal{Y}}$ the respective (flat) completion maps, and f the (proper) map induced by f_0.

Then the map adjoint to the natural composition

$$\mathbf{R}f_*\kappa_{\mathcal{X}}^*f_0^!\kappa_{\mathcal{Y}*} \xrightarrow{3.5.3} \kappa_{\mathcal{Y}}^*\mathbf{R}f_{0*}f_0^!\kappa_{\mathcal{Y}*} \longrightarrow \kappa_{\mathcal{Y}}^*\kappa_{\mathcal{Y}*} \longrightarrow \mathbf{1}$$

is an isomorphism of functors—from $\mathbf{D}(\mathcal{Y})$ to $\mathbf{D}_{\vec{c}}(\mathcal{X})$, see Corollary 4.1.1—

$$\kappa_{\mathcal{X}}^*f_0^!\kappa_{\mathcal{Y}*} \xrightarrow{\sim} f^{\times}.$$

PROOF. For any $\mathcal{E} \in \mathbf{D}_{\vec{c}}(\mathcal{X})$ set $\mathcal{E}_0 := \boldsymbol{j}_X\mathbf{R}Q_X\kappa_{\mathcal{X}*}\mathcal{E} \in \mathbf{D}_{\mathrm{qc}}(X)$ (see Section 3.3). Using Proposition 3.3.1 we have then for any $\mathcal{F} \in \mathbf{D}(\mathcal{Y})$ the natural isomorphisms

$$\begin{aligned} \mathrm{Hom}_{\mathbf{D}(\mathcal{X})}(\mathcal{E}, \kappa_{\mathcal{X}}^*f_0^!\kappa_{\mathcal{Y}*}\mathcal{F}) &\xrightarrow{\sim} \mathrm{Hom}_{\mathbf{D}(X)}(\mathcal{E}_0, f_0^!\kappa_{\mathcal{Y}*}\mathcal{F}) \\ &\xrightarrow{\sim} \mathrm{Hom}_{\mathbf{D}(Y)}(\mathbf{R}f_{0*}\mathcal{E}_0, \kappa_{\mathcal{Y}*}\mathcal{F}) \\ &\xrightarrow{\sim} \mathrm{Hom}_{\mathbf{D}(\mathcal{Y})}(\kappa_{\mathcal{Y}}^*\mathbf{R}f_{0*}\mathcal{E}_0, \mathcal{F}) \\ &\underset{3.5.3}{\xrightarrow{\sim}} \mathrm{Hom}_{\mathbf{D}(\mathcal{Y})}(\mathbf{R}f_*\kappa_{\mathcal{X}}^*\mathcal{E}_0, \mathcal{F}) \xrightarrow{\sim} \mathrm{Hom}_{\mathbf{D}(\mathcal{Y})}(\mathbf{R}f_*\mathcal{E}, \mathcal{F}). \end{aligned}$$

Thus $\kappa_{\mathcal{X}}^*f_0^!\kappa_{\mathcal{Y}*}$ is right-adjoint to $\mathbf{R}f_*|_{\mathbf{D}_{\vec{c}}(\mathcal{X})}$, whence the conclusion. □

PROOF OF THEOREM 4.1. 1. Following Deligne [**H1**, p. 417, top], we begin by considering for $\mathcal{M} \in \mathcal{A}(\mathcal{X})$ the functorial flasque *Godement resolution*

$$0 \to \mathcal{M} \to G^0(\mathcal{M}) \to G^1(\mathcal{M}) \to \cdots.$$

Here, with $G^{-2}(\mathcal{M}) := 0$, $G^{-1}(\mathcal{M}) := \mathcal{M}$, and for $i \ge 0$, $K^i(\mathcal{M})$ the cokernel of $G^{i-2}(\mathcal{M}) \to G^{i-1}(\mathcal{M})$, the sheaf $G^i(\mathcal{M})$ is specified inductively by

$$G^i(\mathcal{M})(\mathcal{U}) := \prod_{x\in\mathcal{U}} K^i(\mathcal{M})_x \qquad (\mathcal{U} \text{ open in } \mathcal{X}).$$

One shows by induction on i that all the functors G^i and K^i (from $\mathcal{A}(\mathcal{X})$ to itself) are *exact*. Moreover, for $i \ge 0$, $G^i(\mathcal{M})$, being flasque, is *f_*-acyclic*, i.e.,

$$R^jf_*G^i(\mathcal{M}) = 0 \quad \text{for all } j > 0.$$

The category $\mathcal{A}_{\vec{c}}(\mathcal{X})$ has small colimits (Proposition 3.2.2), and is generated by its coherent members, of which there exists a small set containing representatives of every isomorphism class. The Special Adjoint Functor Theorem ([**F**, p. 90] or [**M1**, p. 126, Corollary]) guarantees then that a right-exact functor F from $\mathcal{A}_{\vec{c}}$ into an abelian category $\mathcal{A}'$ has a right adjoint iff F is *continuous* in the sense that it commutes with filtered direct limits, i.e., for any small directed system $(\mathcal{M}_\alpha,\, \varphi_{\alpha\beta}\colon \mathcal{M}_\beta \to \mathcal{M}_\alpha)$ in $\mathcal{A}_{\vec{c}}$, with $\varinjlim_\alpha \mathcal{M}_\alpha = (\mathcal{M},\, \varphi_\alpha\colon \mathcal{M}_\alpha \to \mathcal{M})$ it holds that

$$(F(\mathcal{M}), F(\varphi_\alpha)) = \varinjlim_\alpha\, (F(\mathcal{M}_\alpha), F(\varphi_{\alpha\beta})).$$

Accordingly, for constructing right adjoints we need to replace the restrictions of G^i and K^i to $\mathcal{A}_{\vec{c}}(\mathfrak{X})$ by continuous functors.

LEMMA 4.1.3. *Let $\mathfrak{X}$ be a locally noetherian formal scheme and let G be a functor from $\mathcal{A}_{\mathrm{c}}(\mathfrak{X})$ to a category $\mathcal{A}'$ in which direct limits exist for all small directed systems. Let $j\colon \mathcal{A}_{\mathrm{c}}(\mathfrak{X}) \hookrightarrow \mathcal{A}_{\vec{c}}(\mathfrak{X})$ be the inclusion functor. Then:*

(a) *There exists a continuous functor $G_{\vec{c}}\colon \mathcal{A}_{\vec{c}}(\mathfrak{X}) \to \mathcal{A}'$ and an isomorphism of functors $\varepsilon\colon G \xrightarrow{\sim} G_{\vec{c}} \circ j$ such that for any map of functors $\psi\colon G \to F\circ j$ with F continuous, there is a unique map of functors $\psi_{\vec{c}}\colon G_{\vec{c}} \to F$ such that ψ factors as*

$$G \xrightarrow{\varepsilon} G_{\vec{c}}\circ j \xrightarrow{\text{via } \psi_{\vec{c}}} F\circ j.$$

(b) *Assume that $\mathcal{A}'$ is abelian, and has exact filtered direct limits (i.e., satisfies Grothendieck's axiom* AB5*). Then if G is exact, so is $G_{\vec{c}}$.*

PROOF. (a) For $\mathcal{M} \in \mathcal{A}_{\vec{c}}(\mathfrak{X})$, let $(\mathcal{M}_\alpha)$ be the directed system of coherent $\mathcal{O}_{\mathfrak{X}}$-submodules of $\mathcal{M}$, and set

$$G_{\vec{c}}(\mathcal{M}) := \varinjlim_\alpha G(\mathcal{M}_\alpha).$$

For any $\mathcal{A}_{\vec{c}}(\mathfrak{X})$-map $\nu\colon \mathcal{M} \to \mathcal{N}$ and any α, there exists a coherent submodule $\mathcal{N}_\beta \subset \mathcal{N}$ such that $\nu|_{\mathcal{M}_\alpha}$ factors as $\mathcal{M}_\alpha \to \mathcal{N}_\beta \hookrightarrow \mathcal{N}$ (Corollary 3.1.7 and Lemma 3.2.1, with $q=0$); and the resulting composition

$$\nu'_\alpha\colon G(\mathcal{M}_\alpha) \to G(\mathcal{N}_\beta) \to G_{\vec{c}}(\mathcal{N})$$

does not depend on the choice of $\mathcal{N}_\beta$. We define the map

$$G_{\vec{c}}(\nu)\colon G_{\vec{c}}(\mathcal{M}) = \varinjlim_\alpha G(\mathcal{M}_\alpha) \to G_{\vec{c}}(\mathcal{N})$$

to be the unique one whose composition with $G(\mathcal{M}_\alpha) \to G_{\vec{c}}(\mathcal{M})$ is ν'_α for all α. Verification of the rest of assertion (a) is straightforward.

(b) Let $0 \to \mathcal{M} \to \mathcal{N} \xrightarrow{\pi} \mathcal{Q} \to 0$ be an exact sequence in $\mathcal{A}_{\vec{c}}(\mathfrak{X})$. Let $(\mathcal{N}_\beta)$ be the filtered system of coherent submodules of $\mathcal{N}$, so that $\mathcal{N} = \varinjlim \mathcal{N}_\beta$ (Corollary 3.1.7). Then $(\mathcal{M}\cap\mathcal{N}_\beta)$ is a filtered system of coherent $\mathcal{O}_{\mathfrak{X}}$-modules whose $\varinjlim$ is $\mathcal{M}$, and $(\pi\mathcal{N}_\beta)$ is a filtered system of coherent $\mathcal{O}_{\mathfrak{X}}$-modules whose $\varinjlim$ is $\mathcal{Q}$ (see Corollary 3.1.6). The exactness of $G_{\vec{c}}$ is then made apparent by application of $\varinjlim_\beta$ to the system of exact sequences

$$0 \to G(\mathcal{M}\cap\mathcal{N}_\beta) \to G(\mathcal{N}_\beta) \to G(\pi\mathcal{N}_\beta) \to 0. \qquad \square$$

Now for $\mathcal{M} \in \mathcal{A}_{\vec{c}}(\mathfrak{X})$, the $\varinjlim$ of the system of Godement resolutions of all the coherent submodules $\mathcal{M}_\alpha \subset \mathcal{M}$ is a functorial resolution

$$0 \to \mathcal{M} \to G^0_{\vec{c}}(\mathcal{M}) \to G^1_{\vec{c}}(\mathcal{M}) \to \cdots;$$

and the cokernel of $G^{i-2}_{\vec{c}}(\mathcal{M}) \to G^{i-1}_{\vec{c}}(\mathcal{M})$ is $K^i_{\vec{c}}(\mathcal{M}) := \varinjlim K^i(\mathcal{M}_\alpha)$. By (b) above (applied to the exact functors G^i and K^i), the continuous functors $G^i_{\vec{c}}$ and $K^i_{\vec{c}}$ are exact; and $G^i_{\vec{c}}(\mathcal{M}) = \varinjlim G^i(\mathcal{M}_\alpha)$ is f_*-acyclic since $G^i(\mathcal{M}_\alpha)$ is, and—$\mathfrak{X}$ being noetherian—the functors R^jf_* commute with $\varinjlim$. Proposition 3.4.3(b) implies then that there is an integer $e \ge 0$ such that for all $\mathcal{M} \in \mathcal{A}_{\vec{c}}(\mathfrak{X})$, $K^e_{\vec{c}}(\mathcal{M})$ is f_*-acyclic.

So if we define the exact functors $\mathcal{D}^i\colon \mathcal{A}_{\vec{c}}(\mathfrak{X}) \to \mathcal{A}(\mathfrak{X})$ by

$$\mathcal{D}^i(\mathcal{M}) = \begin{cases} G^i_{\vec{c}}(\mathcal{M}) & (0 \le i < e) \\ K^e_{\vec{c}}(\mathcal{M}) & (i = e) \\ 0 & (i > e) \end{cases}$$

then for $\mathcal{M} \in \mathcal{A}_{\vec{c}}(\mathfrak{X})$, each $\mathcal{D}^i(\mathcal{M})$ is f_*-acyclic and the natural sequence

$$0 \longrightarrow \mathcal{M} \xrightarrow{\delta(\mathcal{M})} \mathcal{D}^0(\mathcal{M}) \xrightarrow{\delta^0(\mathcal{M})} \mathcal{D}^1(\mathcal{M}) \xrightarrow{\delta^1(\mathcal{M})} \mathcal{D}^2(\mathcal{M}) \longrightarrow \cdots \longrightarrow \mathcal{D}^e(\mathcal{M}) \longrightarrow 0$$

is exact. In short, the sequence $\mathcal{D}^0 \to \mathcal{D}^1 \to \mathcal{D}^2 \to \cdots \to \mathcal{D}^e \to 0$ is an *exact, continuous, f_*-acyclic, finite resolution of the inclusion functor $\mathcal{A}_{\vec{c}}(\mathfrak{X}) \hookrightarrow \mathcal{A}(\mathfrak{X})$.*

2. We have then a Δ-functor $(\mathcal{D}^\bullet, \mathrm{Id})\colon \mathbf{K}(\mathcal{A}_{\vec{c}}(\mathfrak{X})) \to \mathbf{K}(\mathfrak{X})$ which assigns an f_*-acyclic resolution to each $\mathcal{A}_{\vec{c}}(\mathfrak{X})$-complex $\mathcal{G} = (\mathcal{G}^p)_{p\in\mathbb{Z}}$:

$$(\mathcal{D}^\bullet\mathcal{G})^m := \bigoplus_{p+q=m} \mathcal{D}^q(\mathcal{G}^p) \qquad (m \in \mathbb{Z},\ 0 \le q \le e),$$

the differential $(\mathcal{D}^\bullet\mathcal{G})^m \to (\mathcal{D}^\bullet\mathcal{G})^{m+1}$ being defined on $\mathcal{D}^q(\mathcal{G}^p)$ $(p + q = m)$ to be $d' + (-1)^p d''$ where $d'\colon \mathcal{D}^q(\mathcal{G}^p) \to \mathcal{D}^q(\mathcal{G}^{p+1})$ comes from the differential in $\mathcal{G}$ and $d'' = \delta^q(\mathcal{G}^p)\colon \mathcal{D}^q(\mathcal{G}^p) \to \mathcal{D}^{q+1}(\mathcal{G}^p)$.

It is elementary to check that the natural map $\delta(\mathcal{G})\colon \mathcal{G} \to \mathcal{D}^\bullet\mathcal{G}$ is a *quasi-isomorphism.* The canonical maps are $\mathbf{D}(\mathfrak{Y})$*-isomorphisms*

$$f_*\mathcal{D}^\bullet(\mathcal{G}) \xrightarrow{\sim} \mathbf{R}f_*\mathcal{D}^\bullet(\mathcal{G}) \underset{\mathbf{R}f_*\delta(\mathcal{G})}{\xleftarrow{\sim}} \mathbf{R}f_*\mathcal{G}, \tag{4.1.3.1}$$

i.e., the natural map $\alpha^i\colon H^i(f_*\mathcal{D}^\bullet(\mathcal{G})) \to H^i(\mathbf{R}f_*\mathcal{D}^\bullet(\mathcal{G}))$ is an isomorphism for all $i \in \mathbb{Z}$: this holds for bounded-below $\mathcal{G}$ because $\mathcal{D}^\bullet(\mathcal{G})$ is a complex of f_*-acyclic objects; and for arbitrary $\mathcal{G}$ since for any $n \in \mathbb{Z}$, with $\mathcal{G}^{\ge n}$ denoting the truncation

$$\cdots \to 0 \to 0 \to \operatorname{coker}(\mathcal{G}^{n-1} \to \mathcal{G}^n) \to \mathcal{G}^{n+1} \to \mathcal{G}^{n+2} \to \cdots \tag{4.1.3.2}$$

there is a natural commutative diagram

$$\begin{array}{ccc} H^i(f_*\mathcal{D}^\bullet(\mathcal{G})) & \xrightarrow{\alpha^i} & H^i(\mathbf{R}f_*\mathcal{D}^\bullet(\mathcal{G})) \\ {\scriptstyle\beta^i_n}\big\downarrow & & \big\downarrow{\scriptstyle\gamma^i_n} \\ H^i(f_*\mathcal{D}^\bullet(\mathcal{G}^{\ge n})) & \xrightarrow[\alpha^i_n]{} & H^i(\mathbf{R}f_*\mathcal{D}^\bullet(\mathcal{G}^{\ge n})) \end{array}$$

in which, when $n \ll i$, β^i_n is an isomorphism (since $\mathcal{G}$ and $\mathcal{G}^{\ge n}$ are identical in all degrees $> n$), γ^i_n is an isomorphism (by Proposition 3.4.3(b) applied to the mapping cone of the natural composition $\mathcal{D}^\bullet(\mathcal{G}) \xrightarrow{\sim} \mathcal{G} \longrightarrow \mathcal{G}^{\ge n} \xrightarrow{\sim} \mathcal{D}^\bullet(\mathcal{G}^{\ge n})$), and α^i_n is an isomorphism (since $\mathcal{G}^{\ge n}$ is bounded below).

Thus we have realized $\mathbf{R}f_* \circ \boldsymbol{j}$ at the homotopy level, via the functor $\mathcal{C}^\bullet := f_*\mathcal{D}^\bullet$; and our task is now to find a right adjoint at this level.

3. Each functor $\mathcal{C}^p = f_*\mathcal{D}^p\colon \mathcal{A}_{\vec{c}}(\mathfrak{X}) \to \mathcal{A}(\mathfrak{Y})$ is exact, since $R^1f_*(\mathcal{D}^p(\mathcal{M})) = 0$ for all $\mathcal{M} \in \mathcal{A}_{\vec{c}}(\mathfrak{X})$. $\mathcal{C}^p$ is continuous, since $\mathcal{D}^p$ is and, $\mathfrak{X}$ being noetherian, f_* commutes with $\varinjlim$. As before, the Special Adjoint Functor Theorem yields that *$\mathcal{C}^p$ has a right adjoint $\mathcal{C}_p\colon \mathcal{A}(\mathfrak{Y}) \to \mathcal{A}_{\vec{c}}(\mathfrak{X})$.*

For each $\mathcal{A}(\mathcal{Y})$-complex $\mathcal{F} = (\mathcal{F}^p)_{p\in\mathbb{Z}}$ let $\mathcal{C}_\bullet\mathcal{F}$ be the $\mathcal{A}_{\vec{c}}(\mathcal{X})$-complex with

$$(\mathcal{C}_\bullet\mathcal{F})^m := \prod_{p-q=m} \mathcal{C}_q\mathcal{F}^p \qquad (m \in \mathbb{Z}, 0 \le q \le e),$$

and with differential $(\mathcal{C}_\bullet\mathcal{F})^m \to (\mathcal{C}_\bullet\mathcal{F})^{m+1}$ the unique map making the following diagram commute for all r, s with $r - s = m+1$:

$$\begin{array}{ccc} \prod\limits_{p-q=m} \mathcal{C}_q\mathcal{F}^p & \longrightarrow & \prod\limits_{p-q=m+1} \mathcal{C}_q\mathcal{F}^p \\ \downarrow & & \downarrow \\ \mathcal{C}_s\mathcal{F}^{r-1} \oplus \mathcal{C}_{s+1}\mathcal{F}^r & \xrightarrow[d_{\prime}+(-1)^r d_{\prime\prime}]{} & \mathcal{C}_s\mathcal{F}^r \end{array}$$

where:

(i) the vertical arrows come from projections,
(ii) $d_{\prime}\colon \mathcal{C}_s\mathcal{F}^{r-1} \to \mathcal{C}_s\mathcal{F}^r$ corresponds to the differential in $\mathcal{F}$, and
(iii) with $\delta_s\colon \mathcal{C}_{s+1} \to \mathcal{C}_s$ corresponding by adjunction to $f_*(\delta^s)\colon \mathcal{C}^s \to \mathcal{C}^{s+1}$,

$$d_{\prime\prime} := (-1)^s\delta_s(\mathcal{F}^r)\colon \mathcal{C}_{s+1}\mathcal{F}^r \to \mathcal{C}_s\mathcal{F}^r.$$

This construction leads naturally to a Δ-functor $(\mathcal{C}_\bullet, \mathrm{Id})\colon \mathbf{K}(\mathcal{Y}) \to \mathbf{K}(\mathcal{A}_{\vec{c}}(\mathcal{X}))$. The adjunction isomorphism

$$\mathrm{Hom}_{\mathcal{A}_{\vec{c}}(\mathcal{X})}(\mathcal{M}, \mathcal{C}_p\mathcal{N}) \xrightarrow{\sim} \mathrm{Hom}_{\mathcal{A}(\mathcal{Y})}(\mathcal{C}^p\mathcal{M}, \mathcal{N}) \qquad (\mathcal{M} \in \mathcal{A}_{\vec{c}}(\mathcal{X}),\ \mathcal{N} \in \mathcal{A}(\mathcal{Y}))$$

applied componentwise produces an isomorphism of complexes of abelian groups

$$\mathrm{Hom}^\bullet_{\mathcal{A}_{\vec{c}}(\mathcal{X})}(\mathcal{G}, \mathcal{C}_\bullet\mathcal{F}) \xrightarrow{\sim} \mathrm{Hom}^\bullet_{\mathcal{A}(\mathcal{Y})}(\mathcal{C}^\bullet\mathcal{G}, \mathcal{F}) \tag{4.1.4.1}$$

for all $\mathcal{A}_{\vec{c}}(\mathcal{X})$-complexes $\mathcal{G}$ and $\mathcal{A}(\mathcal{Y})$-complexes $\mathcal{F}$.

4. The isomorphism (4.1.4.1) suggests that we use $\mathcal{C}_\bullet$ to construct $f^\times$, as follows. Recall that a complex $\mathcal{J} \in \mathbf{K}(\mathcal{A}_{\vec{c}}(\mathcal{X}))$ is K-injective iff for each exact complex $\mathcal{G} \in \mathbf{K}(\mathcal{A}_{\vec{c}}(\mathcal{X}))$, the complex $\mathrm{Hom}^\bullet_{\mathcal{A}_{\vec{c}}(\mathcal{X})}(\mathcal{G}, \mathcal{J})$ is exact too. By (4.1.3.1), $\mathcal{C}^\bullet\mathcal{G}$ is exact if $\mathcal{G}$ is; so it follows from (4.1.4.1) that *if $\mathcal{F}$ is K-injective in $\mathbf{K}(\mathcal{Y})$ then $\mathcal{C}_\bullet\mathcal{F}$ is K-injective in $\mathbf{K}(\mathcal{A}_{\vec{c}}(\mathcal{X}))$.* Thus if $\mathbf{K_I}(-) \subset \mathbf{K}(-)$ is the full subcategory of all K-injective complexes, then we have a Δ-functor $(\mathcal{C}_\bullet, \mathrm{Id})\colon \mathbf{K_I}(\mathcal{Y}) \to \mathbf{K_I}(\mathcal{A}_{\vec{c}}(\mathcal{X}))$. Associating a K-injective resolution to each complex in $\mathcal{A}(\mathcal{Y})$ leads to a Δ-functor $(\rho, \Theta)\colon \mathbf{D}(\mathcal{Y}) \to \mathbf{K_I}(\mathcal{Y})$.[16] This ρ is bounded below: an $\mathcal{A}(\mathcal{Y})$-complex $\mathcal{E}$ such that $H^i(\mathcal{E}) = 0$ for all $i < n$ is quasi-isomorphic to its truncation $\mathcal{E}^{\ge n}$ (see (4.1.3.2)), which is quasi-isomorphic to an injective complex $\mathcal{F}$ which vanishes in all degrees below n. (Such an $\mathcal{F}$ is K-injective.)

Finally, one can define $f^\times$ to be the composition of the functors

$$\mathbf{D}(\mathcal{Y}) \xrightarrow{\rho} \mathbf{K_I}(\mathcal{Y}) \xrightarrow{\mathcal{C}_\bullet} \mathbf{K_I}(\mathcal{A}_{\vec{c}}(\mathcal{X})) \xrightarrow{\text{natural}} \mathbf{D}(\mathcal{A}_{\vec{c}}(\mathcal{X})),$$

and check, via (4.1.3.1) and (4.1.4.1) that Theorem 4.1 is satisfied. (This involves some tedium with respect to Δ-details.) □

[16] In fact (ρ, Θ) is an equivalence of Δ-categories, see [**L4**, §1.7]. But note that Θ need not be the identity morphism, i.e., one may not be able to find a complete family of K-injective resolutions commuting with translation. For example, we do not know that every periodic complex has a periodic K-injective resolution.

5. Torsion sheaves.

Refer to §1.2 for notation and first sorites regarding torsion sheaves.

Paragraphs 5.1 and 5.2 develop properties of quasi-coherent torsion sheaves and their derived categories on locally noetherian formal schemes—see e.g., Propositions 5.2.1, 5.2.4, 5.2.6, and Corollary 5.2.11. (There is some overlap here with §4 in [**Y**].) Such properties will be needed throughout the rest of the paper. For instance, Paragraph 5.3 establishes for a noetherian formal scheme $\mathfrak{X}$, either separated or finite-dimensional, an *equivalence of categories* $\mathbf{D}(\mathcal{A}_{\mathrm{qct}}(\mathfrak{X})) \xrightarrow{\approx} \mathbf{D}_{\mathrm{qct}}(\mathfrak{X})$, thereby enabling the use of $\mathbf{D}_{\mathrm{qct}}(\mathfrak{X})$—rather than $\mathbf{D}(\mathcal{A}_{\mathrm{qct}}(\mathfrak{X}))$—in Theorem 6.1 ($\cong$ Theorem 2 of Section 1). Also, Lemma 5.4.1, identifying the derived functor $\mathbf{R}\varGamma_{\mathcal{J}}(-)$ (for any $\mathcal{O}_X$-ideal $\mathcal{J}$, where X is a ringed space) with the homotopy colimit of the functors $\mathbf{R}\mathcal{H}om^{\bullet}(\mathcal{O}_X/\mathcal{J}^n, -)$, plays a key role in the proof of the Base Change Theorem 7.4 ($\cong$ Theorem 3).

5.1. This paragraph deals with categories of quasi-coherent torsion sheaves on locally noetherian formal schemes.

PROPOSITION 5.1.1. *Let $f\colon \mathfrak{X} \to \mathfrak{Y}$ be a map of noetherian formal schemes, and let $\mathcal{M} \in \mathcal{A}_{\mathrm{qct}}(\mathfrak{X})$. Then $f_*\mathcal{M} \in \mathcal{A}_{\mathrm{qct}}(\mathfrak{Y})$. Moreover, if f is pseudo-proper (see §1.2.2) and $\mathcal{M}$ is coherent then $f_*\mathcal{M}$ is coherent.*

PROOF. Let $\mathfrak{J} \subset \mathcal{O}_{\mathfrak{X}}$ and $\mathfrak{I} \subset \mathcal{O}_{\mathfrak{Y}}$ be ideals of definition such that $\mathfrak{I}\mathcal{O}_{\mathfrak{X}} \subset \mathfrak{J}$, and let

$$X_n := (\mathfrak{X}, \mathcal{O}_{\mathfrak{X}}/\mathfrak{J}^n) \xrightarrow{f_n} (\mathfrak{Y}, \mathcal{O}_{\mathfrak{Y}}/\mathfrak{I}^n) =: Y_n \qquad (n > 0)$$

be the scheme-maps induced by f, so that if j_n and i_n are the canonical closed immersions then $fj_n = i_n f_n$. Let $\mathcal{M}_n := \mathcal{H}om(\mathcal{O}_{\mathfrak{X}}/\mathfrak{J}^n, \mathcal{M})$, so that

$$\mathcal{M} = \varGamma'_{\mathfrak{X}}\mathcal{M} = \varinjlim_n \mathcal{M}_n = \varinjlim_n j_{n*}j_n^*\mathcal{M}_n.$$

Since $\mathfrak{J}^n$ is a coherent $\mathcal{O}_{\mathfrak{X}}$-ideal [**GD**, p. 427], therefore $\mathcal{M}_n$ is quasi-coherent (Corollary 3.1.6(d)), and it is straightforward to check that $i_{n*}f_{n*}j_n^*\mathcal{M}_n \in \mathcal{A}_{\mathrm{qct}}(\mathfrak{Y})$. Thus, $\mathfrak{X}$ being noetherian, and by Corollary 5.1.3 below,

$$f_*\mathcal{M} = f_*\varinjlim_n \mathcal{M}_n \cong \varinjlim_n f_*j_{n*}j_n^*\mathcal{M}_n = \varinjlim_n i_{n*}f_{n*}j_n^*\mathcal{M}_n \in \mathcal{A}_{\mathrm{qct}}(\mathfrak{Y}).$$

When f is pseudo-proper then every f_n is proper; and if $\mathcal{M} \in \mathcal{A}_{\mathrm{qct}}(\mathfrak{X})$ is coherent then so is $f_*\mathcal{M}$, because for some n, $f_*\mathcal{M} = f_*j_{n*}j_n^*\mathcal{M}_n = i_{n*}f_{n*}j_n^*\mathcal{M}_n$. □

PROPOSITION 5.1.2. *Let Z be a closed subset of a locally noetherian scheme X, and let $\kappa\colon \mathfrak{X} \to X$ be the completion of X along Z. Then the functors κ^* and κ_* restrict to inverse isomorphisms between the categories $\mathcal{A}_Z(X)$ and $\mathcal{A}_{\mathrm{t}}(\mathfrak{X})$, and between the categories $\mathcal{A}_{\mathrm{qc}Z}(X)$ and $\mathcal{A}_{\mathrm{qct}}(\mathfrak{X})$; and if $\mathcal{M} \in \mathcal{A}_{\mathrm{qct}}(\mathfrak{X})$ is coherent, then so is $\kappa_*\mathcal{M}$.*

PROOF. Let $\mathcal{J}$ be a quasi-coherent $\mathcal{O}_X$-ideal such that the support of $\mathcal{O}_X/\mathcal{J}$ is Z. Applying $\varinjlim_n$ to the natural isomorphisms

$$\kappa^*\mathcal{H}om_X(\mathcal{O}_X/\mathcal{J}^n, \mathcal{N}) \xrightarrow{\sim} \mathcal{H}om_{\mathfrak{X}}(\mathcal{O}_{\mathfrak{X}}/\mathcal{J}^n\mathcal{O}_{\mathfrak{X}}, \kappa^*\mathcal{N}) \qquad (\mathcal{N} \in \mathcal{A}(X),\ n > 0)$$

we get a functorial isomorphism $\kappa^*\varGamma'_Z \xrightarrow{\sim} \varGamma'_{\mathfrak{X}}\kappa^*$, and hence $\kappa^*(\mathcal{A}_Z(X)) \subset \mathcal{A}_t(\mathfrak{X})$. Applying $\varinjlim_n$ to the natural isomorphisms

$$\mathcal{H}om_X(\mathcal{O}_X/\mathcal{J}^n, \kappa_*\mathcal{M}) \xrightarrow{\sim} \kappa_*\mathcal{H}om_{\mathfrak{X}}(\mathcal{O}_{\mathfrak{X}}/\mathcal{J}^n\mathcal{O}_{\mathfrak{X}}, \mathcal{M}) \qquad (\mathcal{M} \in \mathcal{A}(\mathfrak{X}),\ n > 0)$$

we get a functorial isomorphism $\varGamma'_Z\kappa_* \xrightarrow{\sim} \kappa_*\varGamma'_{\mathfrak{X}}$, and hence $\kappa_*(\mathcal{A}_t(\mathfrak{X})) \subset \mathcal{A}_Z(X)$.

As κ is a pseudo-proper map of locally noetherian formal schemes ((0) being an ideal of definition of X), we see as in the proof of Lemma 5.1.1 that for $\mathcal{M} \in \mathcal{A}_{\mathrm{qct}}(\mathfrak{X})$, $\kappa_*\mathcal{M}$ is a $\varinjlim$ of quasi-coherent $\mathcal{O}_X$-modules, so is itself quasi-coherent, and $\kappa_*\mathcal{M}$ is coherent whenever $\mathcal{M}$ is.[17]

Finally, examining stalks (see §1.2) we find that the natural transformations $1 \to \kappa_*\kappa^*$ and $\kappa^*\kappa_* \to 1$ induce isomorphisms

$$\begin{aligned} \varGamma'_Z\mathcal{N} &\xrightarrow{\sim} \kappa_*\kappa^*\varGamma'_Z\mathcal{N} \qquad (\mathcal{N} \in \mathcal{A}(X)),\\ \kappa^*\kappa_*\varGamma'_{\mathfrak{X}}\mathcal{M} &\xrightarrow{\sim} \varGamma'_{\mathfrak{X}}\mathcal{M} \qquad (\mathcal{M} \in \mathcal{A}(\mathfrak{X})). \end{aligned}$$ □

Corollary 5.1.3. *If $\mathfrak{X}$ is a locally noetherian formal scheme then $\mathcal{A}_{\mathrm{qct}}(\mathfrak{X})$ is plump in $\mathcal{A}(\mathfrak{X})$ and closed under small $\mathcal{A}(\mathfrak{X})$-colimits.*

Proof. The assertions are local, and so, since $\mathcal{A}_{\mathrm{t}}(\mathfrak{X})$ is plump (§1.2.1), Proposition 5.1.2 (where κ^* commutes with $\varinjlim$) enables reduction to well-known facts about $\mathcal{A}_{\mathrm{qc}Z}(X) \subset \mathcal{A}(X)$ with X an affine noetherian (ordinary) scheme. □

Lemma 5.1.4. *Let $\mathfrak{X}$ be a locally noetherian formal scheme. If $\mathcal{M}$ is a quasi-coherent $\mathcal{O}_{\mathfrak{X}}$-module then $\varGamma'_{\mathfrak{X}}\mathcal{M} \in \mathcal{A}_{\mathrm{qct}}(\mathfrak{X})$ is the $\varinjlim$ of its coherent submodules. In particular, $\mathcal{A}_{\mathrm{qct}}(\mathfrak{X}) \subset \mathcal{A}_{\vec{\mathrm{c}}}(\mathfrak{X})$.*

Proof. Let $\mathfrak{J}$ be an ideal of definition of $\mathfrak{X}$. For any positive integer n, let X_n be the scheme $(\mathfrak{X}, \mathcal{O}_{\mathfrak{X}}/\mathfrak{J}^n)$, let $j_n\colon X_n \to \mathfrak{X}$ be the canonical closed immersion, and let $\mathcal{M}_n := \mathcal{H}om(\mathcal{O}_{\mathfrak{X}}/\mathfrak{J}^n, \mathcal{M}) \subset \varGamma'_{\mathfrak{X}}(\mathcal{M})$, so that $\mathcal{M}_n \in \mathcal{A}_{\mathrm{qct}}(\mathfrak{X})$ (Corollary 3.1.6(d)). Then the quasi-coherent $\mathcal{O}_{X_n}$-module $j_n^*\mathcal{M}_n$ is the $\varinjlim$ of its coherent submodules [**GD**, p. 319, (6.9.9)], hence so is $\mathcal{M}_n = j_{n*}j_n^*\mathcal{M}_n$ (since j_n^* and j_{n*} preserve both $\varinjlim$ and coherence [**GD**, p. 115, (5.3.13) and (5.3.15)]), and therefore so is $\varGamma'_{\mathfrak{X}}\mathcal{M} = \varinjlim_n \mathcal{M}_n$. That $\varinjlim_n \mathcal{M}_n \in \mathcal{A}_{\mathrm{qct}}(\mathfrak{X})$ results from Corollary 5.1.3. □

Corollary 5.1.5. *For a locally noetherian formal scheme $\mathfrak{X}$, the inclusion functor $j^{\mathrm{t}}_{\mathfrak{X}}\colon \mathcal{A}_{\mathrm{qct}}(\mathfrak{X}) \hookrightarrow \mathcal{A}(\mathfrak{X})$ has a right adjoint $Q^{\mathrm{t}}_{\mathfrak{X}}$. If moreover $\mathfrak{X}$ is noetherian then $Q^{\mathrm{t}}_{\mathfrak{X}}$ commutes with $\varinjlim$.*

Proof. To show that $j^{\mathrm{t}}_{\mathfrak{X}}$ has a right adjoint one can, in view of Corollary 5.1.3 and Lemma 5.1.4, simply apply the Special Adjoint Functor theorem.

More specifically, since $\varGamma'_{\mathfrak{X}}$ is right-adjoint to the inclusion $\mathcal{A}_{\mathrm{t}}(\mathfrak{X}) \hookrightarrow \mathcal{A}(\mathfrak{X})$, and $\mathcal{A}_{\vec{\mathrm{c}}}(\mathfrak{X}) \subset \mathcal{A}_{\mathrm{qc}}(\mathfrak{X})$ (Corollary 3.1.5), it follows from Lemma 5.1.4 that the restriction of $\varGamma'_{\mathfrak{X}}$ to $\mathcal{A}_{\vec{\mathrm{c}}}(\mathfrak{X})$ is right-adjoint to $\mathcal{A}_{\mathrm{qct}}(\mathfrak{X}) \hookrightarrow \mathcal{A}_{\vec{\mathrm{c}}}(\mathfrak{X})$; and by Proposition 3.2.3, $\mathcal{A}_{\vec{\mathrm{c}}}(\mathfrak{X}) \hookrightarrow \mathcal{A}(\mathfrak{X})$ has a right adjoint $Q_{\mathfrak{X}}$; so $Q^{\mathrm{t}}_{\mathfrak{X}} := \varGamma'_{\mathfrak{X}} \circ Q_{\mathfrak{X}}$ is right-adjoint to $j^{\mathrm{t}}_{\mathfrak{X}}$. (Similarly, $Q_{\mathfrak{X}} \circ \varGamma'_{\mathfrak{X}}$ is right-adjoint to $j^{\mathrm{t}}_{\mathfrak{X}}$.)

Commutativity with $\varinjlim$ means that for any small directed system $(\mathcal{G}_\alpha)$ in $\mathcal{A}(\mathfrak{X})$ and any $\mathcal{M} \in \mathcal{A}_{\mathrm{qct}}(\mathfrak{X})$, the natural map

$$\phi\colon \mathrm{Hom}(\mathcal{M}, \varinjlim_\alpha Q^{\mathrm{t}}_{\mathfrak{X}}\mathcal{G}_\alpha) \to \mathrm{Hom}(\mathcal{M}, Q^{\mathrm{t}}_{\mathfrak{X}}\varinjlim_\alpha \mathcal{G}_\alpha)$$

is an *isomorphism.* This follows from Lemma 5.1.4, which allows us to assume that $\mathcal{M}$ is coherent, in which case ϕ is isomorphic to the natural composed isomorphism

$$\varinjlim_\alpha \mathrm{Hom}(\mathcal{M}, Q^{\mathrm{t}}_{\mathfrak{X}}\mathcal{G}_\alpha) \xrightarrow{\sim} \varinjlim_\alpha \mathrm{Hom}(\mathcal{M}, \mathcal{G}_\alpha) \xrightarrow{\sim} \mathrm{Hom}(\mathcal{M}, \varinjlim_\alpha \mathcal{G}_\alpha).$$ □

[17] The noetherian assumption in Lemma 5.1.1 is needed only for commutativity of f_* with $\varinjlim$, a condition clearly satisfied by $f = \kappa$ in the present situation.

Remark. For an ordinary noetherian scheme X we have $Q^{\mathrm{t}}_X = Q_X$ (see §3.1). More generally, if $\kappa\colon \mathfrak{X} \to X$ is as in Proposition 5.1.2, then $Q^{\mathrm{t}}_{\mathfrak{X}} = \kappa^* \varGamma_Z Q_X \kappa_*$. Hence Proposition 5.1.1 (applied to open immersions $\mathfrak{X} \hookrightarrow \mathfrak{Y}$ with $\mathfrak{X}$ affine) lets us construct the functor $Q^{\mathrm{t}}_{\mathfrak{Y}}$ for any noetherian formal scheme $\mathfrak{Y}$ by mimicking the construction for ordinary schemes (cf. [**I**, p. 187, Lemme 3.2].)

5.2. The preceding results carry over to derived categories.

From Corollary 5.1.3 it follows that on a locally noetherian formal scheme $\mathfrak{X}$, $\mathbf{D}_{\mathrm{qct}}(\mathfrak{X})$ is a triangulated subcategory of $\mathbf{D}(\mathfrak{X})$, closed under direct sums.

PROPOSITION 5.2.1. *For a locally noetherian formal scheme $\mathfrak{X}$, set $\mathcal{A}_{\mathrm{t}} := \mathcal{A}_{\mathrm{t}}(\mathfrak{X})$, the category of torsion $\mathcal{O}_{\mathfrak{X}}$-modules, and let $\boldsymbol{i}\colon \mathbf{D}(\mathcal{A}_{\mathrm{t}}) \to \mathbf{D}(\mathfrak{X})$ be the natural functor. Then:*

(a) *An $\mathcal{O}_{\mathfrak{X}}$-complex $\mathcal{E}$ is in $\mathbf{D}_{\mathrm{t}}(\mathfrak{X})$ iff the natural map $\boldsymbol{i}\mathbf{R}\varGamma'_{\mathfrak{X}}\mathcal{E} \to \mathcal{E}$ is a $\mathbf{D}(\mathfrak{X})$-isomorphism.*

(b) *If $\mathcal{E} \in \mathbf{D}_{\mathrm{qc}}(\mathfrak{X})$ then $\mathbf{R}\varGamma'_{\mathfrak{X}}\mathcal{E} \in \mathbf{D}_{\mathrm{qc}}(\mathcal{A}_{\mathrm{t}})$.*

(c) *The functor $\boldsymbol{i}$ and its right adjoint $\mathbf{R}\varGamma'_{\mathfrak{X}}$ induce quasi-inverse equivalences between $\mathbf{D}(\mathcal{A}_{\mathrm{t}})$ and $\mathbf{D}_{\mathrm{t}}(\mathfrak{X})$ and between $\mathbf{D}_{\mathrm{qc}}(\mathcal{A}_{\mathrm{t}})$ and $\mathbf{D}_{\mathrm{qct}}(\mathfrak{X})$.*[18]

PROOF. (a) For $\mathcal{F} \in \mathbf{D}(\mathcal{A}_{\mathrm{t}})$ (e.g., $\mathcal{F} := \mathbf{R}\varGamma'_{\mathfrak{X}}\mathcal{E}$), any complex isomorphic to $\boldsymbol{i}\mathcal{F}$ is clearly in $\mathbf{D}_{\mathrm{t}}(\mathfrak{X})$.

Suppose conversely that $\mathcal{E} \in \mathbf{D}_{\mathrm{t}}(\mathfrak{X})$. The assertion that $\boldsymbol{i}\mathbf{R}\varGamma'_{\mathfrak{X}}\mathcal{E} \cong \mathcal{E}$ is local, so we may assume that $\mathfrak{X} = \mathrm{Spf}(A)$ where $A = \Gamma(\mathfrak{X}, \mathcal{O}_{\mathfrak{X}})$ is a noetherian adic ring, so that any defining ideal $\mathcal{J}$ of $\mathfrak{X}$ is generated by a finite sequence in A. Then $\boldsymbol{i}\mathbf{R}\varGamma'_{\mathfrak{X}}\mathcal{E} \cong \mathcal{K}^\bullet_\infty \otimes \mathcal{E}$, where $\mathcal{K}^\bullet_\infty$ is a bounded flat complex—a $\varinjlim$ of Koszul complexes on powers of the generators of $\mathcal{J}$—see [**AJL**, p. 18, Lemma 3.1.1].

So $\boldsymbol{i}\mathbf{R}\varGamma'_{\mathfrak{X}}$ is a bounded functor, and the usual way-out argument reduces the question to where $\mathcal{E}$ is a single torsion sheaf. But then it is immediate from the construction of $\mathcal{K}^\bullet_\infty$ that $\mathcal{K}^\bullet_\infty \otimes \mathcal{E} = \mathcal{E}$.

(b) Again, we can assume that $\mathfrak{X} = \mathrm{Spf}(A)$ and $\mathbf{R}\varGamma'_{\mathfrak{X}}$ is bounded, and since $\mathcal{A}_{\mathrm{qc}}(\mathfrak{X})$ is plump in $\mathcal{A}(\mathfrak{X})$ (Proposition 3.2.2) we can reduce to where $\mathcal{E}$ is a single quasi-coherent $\mathcal{O}_{\mathfrak{X}}$-module, though it is better to assume only that $\mathcal{E} \in \mathbf{D}^+_{\mathrm{qc}}(\mathfrak{X})$, for then we may also assume $\mathcal{E}$ injective, so that

$$\mathbf{R}\varGamma'_{\mathfrak{X}}\mathcal{E} \cong \varGamma'_{\mathfrak{X}}\mathcal{E} = \varinjlim_{n>0} \mathcal{H}om(\mathcal{O}/\mathcal{J}^n, \mathcal{E}).$$

From Corollary 3.1.6(d) it follows that $\mathcal{H}om(\mathcal{O}/\mathcal{J}^n, \mathcal{E}) \in \mathbf{D}_{\mathrm{qct}}(\mathfrak{X})$—for this assertion another way-out argument reduces us again to where $\mathcal{E}$ is a single quasi-coherent $\mathcal{O}_{\mathfrak{X}}$-module—and since homology commutes with $\varinjlim$ and $\mathcal{A}_{\mathrm{qct}}$ is closed under $\varinjlim$ (Corollary 5.1.3), therefore $\mathbf{R}\varGamma'_{\mathfrak{X}}\mathcal{E}$ has quasi-coherent homology.

Assertion (c) results now from the following simple lemma. □

LEMMA 5.2.2. *Let $\mathcal{A}$ be an abelian category, let $j\colon \mathcal{A}_\flat \to \mathcal{A}$ be the inclusion of a plump subcategory such that j has a right adjoint $\varGamma$, and let derived-category extension of j. Suppose that every $\mathcal{A}$-complex has a K-injective resolution, so that the derived functor $\mathbf{R}\varGamma\colon \mathbf{D}(\mathcal{A}) \to \mathbf{D}(\mathcal{A}_\flat)$ exists. Then $\mathbf{R}\varGamma$ is right-adjoint to $\boldsymbol{j}$. Furthermore, the following conditions are equivalent.*

(1) *$\boldsymbol{j}$ induces an equivalence of categories from $\mathbf{D}(\mathcal{A}_\flat)$ to $\mathbf{D}_\flat(\mathcal{A})$, with quasi-inverse $\mathbf{R}_\flat\varGamma := \mathbf{R}\varGamma|_{\mathbf{D}_\flat(\mathcal{A})}$.*

---

[18] We may therefore sometimes abuse notation and write $\mathbf{R}\varGamma'_{\mathfrak{X}}$ instead of $\boldsymbol{i}\mathbf{R}\varGamma'_{\mathfrak{X}}$; but the meaning should be clear from the context.

(2) *For every* $\mathcal{E} \in \mathbf{D}_\flat(\mathcal{A})$ *the natural map* $\boldsymbol{j}\mathbf{R}\varGamma\mathcal{E} \to \mathcal{E}$ *is an isomorphism.*

(3) *The functor* $\mathbf{R}_\flat\varGamma$ *is bounded, and for* $\mathcal{E}_0 \in \mathcal{A}_\flat$ *the natural map* $\boldsymbol{j}\mathbf{R}\varGamma\mathcal{E}_0 \to \mathcal{E}_0$ *is a* $\mathbf{D}(\mathcal{A})$*-isomorphism.*

When these conditions hold, every $\mathcal{A}_\flat$*-complex has a K-injective resolution.*

PROOF. Since $\varGamma$ has an exact left adjoint, it takes K-injective $\mathcal{A}$-complexes to K-injective $\mathcal{A}_\flat$-complexes, whence there is a bifunctorial isomorphism in the derived category of abelian groups

$$\mathbf{R}\mathrm{Hom}^\bullet_{\mathcal{A}}(\boldsymbol{j}\mathcal{G},\mathcal{E}) \xrightarrow{\ \sim\ } \mathbf{R}\mathrm{Hom}^\bullet_{\mathcal{A}_\flat}(\mathcal{G},\mathbf{R}\varGamma\mathcal{E}) \qquad (\mathcal{G}\in\mathbf{D}(\mathcal{A}_\flat),\ \mathcal{E}\in\mathbf{D}(\mathcal{A})).$$

(To see this, one can assume $\mathcal{E}$ to be K-injective, and then drop the $\mathbf{R}$'s ...) Apply homology H^0 to this isomorphism to get adjointness of $\boldsymbol{j}$ and $\mathbf{R}\varGamma$.

The implications (1) ⇒ (3) ⇒ (2) are straightforward. For (2) ⇒ (1), one needs that for $\mathcal{G} \in \mathbf{D}(\mathcal{A}_\flat)$ the natural map $\mathcal{G} \to \mathbf{R}\varGamma\boldsymbol{j}\mathcal{G}$ is an isomorphism, or equivalently (look at homology), that the corresponding map $\boldsymbol{j}\mathcal{G} \to \boldsymbol{j}\mathbf{R}\varGamma\boldsymbol{j}\mathcal{G}$ is an isomorphism. But the composition of this last map with the isomorphism $\boldsymbol{j}\mathbf{R}\varGamma\boldsymbol{j}\mathcal{G} \xrightarrow{\ \sim\ } \boldsymbol{j}\mathcal{G}$ (given by (2)) is the identity, whence the conclusion.

Finally, if $\mathcal{G}$ is an $\mathcal{A}_\flat$-complex and $j\mathcal{G} \to \mathcal{J}$ is a K-injective $\mathcal{A}$-resolution, then as before $\varGamma\mathcal{J}$ is a K-injective $\mathcal{A}_\flat$-complex; and (1) implies that the natural composition

$$\mathcal{G} \to \varGamma j\mathcal{G} \to \varGamma\mathcal{J}\ \ (\cong \mathbf{R}\varGamma\boldsymbol{j}\mathcal{G})$$

is a $\mathbf{D}(\mathcal{A}_\flat)$-isomorphism, hence an $\mathcal{A}_\flat$-K-injective resolution. □

COROLLARY 5.2.3. *For any complexes* $\mathcal{E} \in \mathbf{D}_{\mathrm{t}}(\mathfrak{X})$ *and* $\mathcal{F} \in \mathbf{D}(\mathfrak{X})$ *the natural map* $\mathbf{R}\varGamma'_{\mathfrak{X}}\mathcal{F} \to \mathcal{F}$ *induces an isomorphism*

$$\mathbf{R}\mathcal{H}om^\bullet(\mathcal{E},\mathbf{R}\varGamma'_{\mathfrak{X}}\mathcal{F}) \xrightarrow{\ \sim\ } \mathbf{R}\mathcal{H}om^\bullet(\mathcal{E},\mathcal{F}).$$

PROOF. Consideration of homology presheaves shows it sufficient that for each affine open $\mathfrak{U} \subset \mathfrak{X}$, the natural map

$$\mathrm{Hom}_{\mathbf{D}(\mathfrak{U})}(\mathcal{E}|_{\mathfrak{U}}, (\mathbf{R}\varGamma'_{\mathfrak{X}}\mathcal{F})|_{\mathfrak{U}}) \to \mathrm{Hom}_{\mathbf{D}(\mathfrak{U})}(\mathcal{E}|_{\mathfrak{U}}, \mathcal{F}|_{\mathfrak{U}})$$

be an isomorphism. But since $\mathbf{R}\varGamma'_{\mathfrak{X}}$ commutes with restriction to $\mathfrak{U}$, that is a direct consequence of Proposition 5.2.1(c) (with $\mathfrak{X}$ replaced by $\mathfrak{U}$). □

Parts (b) and (c) of the following Proposition will be generalized in parts (d) and (b), respectively, of Proposition 5.2.8.

PROPOSITION 5.2.4. *Let* Z *be a closed subset of a locally noetherian scheme* X, *and let* $\kappa\colon \mathfrak{X} \to X$ *be the completion of* X *along* Z. *Then:*

(a) *The exact functors* κ^* *and* κ_* *restrict to inverse isomorphisms between the categories* $\mathbf{D}_Z(X)$ *and* $\mathbf{D}_{\mathrm{t}}(\mathfrak{X})$, *and between the categories* $\mathbf{D}_{\mathrm{qc}Z}(X)$ *and* $\mathbf{D}_{\mathrm{qct}}(\mathfrak{X})$; *and if* $\mathcal{M} \in \mathbf{D}_{\mathrm{qct}}(\mathfrak{X})$ *has coherent homology, then so does* $\kappa_*\mathcal{M}$.

(b) *There is a unique derived-category isomorphism*

$$\mathbf{R}\varGamma'_Z\kappa_*\mathcal{E} \xrightarrow{\ \sim\ } \kappa_*\mathbf{R}\varGamma'_{\mathfrak{X}}\mathcal{E} \qquad (\mathcal{E}\in\mathbf{D}(\mathfrak{X}))$$

whose composition with the natural map $\kappa_*\mathbf{R}\varGamma'_{\mathfrak{X}}\mathcal{E} \to \kappa_*\mathcal{E}$ *is just the natural map* $\mathbf{R}\varGamma'_Z\kappa_*\mathcal{E} \to \kappa_*\mathcal{E}$.

(c) *There is a unique derived-category isomorphism*

$$\kappa^*\mathbf{R}\varGamma'_Z\mathcal{F} \xrightarrow{\ \sim\ } \mathbf{R}\varGamma'_{\mathfrak{X}}\kappa^*\mathcal{F} \qquad (\mathcal{F}\in\mathbf{D}(X))$$

whose composition with the natural map $\mathbf{R}\varGamma'_{\mathfrak{X}}\kappa^*\mathcal{F} \to \kappa^*\mathcal{F}$ *is just the natural map* $\kappa^*\mathbf{R}\varGamma'_Z\mathcal{F} \to \kappa^*\mathcal{F}$.

PROOF. The assertions in (a) follow at once from Proposition 5.1.2.

(b) Since κ_* has an exact left adjoint (namely κ^*), therefore κ_* transforms K-injective $\mathcal{A}(\mathfrak{X})$-complexes into K-injective $\mathcal{A}(X)$-complexes, and consequently the isomorphism in (b) results from the isomorphism $\Gamma'_Z\kappa_* \xrightarrow{\sim} \kappa_*\Gamma'_{\mathfrak{X}}$ in the proof of Proposition 5.1.2. That the composition in (b) is as asserted comes down then to the elementary fact that the natural composition

$$\mathcal{H}om_X(\mathcal{O}_X/\mathcal{J}^n, \kappa_*\mathcal{M}) \xrightarrow{\sim} \kappa_*\mathcal{H}om_{\mathfrak{X}}(\mathcal{O}_{\mathfrak{X}}/\mathcal{J}^n\mathcal{O}_{\mathfrak{X}}, \mathcal{M}) \longrightarrow \kappa_*\mathcal{M}$$

(see proof of Proposition 5.1.2) is just the obvious map. Since $\kappa_*\mathbf{R}\Gamma'_{\mathfrak{X}}\mathcal{E} \in \mathbf{D}_Z(X)$ (by (a) and Proposition 5.2.1(a)), the uniqueness assertion (for the inverse isomorphism) results from adjointness of $\mathbf{R}\Gamma'_Z$ and the inclusion $\mathbf{D}_Z(X) \hookrightarrow \mathbf{D}(X)$. (The proof is similar to that of Proposition 5.2.1(c)).

(c) Using (b), we have the natural composed map

$$\kappa^*\mathbf{R}\Gamma'_Z\mathcal{F} \to \kappa^*\mathbf{R}\Gamma'_Z\kappa_*\kappa^*\mathcal{F} \xrightarrow{\sim} \kappa^*\kappa_*\mathbf{R}\Gamma'_{\mathfrak{X}}\kappa^*\mathcal{F} \to \mathbf{R}\Gamma'_{\mathfrak{X}}\kappa^*\mathcal{F}.$$

Showing this to be an isomorphism is a local problem, so assume $X = \mathrm{Spec}(A)$ with A a noetherian adic ring. Let $K^\bullet_\infty$ be the usual $\varinjlim$ of Koszul complexes on powers of a finite system of generators of an ideal of definition of A ([**AJL**, §3.1]); and let $\widetilde{K}^\bullet_\infty$ be the corresponding quasi-coherent complex on Spec(A), so that the complex $\mathcal{K}^\bullet_\infty$ in the proof of Proposition 5.2.1(a) is just $\kappa^*\widetilde{K}^\bullet_\infty$. Then one checks via [**AJL**, p. 18, Lemma (3.1.1)] that the map in question is isomorphic to the natural isomorphism of complexes

$$\kappa^*(\widetilde{K}^\bullet_\infty \otimes_{\mathcal{O}_X} \mathcal{F}) \xrightarrow{\sim} \kappa^*\widetilde{K}^\bullet_\infty \otimes_{\mathcal{O}_{\mathfrak{X}}} \kappa^*\mathcal{F}.$$

That the composition in (c) is as asserted results from the following natural commutative diagram, whose bottom row composes to the identity:

$$\begin{array}{ccccccc}
\kappa^*\mathbf{R}\Gamma'_Z\mathcal{F} & \longrightarrow & \kappa^*\mathbf{R}\Gamma'_Z\kappa_*\kappa^*\mathcal{F} & \xrightarrow{\sim} & \kappa^*\kappa_*\mathbf{R}\Gamma'_{\mathfrak{X}}\kappa^*\mathcal{F} & \longrightarrow & \mathbf{R}\Gamma'_{\mathfrak{X}}\kappa^*\mathcal{F} \\
\downarrow & & \downarrow & \text{(b)} & \downarrow & & \downarrow \\
\kappa^*\mathcal{F} & \longrightarrow & \kappa^*\kappa_*\kappa^*\mathcal{F} & = & \kappa^*\kappa_*\kappa^*\mathcal{F} & \longrightarrow & \kappa^*\mathcal{F}
\end{array}$$

Uniqueness is shown as in (b). □

COROLLARY 5.2.5. *The natural maps are isomorphisms*

$$\begin{aligned}
&\mathrm{Hom}_X(\mathcal{E},\mathcal{F}) \cong \mathrm{Hom}_X(\mathcal{E},\kappa_*\kappa^*\mathcal{F}) \cong \mathrm{Hom}_{\mathfrak{X}}(\kappa^*\mathcal{E},\kappa^*\mathcal{F}) \quad (\mathcal{E}\in\mathbf{D}_Z(X),\ \mathcal{F}\in\mathbf{D}(X)),\\
&\mathrm{Hom}_X(\mathcal{E},\mathcal{F}) \cong \mathrm{Hom}_X(\mathcal{E},\kappa_*\kappa^*\mathcal{F}) \cong \mathrm{Hom}_{\mathfrak{X}}(\kappa^*\mathcal{E},\kappa^*\mathcal{F}) \quad (\mathcal{E}\in\mathbf{D}(X),\ \mathcal{F}\in\mathbf{D}_Z(X)),\\
&\mathrm{Hom}_{\mathfrak{X}}(\mathcal{G},\mathcal{H}) \cong \mathrm{Hom}_{\mathfrak{X}}(\kappa^*\kappa_*\mathcal{G},\mathcal{H}) \cong \mathrm{Hom}_X(\kappa_*\mathcal{G},\kappa_*\mathcal{H}) \quad (\mathcal{G}\in\mathbf{D}_{\mathrm{t}}(\mathfrak{X}),\ \mathcal{H}\in\mathbf{D}(\mathfrak{X})).
\end{aligned}$$

PROOF. For the first line, use Proposition 5.2.1 and its analogue for $\mathbf{D}_Z(X)$, Lemma 5.2.2, and Proposition 5.2.4 to get the equivalent sequence of natural isomorphisms

$$\begin{aligned}
\mathrm{Hom}_X(\mathcal{E},\mathcal{F}) &\cong \mathrm{Hom}_X(\mathcal{E},\mathbf{R}\Gamma'_Z\mathcal{F})\\
&\cong \mathrm{Hom}_{\mathfrak{X}}(\kappa^*\mathcal{E},\kappa^*\mathbf{R}\Gamma'_Z\mathcal{F})\\
&\cong \mathrm{Hom}_{\mathfrak{X}}(\kappa^*\mathcal{E},\mathbf{R}\Gamma'_{\mathfrak{X}}\kappa^*\mathcal{F})\\
&\cong \mathrm{Hom}_{\mathfrak{X}}(\kappa^*\mathcal{E},\kappa^*\mathcal{F})\\
&\cong \mathrm{Hom}_X(\mathcal{E},\kappa_*\kappa^*\mathcal{F}).
\end{aligned}$$

The rest is immediate from Proposition 5.2.4(a). □

The next series of results concerns the behavior of $\mathbf{D}_{\mathrm{qct}}$ with respect to maps of formal schemes.

PROPOSITION 5.2.6. *Let $f\colon \mathcal{X} \to \mathcal{Y}$ be a map of noetherian formal schemes. Then $\mathbf{R}f_*|_{\mathbf{D}_{\mathrm{qct}}(\mathcal{X})}$ is bounded, and*

$$\mathbf{R}f_*(\mathbf{D}_{\mathrm{qct}}(\mathcal{X})) \subset \mathbf{D}_{\mathrm{qct}}(\mathcal{Y}).$$

Moreover, if f is pseudo-proper and $\mathcal{F} \in \mathbf{D}_{\mathrm{t}}(\mathcal{X})$ has coherent homology, then so does $\mathbf{R}f_*\mathcal{F} \in \mathbf{D}_{\mathrm{t}}(\mathcal{Y})$.

PROOF. Since $\mathbf{D}_{\mathrm{qct}}(\mathcal{X}) \subset \mathbf{D}_{\vec{\mathrm{c}}}(\mathcal{X})$ (Lemma 5.1.4), the boundedness assertion is given by Proposition 3.4.3(b). (Clearly, $\mathbf{R}f_*$ is bounded-below.) It suffices then for the next assertion (by the usual way-out arguments [**H1**, p. 73, Proposition 7.3]) to show for any $\mathcal{M} \in \mathcal{A}_{\mathrm{qct}}(\mathcal{X})$ that $\mathbf{R}f_*\mathcal{M} \in \mathbf{D}_{\mathrm{qct}}(\mathcal{Y})$.

Let $\mathcal{E}$ be an injective resolution of $\mathcal{M}$, let $\mathcal{J}$ be an ideal of definition of $\mathcal{X}$, and let $\mathcal{E}_n$ be the flasque complex $\mathcal{E}_n := \mathcal{H}om(\mathcal{O}/\mathcal{J}^n, \mathcal{E})$. Then by Proposition 5.2.1(a), $\mathcal{M} \cong \mathbf{R}\varGamma'_{\mathcal{X}}\mathcal{M} \cong \varinjlim_n \mathcal{E}_n$. Since $\mathcal{X}$ is noetherian, $\varinjlim$'s of flasque sheaves are f_*-acyclic and $\varinjlim$ commutes with f_*; so with notation as in the proof of Proposition 5.1.1,

$$\mathbf{R}f_*\mathcal{M} \cong \mathbf{R}f_*\mathbf{R}\varGamma'_{\mathcal{X}}\mathcal{M} \cong f_*\varinjlim_n \mathcal{E}_n \cong \varinjlim_n f_*j_{n*}j_n^*\mathcal{E}_n \cong \varinjlim_n i_{n*}f_{n*}j_n^*\mathcal{E}_n.$$

Since $\mathcal{E} \in \mathbf{D}^+_{\mathrm{qc}}(\mathcal{X})$, therefore

$$j_{n*}j_n^*\mathcal{E}_n = \mathcal{H}om(\mathcal{O}/\mathcal{J}^n, \mathcal{E}) \in \mathbf{D}_{\mathrm{qc}}(\mathcal{X}),$$

as we see by way-out reduction to where $\mathcal{E}$ is a single quasi-coherent sheaf and then by Corollary 3.1.6(d); and hence $j_n^*\mathcal{E}_n \in \mathbf{D}_{\mathrm{qc}}(X_n)$ (see [**GD**, p. 115, (5.3.15)]). Now $j_n^*\mathcal{E}_n$ is a flasque bounded-below $\mathcal{O}_{X_n}$-complex, so by way-out reduction to (for example) [**Ke**, p. 643, corollary 11],

$$f_{n*}j_n^*\mathcal{E}_n \cong \mathbf{R}f_{n*}j_n^*\mathcal{E}_n \in \mathbf{D}_{\mathrm{qc}}(Y_n);$$

and finally, in view of Corollary 5.1.3,

$$\mathbf{R}f_*\mathcal{M} \cong i_{n*}\varinjlim_n f_{n*}j_n^*\mathcal{E}_n \in \mathbf{D}_{\mathrm{qct}}(\mathcal{Y}).$$

For the last assertion, we reduce as before to showing for each coherent torsion $\mathcal{O}_{\mathcal{X}}$-module $\mathcal{M}$ and each $p \geq 0$ that $R^pf_*\mathcal{M} := H^p\mathbf{R}f_*\mathcal{M}$ is a coherent $\mathcal{O}_{\mathcal{Y}}$-module. With notation remaining as in Proposition 5.1.1, the maps i_n and j_n are exact, and for some n, $\mathcal{M} = j_{n*}j_n^*\mathcal{M}_n$. So

$$R^pf_*\mathcal{M} = R^pf_*j_{n*}j_n^*\mathcal{M}_n = i_{n*}R^pf_{n*}j_n^*\mathcal{M}_n,$$

which is coherent since $j_n^*\mathcal{M}_n$ is a coherent $\mathcal{O}_{X_n}$-module and $f_n\colon X_n \to Y_n$ is a proper scheme-map. □

COROLLARY 5.2.7 (cf. Corollary 3.5.3). *Let $f_0\colon X \to Y$ be a map of locally noetherian schemes, let $W \subset Y$ and $Z \subset f_0^{-1}W$ be closed subsets, with associated (flat) completion maps $\kappa_{\mathcal{Y}}\colon \mathcal{Y} = Y_{/W} \to Y$, $\kappa_{\mathcal{X}}\colon \mathcal{X} = X_{/Z} \to X$, and let $f\colon \mathcal{X} \to \mathcal{Y}$ be the map induced by f_0. For $\mathcal{E} \in \mathbf{D}(X)$ let*

$$\theta_{\mathcal{E}}\colon \kappa_{\mathcal{Y}}^*\mathbf{R}f_{0*}\mathcal{E} \to \mathbf{R}f_*\kappa_{\mathcal{X}}^*\mathcal{E}$$

be the map adjoint to the natural composition

$$\mathbf{R}f_{0*}\mathcal{E} \longrightarrow \mathbf{R}f_{0*}\kappa_{\mathcal{X}*}\kappa_{\mathcal{X}}^*\mathcal{E} \xrightarrow{\ \sim\ } \kappa_{\mathcal{Y}*}\mathbf{R}f_*\kappa_{\mathcal{X}}^*\mathcal{E}.$$

Then $\theta_{\mathcal{E}}$ is an isomorphism for all $\mathcal{E} \in \mathbf{D}_{\mathrm{qc}Z}(X)$.

PROOF. $\theta_{\mathcal{E}}$ is the composition of the natural maps

$$\kappa_{\mathcal{Y}}^{*}\mathbf{R}f_{0*}\mathcal{E} \to \kappa_{\mathcal{Y}}^{*}\mathbf{R}f_{0*}\kappa_{\mathcal{X}*}\kappa_{\mathcal{X}}^{*}\mathcal{E} \xrightarrow{\sim} \kappa_{\mathcal{Y}}^{*}\kappa_{\mathcal{Y}*}\mathbf{R}f_{*}\kappa_{\mathcal{X}}^{*}\mathcal{E} \to \mathbf{R}f_{*}\kappa_{\mathcal{X}}^{*}\mathcal{E}.$$

By Proposition 5.2.4, the first map and (in view of Proposition 5.2.6) the third map are both isomorphisms. □

PROPOSITION 5.2.8. *Let $f\colon \mathcal{X} \to \mathcal{Y}$ be a map of locally noetherian formal schemes. Let $\mathcal{I}$ be a coherent $\mathcal{O}_{\mathcal{Y}}$-ideal, and let $\mathbf{D}_{\mathcal{I}}(\mathcal{Y})$ be the triangulated subcategory of $\mathbf{D}(\mathcal{Y})$ whose objects are the complexes $\mathcal{F}$ with $\mathcal{I}$-torsion homology (i.e., $\Gamma_{\mathcal{I}} H^i\mathcal{F} = H^i\mathcal{F}$ for all $i \in \mathbb{Z}$—see §§1 and 1.2.1). Then:*

(a) $\mathbf{L}f^{*}(\mathbf{D}_{\mathcal{I}}(\mathcal{Y})) \subset \mathbf{D}_{\mathcal{I}\mathcal{O}_{\mathcal{X}}}(\mathcal{X})$.

(b) *There is a unique functorial isomorphism*

$$\xi(\mathcal{E})\colon \mathbf{L}f^{*}\mathbf{R}\Gamma_{\mathcal{I}}\,\mathcal{E} \xrightarrow{\sim} \mathbf{R}\Gamma_{\mathcal{I}\mathcal{O}_{\mathcal{X}}}\mathbf{L}f^{*}\mathcal{E} \qquad (\mathcal{E} \in \mathbf{D}(\mathcal{Y}))$$

whose composition with the natural map $\mathbf{R}\Gamma_{\mathcal{I}\mathcal{O}_{\mathcal{X}}}\mathbf{L}f^{*}\mathcal{E} \to \mathbf{L}f^{*}\mathcal{E}$ is the natural map $\mathbf{L}f^{*}\mathbf{R}\Gamma_{\mathcal{I}}\,\mathcal{E} \to \mathbf{L}f^{*}\mathcal{E}$.

(c) *The natural map is an isomorphism*

$$\mathbf{R}\Gamma'_{\mathcal{X}}\mathbf{L}f^{*}\mathbf{R}\Gamma'_{\mathcal{Y}}\mathcal{E} \xrightarrow{\sim} \mathbf{R}\Gamma'_{\mathcal{X}}\mathbf{L}f^{*}\mathcal{E} \qquad (\mathcal{E} \in \mathbf{D}(\mathcal{Y})).$$

(d) *If $\mathcal{X}$ is noetherian, there is a unique functorial isomorphism*

$$\mathbf{R}\Gamma_{\mathcal{I}}\,\mathbf{R}f_{*}\mathcal{G} \xrightarrow{\sim} \mathbf{R}f_{*}\mathbf{R}\Gamma_{\mathcal{I}\mathcal{O}_{\mathcal{X}}}\mathcal{G} \qquad (\mathcal{G} \in \mathbf{D}^{+}(\mathcal{X}))$$

whose composition with the natural map $\mathbf{R}f_{*}\mathbf{R}\Gamma_{\mathcal{I}\mathcal{O}_{\mathcal{X}}}\mathcal{G} \to \mathbf{R}f_{*}\mathcal{G}$ is the natural map $\mathbf{R}\Gamma_{\mathcal{I}}\,\mathbf{R}f_{*}\mathcal{G} \to \mathbf{R}f_{*}\mathcal{G}$.

PROOF. (a) Let $\mathcal{F} \in \mathbf{D}_{\mathcal{I}}(\mathcal{Y})$. To show that $\mathbf{L}f^{*}\mathcal{F} \in \mathbf{D}_{\mathcal{I}\mathcal{O}_{\mathcal{X}}}(\mathcal{X})$ we may assume that $\mathcal{F}$ is K-injective. Let $x \in \mathcal{X}$, set $y := f(x)$, and let $P_x^{\bullet}$ be a flat resolution of the $\mathcal{O}_{\mathcal{Y},y}$-module $\mathcal{O}_{\mathcal{X},x}$. Then, as in the proof of Proposition 5.2.1(a), there is a canonical $\mathbf{D}(\mathcal{Y})$-isomorphism

$$\varinjlim_{n} \mathcal{H}om^{\bullet}(\mathcal{O}_{\mathcal{Y}}/\mathcal{I}^{n}, \mathcal{F}) = \Gamma_{\mathcal{I}}\,\mathcal{F} = \mathbf{R}\Gamma_{\mathcal{I}}\,\mathcal{F} \xrightarrow{\sim} \mathcal{F},$$

and it follows that for any i the stalk at x of the homology $H^i\mathbf{L}f^{*}\mathcal{F}$ is

$$H^i(P_x^{\bullet} \otimes_{\mathcal{O}_{\mathcal{Y},y}} \mathcal{F}_y) = \varinjlim_{n} H^i(P_x^{\bullet} \otimes_{\mathcal{O}_{\mathcal{Y},y}} \mathrm{Hom}^{\bullet}_{\mathcal{O}_{\mathcal{Y},y}}(\mathcal{O}_{\mathcal{Y},y}/\mathcal{I}_y^{n}, \mathcal{F}_y)).$$

Hence each element of the stalk is annihilated by a power of $\mathcal{I}\mathcal{O}_{\mathcal{X},x}$, and (a) results.

(b) The existence and uniqueness of a functorial map $\xi(\mathcal{E})$ satisfying everything except the isomorphism property result from (a) and the fact that $\mathbf{R}\Gamma_{\mathcal{I}\mathcal{O}_{\mathcal{X}}}$ is right-adjoint to the inclusion $\mathbf{D}_{\mathcal{I}\mathcal{O}_{\mathcal{X}}}(\mathcal{X}) \hookrightarrow \mathbf{D}(\mathcal{X})$.

To show that $\xi(\mathcal{E})$ is an isomorphism we may assume that $\mathcal{Y}$ is affine and that $\mathcal{E}$ is K-flat, and then proceed as in the proof of (the special case) Proposition 5.2.4(c), via the bounded flat complex $K_{\infty}^{\bullet}$.

(c) Let $\mathcal{I}$, $\mathcal{J}$ be defining ideals of $\mathcal{Y}$ and $\mathcal{X}$ respectively, so that $\mathcal{K} := \mathcal{I}\mathcal{O}_{\mathcal{X}} \subset \mathcal{J}$. The natural map $\mathbf{R}\Gamma'_{\mathcal{X}}\mathbf{R}\Gamma_{\mathcal{K}} := \mathbf{R}\Gamma_{\mathcal{J}}\mathbf{R}\Gamma_{\mathcal{K}} \to \mathbf{R}\Gamma_{\mathcal{J}} =: \mathbf{R}\Gamma'_{\mathcal{X}}$ is an *isomorphism,* as one checks locally via [**AJL**, p. 20, Corollary (3.1.3)]. So for any $\mathcal{E} \in \mathbf{D}(\mathcal{Y})$, (b) gives

$$\mathbf{R}\Gamma'_{\mathcal{X}}\mathbf{L}f^{*}\mathcal{E} \cong \mathbf{R}\Gamma'_{\mathcal{X}}\mathbf{R}\Gamma_{\mathcal{K}}\mathbf{L}f^{*}\mathcal{E} \cong \mathbf{R}\Gamma'_{\mathcal{X}}\mathbf{L}f^{*}\mathbf{R}\Gamma'_{\mathcal{Y}}\mathcal{E}.$$

(d) $\mathcal{G}$ may be assumed bounded-below and injective, so that

$$\mathcal{G}_n := \mathcal{H}om^{\bullet}(\mathcal{O}_{\mathcal{X}}/\mathcal{I}^{n}\mathcal{O}_{\mathcal{X}}, \mathcal{G})$$

is flasque.

Then, since $\mathcal{X}$ is noetherian, $\varGamma_{\mathcal{J}\mathcal{O}_{\mathcal{X}}}\mathcal{G} = \varinjlim_n \mathcal{G}_n$ is flasque too, and

$$\mathbf{R}f_*\mathbf{R}\varGamma_{\mathcal{J}\mathcal{O}_{\mathcal{X}}}\mathcal{G} \cong \mathbf{R}f_*\varGamma_{\mathcal{J}\mathcal{O}_{\mathcal{X}}}\mathcal{G} \cong f_*\varinjlim_n \mathcal{G}_n \cong \varinjlim_n f_*\mathcal{G}_n \in \mathbf{D}_{\mathcal{J}}(\mathcal{Y}).$$

By Lemma 5.2.2, $\mathbf{R}\varGamma_{\mathcal{J}}$ (resp. $\mathbf{R}\varGamma_{\mathcal{J}\mathcal{O}_{\mathcal{X}}}$) is right-adjoint to the inclusion $\mathbf{D}_{\mathcal{J}}(\mathcal{Y}) \hookrightarrow \mathbf{D}(\mathcal{Y})$ (resp. $\mathbf{D}_{\mathcal{J}\mathcal{O}_{\mathcal{X}}}(\mathcal{X}) \hookrightarrow \mathbf{D}(\mathcal{X})$), whence, in particular, the uniqueness in (d). Moreover, in view of (a), for any $\mathcal{E} \in \mathbf{D}_{\mathcal{J}}(\mathcal{Y})$ the natural maps are isomorphisms

$$\begin{aligned} \mathrm{Hom}_{\mathcal{Y}}(\mathcal{E}, \mathbf{R}\varGamma_{\mathcal{J}}\mathbf{R}f_*\mathcal{G}) &\xrightarrow{\sim} \mathrm{Hom}_{\mathcal{Y}}(\mathcal{E}, \mathbf{R}f_*\mathcal{G}) \xrightarrow{\sim} \mathrm{Hom}_{\mathcal{X}}(\mathbf{L}f^*\mathcal{E}, \mathcal{G}) \\ &\xrightarrow{\sim} \mathrm{Hom}_{\mathcal{X}}(\mathbf{L}f^*\mathcal{E}, \mathbf{R}\varGamma_{\mathcal{J}\mathcal{O}_{\mathcal{X}}}\mathcal{G}) \xrightarrow{\sim} \mathrm{Hom}_{\mathcal{Y}}(\mathcal{E}, \mathbf{R}f_*\mathbf{R}\varGamma_{\mathcal{J}\mathcal{O}_{\mathcal{X}}}\mathcal{G}). \end{aligned}$$

It follows formally that the image under this composed isomorphism of the identity map of $\mathbf{R}\varGamma_{\mathcal{J}}\mathbf{R}f_*\mathcal{G}$ is an isomorphism as asserted. (In fact this isomorphism is adjoint to the composition $\mathbf{L}f^*\mathbf{R}\varGamma_{\mathcal{J}}\mathbf{R}f_*\mathcal{G} \xrightarrow[\xi(\mathbf{R}f_*\mathcal{G})]{} \mathbf{R}\varGamma_{\mathcal{J}\mathcal{O}_{\mathcal{X}}}\mathbf{L}f^*\mathbf{R}f_*\mathcal{G} \xrightarrow[\text{nat'l}]{} \mathbf{R}\varGamma_{\mathcal{J}\mathcal{O}_{\mathcal{X}}}\mathcal{G}$.) □

DEFINITION 5.2.9. For a locally noetherian formal scheme $\mathcal{X}$,

$$\widetilde{\mathbf{D}}_{\mathrm{qc}}(\mathcal{X}) := \mathbf{R}\varGamma_{\mathcal{X}}'^{-1}(\mathbf{D}_{\mathrm{qc}}(\mathcal{X}))$$

is the Δ-subcategory of $\mathbf{D}(\mathcal{X})$ whose objects are those complexes $\mathcal{F}$ such that $\mathbf{R}\varGamma_{\mathcal{X}}'\mathcal{F} \in \mathbf{D}_{\mathrm{qc}}(\mathcal{X})$—or equivalently, $\mathbf{R}\varGamma_{\mathcal{X}}'\mathcal{F} \in \mathbf{D}_{\mathrm{qct}}(\mathcal{X})$.

REMARKS 5.2.10. (1) By Proposition 5.2.1(b), $\mathbf{D}_{\mathrm{qc}}(\mathcal{X}) \subset \widetilde{\mathbf{D}}_{\mathrm{qc}}(\mathcal{X})$. Hence

$$\mathbf{R}\varGamma_{\mathcal{X}}'(\widetilde{\mathbf{D}}_{\mathrm{qc}}(\mathcal{X})) \subset \widetilde{\mathbf{D}}_{\mathrm{qc}}(\mathcal{X}).$$

(2) Since $\mathbf{R}\varGamma_{\mathcal{X}}'$ is idempotent (see Proposition 5.2.1), the vertex of any triangle based on the canonical map $\mathbf{R}\varGamma_{\mathcal{X}}'\mathcal{E} \to \mathcal{E}$ ($\mathcal{E} \in \mathbf{D}(\mathcal{X})$) is annihilated by $\mathbf{R}\varGamma_{\mathcal{X}}'$. It follows that $\widetilde{\mathbf{D}}_{\mathrm{qc}}(\mathcal{X})$ is the smallest Δ-subcategory of $\mathbf{D}(\mathcal{X})$ containing $\mathbf{D}_{\mathrm{qct}}(\mathcal{X})$ and all complexes $\mathcal{F}$ such that $\mathbf{R}\varGamma_{\mathcal{X}}'\mathcal{F} = 0$.

(3) The functor $\mathbf{R}\varGamma_{\mathcal{X}}' \colon \mathbf{D}(\mathcal{X}) \to \mathbf{D}(\mathcal{X})$ has a right adjoint

$$\mathbf{\Lambda}_{\mathcal{X}}(-) := \mathbf{R}\mathcal{H}om^\bullet(\mathbf{R}\varGamma_{\mathcal{X}}'\mathcal{O}_{\mathcal{X}}, -).$$

Indeed, there are natural functorial isomorphisms for $\mathcal{E}, \mathcal{F} \in \mathbf{D}(\mathcal{X})$,

$$\begin{aligned} \mathrm{Hom}_{\mathbf{D}(\mathcal{X})}(\mathbf{R}\varGamma_{\mathcal{X}}'\mathcal{E}, \mathcal{F}) &\xrightarrow{\sim} \mathrm{Hom}_{\mathbf{D}(\mathcal{X})}(\mathcal{E} \underline{\underline{\otimes}} \mathbf{R}\varGamma_{\mathcal{X}}'\mathcal{O}_{\mathcal{X}}, \mathcal{F}) \\ &\xrightarrow{\sim} \mathrm{Hom}_{\mathbf{D}(\mathcal{X})}(\mathcal{E}, \mathbf{R}\mathcal{H}om^\bullet(\mathbf{R}\varGamma_{\mathcal{X}}'\mathcal{O}_{\mathcal{X}}, \mathcal{F})). \end{aligned} \tag{5.2.10.1}$$

(Whether the natural map $\mathcal{E} \underline{\underline{\otimes}} \mathbf{R}\varGamma_{\mathcal{X}}'\mathcal{O}_{\mathcal{X}} \xrightarrow{\sim} \mathbf{R}\varGamma_{\mathcal{X}}'\mathcal{E}$ is an isomorphism is a local question, dealt with e.g., in [**AJL**, p. 20, Corollary (3.1.2)]. The second isomorphism is given, e.g., by [**Sp**, p. 147, Proposition 6.6 (1)].)

There is a natural isomorphism $\mathbf{R}\varGamma_{\mathcal{X}}' \xrightarrow{\sim} \mathbf{R}\varGamma_{\mathcal{X}}'\mathbf{\Lambda}_{\mathcal{X}}$ (see (d) in Remark 6.3.1 below), and consequently

$$\mathbf{\Lambda}_{\mathcal{X}}(\widetilde{\mathbf{D}}_{\mathrm{qc}}(\mathcal{X})) \subset \widetilde{\mathbf{D}}_{\mathrm{qc}}(\mathcal{X}).$$

(4) *If* $\mathcal{E} \in \mathbf{D}_{\mathrm{c}}^-(\mathcal{X})$ *and* $\mathcal{F} \in \widetilde{\mathbf{D}}_{\mathrm{qc}}(\mathcal{X})$ *then* $\mathbf{R}\mathcal{H}om^\bullet(\mathcal{E}, \mathcal{F}) \in \widetilde{\mathbf{D}}_{\mathrm{qc}}(\mathcal{X})$, *and hence* $\mathbf{R}\mathcal{H}om^\bullet(\mathbf{R}\varGamma_{\mathcal{X}}'\mathcal{E}, \mathcal{F}) \in \widetilde{\mathbf{D}}_{\mathrm{qc}}(\mathcal{X})$. Indeed, the natural map

$$\mathbf{R}\varGamma_{\mathcal{X}}'\mathbf{R}\mathcal{H}om^\bullet(\mathcal{E}, \mathbf{R}\varGamma_{\mathcal{X}}'\mathcal{F}) \to \mathbf{R}\varGamma_{\mathcal{X}}'\mathbf{R}\mathcal{H}om^\bullet(\mathcal{E}, \mathcal{F})$$

is an *isomorphism,* since for any $\mathcal{G}$ in $\mathbf{D}_{\mathrm{t}}(\mathcal{X})$, $\mathcal{G} \underline{\underline{\otimes}} \mathcal{E} \in \mathbf{D}_{\mathrm{t}}(X)$ (an assertion which can be checked locally, using Proposition 5.2.1(a) and the complex $\mathcal{K}_\infty^\bullet$ in its proof), so that there is a sequence of natural isomorphisms (see Proposition 5.2.1(c)):

$$\begin{aligned}\mathrm{Hom}(\mathcal{G}, \mathbf{R}\varGamma'_{\mathfrak{X}}\mathbf{R}\mathcal{H}om^\bullet(\mathcal{E}, \mathbf{R}\varGamma'_{\mathfrak{X}}\mathcal{F})) &\xrightarrow{\sim} \mathrm{Hom}(\mathcal{G}, \mathbf{R}\mathcal{H}om^\bullet(\mathcal{E}, \mathbf{R}\varGamma'_{\mathfrak{X}}\mathcal{F}))\\ &\xrightarrow{\sim} \mathrm{Hom}(\mathcal{G} \underline{\underline{\otimes}} \mathcal{E}, \mathbf{R}\varGamma'_{\mathfrak{X}}\mathcal{F})\\ &\xrightarrow{\sim} \mathrm{Hom}(\mathcal{G} \underline{\underline{\otimes}} \mathcal{E}, \mathcal{F})\\ &\xrightarrow{\sim} \mathrm{Hom}(\mathcal{G}, \mathbf{R}\mathcal{H}om^\bullet(\mathcal{E}, \mathcal{F}))\\ &\xrightarrow{\sim} \mathrm{Hom}(\mathcal{G}, \mathbf{R}\varGamma'_{\mathfrak{X}}\mathbf{R}\mathcal{H}om^\bullet(\mathcal{E}, \mathcal{F})).\end{aligned}$$

Since $\mathcal{A}_{\mathrm{qct}}(\mathfrak{X})$ is plump in $\mathcal{A}(\mathfrak{X})$ (Corollary 5.1.3), Proposition 3.2.4 shows that $\mathbf{R}\varGamma'_{\mathfrak{X}}\mathbf{R}\mathcal{H}om^\bullet(\mathcal{E}, \mathbf{R}\varGamma'_{\mathfrak{X}}\mathcal{F}) \in \mathbf{D}_{\mathrm{qct}}(\mathfrak{X})$, whence $\mathbf{R}\mathcal{H}om^\bullet(\mathcal{E}, \mathcal{F}) \in \widetilde{\mathbf{D}}_{\mathrm{qc}}(\mathfrak{X})$.

From (3) and the natural isomorphisms

$$\mathbf{R}\mathcal{H}om^\bullet(\mathbf{R}\varGamma'_{\mathfrak{X}}\mathcal{E}, \mathcal{F}) \cong \mathbf{R}\mathcal{H}om^\bullet(\mathbf{R}\varGamma'_{\mathfrak{X}}\mathcal{O}_{\mathfrak{X}} \underline{\underline{\otimes}} \mathcal{E}, \mathcal{F}) \cong \mathbf{\Lambda}_{\mathfrak{X}}\mathbf{R}\mathcal{H}om^\bullet(\mathcal{E}, \mathcal{F})$$

we see then that

$$\mathbf{R}\mathcal{H}om^\bullet(\mathbf{R}\varGamma'_{\mathfrak{X}}\mathcal{E}, \mathcal{F}) \in \widetilde{\mathbf{D}}_{\mathrm{qc}}(\mathfrak{X}).$$

(5) For $\mathcal{F} \in \mathbf{D}(\mathfrak{X})$ it holds that

$$\mathcal{F} \in \widetilde{\mathbf{D}}_{\mathrm{qc}}(\mathfrak{X}) \iff \mathbf{R}\mathcal{H}om^\bullet(\mathcal{O}_{\mathfrak{X}}/\mathcal{J}, \mathcal{F}) \in \mathbf{D}_{\mathrm{qct}}(\mathfrak{X}) \text{ for all defining ideals } \mathcal{J} \text{ of } \mathfrak{X}.$$

The implication $\Longrightarrow$ is given, in view of Corollary 5.2.3, by Proposition 3.2.4; and the converse is given by Lemma 5.4.1, since Corollary 5.1.3 implies that $\mathbf{D}_{\mathrm{qct}}(\mathfrak{X})$ is a Δ-subcategory of $\mathbf{D}(\mathfrak{X})$ closed under direct sums.

(6) Let $f\colon \mathfrak{X} \to \mathfrak{Y}$ be a map of locally noetherian formal schemes. For any $\mathcal{F} \in \widetilde{\mathbf{D}}_{\mathrm{qc}}(\mathfrak{Y})$, Lemma 5.1.4 and Proposition 3.3.5 give

$$\mathbf{L}f^*\mathbf{R}\varGamma'_{\mathfrak{Y}}\mathcal{F} \in \mathbf{L}f^*(\mathbf{D}_{\mathrm{qct}}(\mathfrak{Y})) \subset \mathbf{L}f^*(\mathbf{D}_{\vec{\mathrm{c}}}(\mathfrak{Y})) \subset \mathbf{D}_{\mathrm{qc}}(\mathfrak{X}) \subset \widetilde{\mathbf{D}}_{\mathrm{qc}}(\mathfrak{X}),$$

and so $\mathbf{R}\varGamma'_{\mathfrak{X}}\mathbf{L}f^*\mathcal{F} \underset{5.2.8(\mathrm{c})}{\cong} \mathbf{R}\varGamma'_{\mathfrak{X}}\mathbf{L}f^*\mathbf{R}\varGamma'_{\mathfrak{Y}}\mathcal{F} \in \mathbf{D}_{\mathrm{qct}}(\mathfrak{X})$. Thus

$$\mathbf{L}f^*(\widetilde{\mathbf{D}}_{\mathrm{qc}}(\mathfrak{Y})) \subset \widetilde{\mathbf{D}}_{\mathrm{qc}}(\mathfrak{X}).$$

COROLLARY 5.2.11. *Let $f\colon \mathfrak{X} \to \mathfrak{Y}$ be an adic map of locally noetherian formal schemes. Then:*

(a) $\mathbf{L}f^*(\mathbf{D}_{\mathrm{t}}(\mathfrak{Y})) \subset \mathbf{D}_{\mathrm{t}}(\mathfrak{X})$.

(b) $\mathbf{L}f^*(\mathbf{D}_{\mathrm{qct}}(\mathfrak{Y})) \subset \mathbf{D}_{\mathrm{qct}}(\mathfrak{X})$.

(c) *There is a unique functorial isomorphism*

$$\mathbf{L}f^*\mathbf{R}\varGamma'_{\mathfrak{Y}}\mathcal{E} \xrightarrow{\sim} \mathbf{R}\varGamma'_{\mathfrak{X}}\mathbf{L}f^*\mathcal{E} \qquad (\mathcal{E} \in \mathbf{D}(\mathfrak{Y}))$$

whose composition with the natural map $\mathbf{R}\varGamma'_{\mathfrak{X}}\mathbf{L}f^*\mathcal{E} \to \mathbf{L}f^*\mathcal{E}$ is the natural map $\mathbf{L}f^*\mathbf{R}\varGamma'_{\mathfrak{Y}}\mathcal{E} \to \mathbf{L}f^*\mathcal{E}$. There results a conjugate isomorphism of right-adjoint functors

$$\mathbf{R}f_*\mathbf{\Lambda}_{\mathfrak{X}}\mathcal{G} \xrightarrow{\sim} \mathbf{\Lambda}_{\mathfrak{Y}}\mathbf{R}f_*\mathcal{G} \qquad (\mathcal{G} \in \mathbf{D}(\mathfrak{X})).$$

whose composition with the natural map $\mathbf{R}f_*\mathcal{G} \to \mathbf{R}f_*\mathbf{\Lambda}_{\mathfrak{X}}\mathcal{G}$ is the natural map $\mathbf{R}f_*\mathcal{G} \to \mathbf{\Lambda}_{\mathfrak{Y}}\mathbf{R}f_*\mathcal{G}$.

(d) *If $\mathscr{X}$ is noetherian then there is a unique functorial isomorphism*

$$\mathbf{R}\varGamma'_{\mathscr{Y}}\mathbf{R}f_*\mathcal{G} \xrightarrow{\sim} \mathbf{R}f_*\mathbf{R}\varGamma'_{\mathscr{X}}\mathcal{G} \qquad (\mathcal{G}\in\mathbf{D}^+(\mathscr{X}) \text{ or } \mathcal{G}\in\widetilde{\mathbf{D}}_{\mathrm{qc}}(\mathscr{X}))$$

whose composition with the natural map $\mathbf{R}f_*\mathbf{R}\varGamma'_{\mathscr{X}}\mathcal{G}\to\mathbf{R}f_*\mathcal{G}$ is the natural map $\mathbf{R}\varGamma'_{\mathscr{Y}}\mathbf{R}f_*\mathcal{G}\to\mathbf{R}f_*\mathcal{G}$.

(e) *If $\mathscr{X}$ is noetherian then $\mathbf{R}f_*(\widetilde{\mathbf{D}}_{\mathrm{qc}}(\mathscr{Y}))\subset\widetilde{\mathbf{D}}_{\mathrm{qc}}(\mathscr{X})$.*

PROOF. To get (a) and (c) take $\mathcal{J}$ in Proposition 5.2.8 to be an ideal of definition of $\mathscr{Y}$. (The second assertion in (c) is left to the reader.) As $\mathbf{D}_{\mathrm{qct}}(\mathscr{Y})=\mathbf{D}_{\vec{\mathrm{c}}}(\mathscr{Y})\cap\mathbf{D}_{\mathrm{t}}(\mathscr{Y})$ (Corollary 3.1.5 and Lemma 5.1.4), (b) follows from (a) and Proposition 3.3.5. The same choice of $\mathcal{J}$ gives (d) for $\mathcal{G}\in\mathbf{D}^+(\mathscr{X})$—and the argument also works for $\mathcal{G}\in\widetilde{\mathbf{D}}_{\mathrm{qc}}(\mathscr{X})$ once one notes that

$$\mathbf{R}f_*\mathbf{R}\varGamma'_{\mathscr{X}}(\widetilde{\mathbf{D}}_{\mathrm{qc}}(\mathscr{X}))\subset\mathbf{R}f_*(\mathbf{D}_{\mathrm{qct}}(\mathscr{X}))\underset{5.2.6}{\subset}\mathbf{D}_{\mathrm{qct}}(\mathscr{Y})\subset\mathbf{D}_{\mathrm{t}}(\mathscr{Y}).$$

The isomorphism in (d) gives (e) via Proposition 5.2.6. □

COROLLARY 5.2.12. *In Corollary 5.2.7, if $\mathscr{X}$ is noetherian and $Z=f_0^{-1}W$ then for all $\mathcal{F}\in\mathbf{D}_{\mathrm{qc}}(X)$ the map $\theta'_{\mathcal{F}}:=\mathbf{R}\varGamma'_{\mathscr{Y}}(\theta_{\mathcal{F}})$ is an isomorphism*

$$\theta'_{\mathcal{F}}\colon\mathbf{R}\varGamma'_{\mathscr{Y}}\kappa^*_{\mathscr{Y}}\mathbf{R}f_{0*}\mathcal{F}\xrightarrow{\sim}\mathbf{R}\varGamma'_{\mathscr{Y}}\mathbf{R}f_*\kappa^*_{\mathscr{X}}\mathcal{F}.$$

PROOF. Arguing as in Proposition 5.2.1, we find that $\mathbf{R}\varGamma_Z\mathcal{F}\in\mathbf{D}_{\mathrm{qc}Z}(X)$, so that we have the isomorphism $\theta_{\mathbf{R}\varGamma_Z\mathcal{F}}$ of Corollary 5.2.7.

Imitating the proof of Corollary 5.2.11, we get an isomorphism

$$\alpha_{\mathcal{F}}\colon\mathbf{R}f_{0*}\mathbf{R}\varGamma_Z\mathcal{F}\xrightarrow{\sim}\mathbf{R}\varGamma_W\mathbf{R}f_{0*}\mathcal{F}$$

whose composition with the natural map $\mathbf{R}\varGamma_W\mathbf{R}f_{0*}\mathcal{F}\to\mathbf{R}f_{0*}\mathcal{F}$ is the natural map $\mathbf{R}f_{0*}\mathbf{R}\varGamma_Z\mathcal{F}\to\mathbf{R}f_{0*}\mathcal{F}$.

Consider then the diagram

$$\begin{array}{ccccccc}
\kappa^*_{\mathscr{Y}}\mathbf{R}f_{0*}\mathbf{R}\varGamma_Z\mathcal{F} & \xrightarrow[\kappa^*_{\mathscr{Y}}(\alpha_{\mathcal{F}})]{\sim} & \kappa^*_{\mathscr{Y}}\mathbf{R}\varGamma_W\mathbf{R}f_{0*}\mathcal{F} & \xrightarrow[5.2.4(\mathrm{c})]{\sim} & \mathbf{R}\varGamma'_{\mathscr{Y}}\kappa^*_{\mathscr{Y}}\mathbf{R}f_{0*}\mathcal{F} & \xrightarrow{\text{nat'l}} & \kappa^*_{\mathscr{Y}}\mathbf{R}f_{0*}\mathcal{F}\\
{\scriptstyle\theta_{\mathbf{R}\varGamma_Z\mathcal{F}}}\downarrow\simeq & & (1) & & \downarrow{\scriptstyle\theta'_{\mathcal{F}}} & & \downarrow{\scriptstyle\theta_{\mathcal{F}}}\\
\mathbf{R}f_*\kappa^*_{\mathscr{X}}\mathbf{R}\varGamma_Z\mathcal{F} & \xrightarrow[5.2.4(\mathrm{c})]{\sim} & \mathbf{R}f_*\mathbf{R}\varGamma'_{\mathscr{X}}\kappa^*_{\mathscr{X}}\mathcal{F} & \xrightarrow[5.2.11(\mathrm{d})]{\sim} & \mathbf{R}\varGamma'_{\mathscr{Y}}\mathbf{R}f_*\kappa^*_{\mathscr{X}}\mathcal{F} & \xrightarrow[\text{nat'l}]{} & \mathbf{R}f_*\kappa^*_{\mathscr{X}}\mathcal{F}
\end{array}$$

It suffices to show that subdiagram (1) commutes; and since $\mathbf{R}\varGamma'_{\mathscr{Y}}$ is right-adjoint to the inclusion $\mathbf{D}_{\mathrm{t}}(\mathscr{Y})\hookrightarrow\mathbf{D}(\mathscr{Y})$ it follows that it's enough to show that the outer border of the diagram commutes. But it is straightforward to check that the top and bottom rows compose to the maps induced by the natural map $\mathbf{R}\varGamma_Z\to\mathbf{1}$, whence the conclusion. □

5.3. From the following key Proposition 5.3.1—generalizing the noetherian case of [**AJL**, p. 12, Proposition (1.3)]—there will result, for complexes with quasi-coherent torsion homology, a stronger version of the Duality Theorem 4.1, see Section 6.

Recall what it means for a noetherian formal scheme $\mathscr{X}$ to be *separated* (§3.4.1). Recall also from Corollary 5.1.5 that the inclusion functor $j^{\mathrm{t}}_{\mathscr{X}}\colon\mathcal{A}_{\mathrm{qct}}(\mathscr{X})\hookrightarrow\mathcal{A}(\mathscr{X})$ has a right adjoint $Q^{\mathrm{t}}_{\mathscr{X}}$.

PROPOSITION 5.3.1. *Let $\mathcal{X}$ be a noetherian formal scheme.*

(a) *The extension of $j^{\mathrm{t}}_{\mathcal{X}}$ induces an* equivalence of categories

$$\boldsymbol{j}^{\mathrm{t}}_{\mathcal{X}}\colon \mathbf{D}^+(\mathcal{A}_{\mathrm{qct}}(\mathcal{X})) \xrightarrow{\approx} \mathbf{D}^+_{\mathrm{qct}}(\mathcal{X}),$$

with bounded quasi-inverse $\mathbf{R}Q^{\mathrm{t}}_{\mathcal{X}}|_{\mathbf{D}^+_{\mathrm{qct}}(\mathcal{X})}$.

(b) *If $\mathcal{X}$ is separated, or of finite Krull dimension, then the extension of $j^{\mathrm{t}}_{\mathcal{X}}$ induces an* equivalence of categories

$$\boldsymbol{j}^{\mathrm{t}}_{\mathcal{X}}\colon \mathbf{D}(\mathcal{A}_{\mathrm{qct}}(\mathcal{X})) \xrightarrow{\approx} \mathbf{D}_{\mathrm{qct}}(\mathcal{X}),$$

with bounded quasi-inverse $\mathbf{R}Q^{\mathrm{t}}_{\mathcal{X}}|_{\mathbf{D}_{\mathrm{qct}}(\mathcal{X})}$.

PROOF. (a) The asserted equivalence is given by [**Y**, Theorem 4.8]. The idea is that $\mathcal{A}_{\mathrm{qct}}(\mathcal{X})$ contains enough $\mathcal{A}_{\mathrm{t}}(\mathcal{X})$-injectives [**Y**, Proposition 4.2], so by [**H1**, p. 47, Proposition 4.8], $\mathbf{D}^+(\mathcal{A}_{\mathrm{qct}}(\mathcal{X}))$ is equivalent to $\mathbf{D}^+(\mathcal{A}_{\mathrm{t}}(\mathcal{X}))$, which is equivalent to $\mathbf{D}^+_{\mathrm{qct}}(\mathcal{X})$ (Proposition 5.2.1(c)).

Since $\mathbf{R}Q^{\mathrm{t}}_{\mathcal{X}}$ is right-adjoint to $\boldsymbol{j}^{\mathrm{t}}_{\mathcal{X}}$ (Lemma 5.2.2), its restriction to $\mathbf{D}^+_{\mathrm{qct}}(\mathcal{X})$ is quasi-inverse to $\boldsymbol{j}^{\mathrm{t}}_{\mathcal{X}}|_{\mathbf{D}^+(\mathcal{A}_{\mathrm{qct}}(\mathcal{X}))}$. From the resulting isomorphism

$$\iota_{\mathcal{E}}\colon \boldsymbol{j}^{\mathrm{t}}_{\mathcal{X}}\mathbf{R}Q^{\mathrm{t}}_{\mathcal{X}}\mathcal{E} \xrightarrow{\sim} \mathcal{E} \qquad (\mathcal{E}\in \mathbf{D}^+_{\mathrm{qct}}(\mathcal{X}))$$

we see that if $H^i\mathcal{E}=0$ then $H^i\mathbf{R}Q^{\mathrm{t}}_{\mathcal{X}}\mathcal{E}=0$, so that $\mathbf{R}Q^{\mathrm{t}}_{\mathcal{X}}|_{\mathbf{D}^+_{\mathrm{qct}}(\mathcal{X})}$ is bounded.

(b) By Lemma 5.2.2, and having the isomorphism $\iota_{\mathcal{E}}$, we need only show that $\mathbf{R}Q^{\mathrm{t}}_{\mathcal{X}}$ is bounded on $\mathbf{D}_{\mathrm{qct}}(\mathcal{X})$.

Suppose that $\mathcal{X}$ is the completion of a separated ordinary noetherian scheme X along some closed subscheme, and let $\kappa\colon \mathcal{X}\to X$ be the completion map, so that $Q^{\mathrm{t}}_{\mathcal{X}} = \kappa^*\varGamma_Z Q_X\kappa_*$ (see remark following Corollary 5.1.5). The exact functor κ_* preserves K-injectivity, since it has an exact left adjoint, namely κ^*. Similarly Q_X transforms K-injective $\mathcal{A}(X)$-complexes into K-injective $\mathcal{A}_{\mathrm{qc}}(X)$-complexes. Hence $\mathbf{R}Q^{\mathrm{t}}_{\mathcal{X}} \cong \kappa^*\mathbf{R}\varGamma^{\mathrm{qc}}_Z\mathbf{R}Q_X\kappa_*$, where $\varGamma^{\mathrm{qc}}_Z\colon \mathcal{A}_{\mathrm{qc}}(X)\to\mathcal{A}_{\mathrm{qc}Z}(X)$ is the restriction of $\varGamma_Z$. Now by the proof of [**AJL**, p. 12, Proposition (1.3)], $\mathbf{R}Q_X$ is bounded on $\mathbf{D}_{\mathrm{qc}}(X)\supset\kappa_*\mathbf{D}_{\mathrm{qct}}(\mathcal{X})$ (Proposition 5.2.4). Also, by [**AJL**, p. 24, Lemma (3.2.3)], $\mathbf{R}\varGamma_Z$ is bounded; and hence by [**AJL**, p. 26, Proposition (3.2.6)], so is $\varGamma^{\mathrm{qc}}_Z$. Thus $\mathbf{R}Q^{\mathrm{t}}_{\mathcal{X}}$ is bounded on $\mathbf{D}_{\mathrm{qct}}(\mathcal{X})$.

In the general separated case, one proceeds by induction on the least number of affine open subsets covering $\mathcal{X}$, as in the proof of [**AJL**, p. 12, Proposition (1.3)] (which is Proposition 5.3.1 for $\mathcal{X}$ an ordinary scheme), *mutatis mutandis*—namely, substitute "$\mathcal{X}$" for "X," "qct" for "qc," "Q^{t}" for "Q," and recall for a map $v\colon \mathcal{V}\to\mathcal{X}$ of noetherian formal schemes that $v_*(\mathcal{A}_{\mathrm{qct}}(\mathcal{V}))\subset\mathcal{A}_{\mathrm{qct}}(\mathcal{X})$ (Proposition 5.1.1), and furthermore that if v is affine then $v_*|_{\mathcal{A}_{\mathrm{qct}}(\mathcal{V})}$ is *exact* (Lemmas 5.1.4 and 3.4.2).

A similar procedure works when the Krull dimension $\dim\mathcal{X}$ is finite, but now the induction is on $n(\mathcal{X}) :=$ least n such that $\mathcal{X}$ has an open covering $\mathcal{X}=\cup_{i=1}^n\mathcal{U}_i$ where for each i there is a separated ordinary noetherian scheme U_i such that $\mathcal{U}_i$ is isomorphic to the completion of U_i along one of its closed subschemes. (This property of $\mathcal{U}_i$ is inherited by any of its open subsets).

The case $n(\mathcal{X})=1$ has just been done. Consider, for any open immersion $v\colon \mathcal{V}\hookrightarrow\mathcal{X}$, the functor $v^{\mathrm{qct}}_* := v_*|_{\mathcal{A}_{\mathrm{qct}}(\mathcal{V})}$. To complete the induction as in the proof of [**AJL**, p. 12, Proposition (1.3)], one needs to show that *the derived functor* $\mathbf{R}v^{\mathrm{qct}}_*\colon \mathbf{D}(\mathcal{A}_{\mathrm{qct}}(\mathcal{V}))\to\mathbf{D}(\mathcal{X})$ *is bounded above.*

For $\mathcal{N} \in \mathcal{A}_{\mathrm{qct}}(\mathcal{V})$, let $\mathcal{N} \to \mathcal{J}^\bullet$ be an $\mathcal{A}_{\mathrm{qct}}$-injective—hence flasque—resolution [**Y**, Proposition 4.2]. Now $H^i\mathbf{R}v_*^{\mathrm{qct}}(\mathcal{N})$ is the sheafification of the presheaf sending an open $\mathcal{W} \subset \mathcal{X}$ to $\mathrm{H}^i\Gamma(\mathcal{W} \cap \mathcal{V}, \mathcal{J}^\bullet) = \mathrm{H}^i(\mathcal{W} \cap \mathcal{V}, \mathcal{N})$, which vanishes when $i > \dim \mathcal{X}$, whence the conclusion ([**L4**, Proposition (2.7.5)]). □

5.4. Let $(X, \mathcal{O}_X)$ be a ringed space, and let $\mathcal{J}$ be an $\mathcal{O}_X$-ideal. The next Lemma, expressing $\mathbf{R}\varGamma_{\mathcal{J}}$ as a "homotopy colimit," lifts back to $\mathbf{D}(X)$ the well-known relation

$$H^i\mathbf{R}\varGamma_{\mathcal{J}}\mathcal{G} = \varinjlim_n \mathcal{E}xt^i_{\mathcal{O}_X}(\mathcal{O}_X/\mathcal{J}^n, \mathcal{G}) \qquad (\mathcal{G} \in \mathbf{D}(X)).$$

Define $\boldsymbol{h}_n \colon \mathbf{D}(X) \to \mathbf{D}(X)$ by

$$\boldsymbol{h}_n(\mathcal{G}) := \mathbf{R}\mathcal{H}om^\bullet(\mathcal{O}_X/\mathcal{J}^n, \mathcal{G}) \qquad (n \geq 1,\ \mathcal{G} \in \mathbf{D}(X)).$$

There are natural functorial maps $s_n \colon \boldsymbol{h}_n \to \boldsymbol{h}_{n+1}$ and $\varepsilon_n \colon \boldsymbol{h}_n \to \mathbf{R}\varGamma_{\mathcal{J}}$, satisfying $\varepsilon_{n+1}s_n = \varepsilon_n$. The family

$$(1, -s_m) \colon \boldsymbol{h}_m \to \boldsymbol{h}_m \oplus \boldsymbol{h}_{m+1} \subset \oplus_{n\geq 1}\boldsymbol{h}_n \qquad (m \geq 1)$$

defines a natural map $s \colon \oplus_{n\geq 1}\boldsymbol{h}_n \to \oplus_{n\geq 1}\boldsymbol{h}_n$. There results, for each $\mathcal{G} \in \mathbf{D}(X)$, a map of triangles

$$\begin{array}{ccccccc}
\oplus_{n\geq 1}\boldsymbol{h}_n\mathcal{G} & \xrightarrow{\ s\ } & \oplus_{n\geq 1}\boldsymbol{h}_n\mathcal{G} & \longrightarrow & ?? & \xrightarrow{\ +\ } & \\
\downarrow & & \downarrow{\scriptstyle \sum\varepsilon_n} & & \downarrow{\scriptstyle \bar{\varepsilon}} & & \\
0 & \longrightarrow & \mathbf{R}\varGamma_{\mathcal{J}}\mathcal{G} & =\!= & \mathbf{R}\varGamma_{\mathcal{J}}\mathcal{G} & \xrightarrow{\ +\ } &
\end{array}$$

LEMMA 5.4.1. *The map $\bar{\varepsilon}$ is a $\mathbf{D}(X)$-isomorphism, and so we have a triangle*

$$\oplus_{n\geq 1}\boldsymbol{h}_n\mathcal{G} \xrightarrow{\ s\ } \oplus_{n\geq 1}\boldsymbol{h}_n\mathcal{G} \xrightarrow{\ \sum\varepsilon_n\ } \mathbf{R}\varGamma_{\mathcal{J}}\mathcal{G} \xrightarrow{\ +\ }$$

PROOF. In the exact homology sequence

$$\cdots \to H^i(\oplus_{n\geq 1}\boldsymbol{h}_n\mathcal{G}) \xrightarrow{\ \sigma^i\ } H^i(\oplus_{n\geq 1}\boldsymbol{h}_n\mathcal{G}) \longrightarrow H^i(??) \longrightarrow H^{i+1}(\oplus_{n\geq 1}\boldsymbol{h}_n\mathcal{G}) \to \cdots$$

the map σ^i is injective, as can be verified stalkwise at each $x \in X$. Assuming, as one may, that $\mathcal{G}$ is K-injective, one deduces that

$$H^i(??) = \varinjlim_n H^i(\boldsymbol{h}_n\mathcal{G}) = H^i \varinjlim_n (\boldsymbol{h}_n\mathcal{G}) = H^i \varinjlim_n \mathcal{H}om^\bullet(\mathcal{O}_X/\mathcal{J}^n, \mathcal{G}) = H^i(\mathbf{R}\varGamma_{\mathcal{J}}\mathcal{G}),$$

whence the assertion. □

6. Duality for torsion sheaves.

Paragraph 6.1 contains the proof of Theorem 2 (section 1), that is, of two essentially equivalent forms of Torsion Duality on formal schemes—Theorem 6.1 and Corollary 6.1.4. The rest of the paragraph deals with numerous relations among the functors which have been introduced, and with compatibilities among dualizing functors occurring before and after completion of maps of ordinary schemes.

More can be said for complexes with coherent homology, thanks to Greenlees-May duality. This is done in paragraph 6.2.

Paragraph 6.3 discusses additional relations involving $\mathbf{R}\mathit{\Gamma}'_{\mathcal{X}}\colon \mathbf{D}(\mathcal{X}) \to \mathbf{D}(\mathcal{X})$ and its right adjoint $\mathbf{R}\mathcal{H}om^\bullet(\mathbf{R}\mathit{\Gamma}'_{\mathcal{X}}\mathcal{O}_{\mathcal{X}}, -)$ on a locally noetherian formal scheme $\mathcal{X}$.

THEOREM 6.1. (a) *Let $f\colon \mathcal{X} \to \mathcal{Y}$ be a map of noetherian formal schemes. Assume that f is separated, or $\mathcal{X}$ has finite Krull dimension, or else restrict to bounded-below complexes. Then the Δ-functor $\mathbf{R}f_*\colon \mathbf{D}_{\mathrm{qct}}(\mathcal{X}) \xrightarrow{5.2.6} \mathbf{D}_{\mathrm{qct}}(\mathcal{Y}) \hookrightarrow \mathbf{D}(\mathcal{Y})$ has a right Δ-adjoint.*

In fact there is a bounded-below Δ-functor $f_{\mathrm{t}}^\times\colon \mathbf{D}(\mathcal{Y}) \to \mathbf{D}_{\mathrm{qct}}(\mathcal{X})$ and a map of Δ-functors $\tau_{\mathrm{t}}\colon \mathbf{R}f_*f_{\mathrm{t}}^\times \to \mathbf{1}$ such that for all $\mathcal{G} \in \mathbf{D}_{\mathrm{qct}}(\mathcal{X})$ and $\mathcal{F} \in \mathbf{D}(\mathcal{Y})$, the composed map (in the derived category of abelian groups)

$$\begin{aligned}\mathbf{R}\mathrm{Hom}^\bullet_{\mathcal{X}}(\mathcal{G}, f_{\mathrm{t}}^\times\mathcal{F}) &\xrightarrow{\text{natural}} \mathbf{R}\mathrm{Hom}^\bullet_{\mathcal{Y}}(\mathbf{R}f_*\mathcal{G}, \mathbf{R}f_*f_{\mathrm{t}}^\times\mathcal{F})\\ &\xrightarrow{\text{via } \tau_{\mathrm{t}}} \mathbf{R}\mathrm{Hom}^\bullet_{\mathcal{Y}}(\mathbf{R}f_*\mathcal{G}, \mathcal{F})\end{aligned}$$

is an isomorphism.

(b) *If $g\colon \mathcal{Y} \to \mathcal{Z}$ is another such map then there is a natural isomorphism*

$$(gf)_{\mathrm{t}}^\times \xrightarrow{\sim} f_{\mathrm{t}}^\times g_{\mathrm{t}}^\times.$$

Proof. Assertion (b) follows from (a), which easily implies that $(gf)_{\mathrm{t}}^\times$ and $f_{\mathrm{t}}^\times g_{\mathrm{t}}^\times$ are both right-adjoint to the restriction of $\mathbf{R}(gf)_* = \mathbf{R}g_*\mathbf{R}f_*$ to $\mathbf{D}_{\mathrm{qct}}(\mathcal{X})$.

As for (a), assuming first that $\mathcal{X}$ is separated or finite-dimensional, or that only bounded-below complexes are considered, we can replace $\mathbf{D}_{\mathrm{qct}}(\mathcal{X})$ by the *equivalent* category $\mathbf{D}(\mathcal{A}_{\mathrm{qct}}(\mathcal{X}))$ (Proposition 5.3.1). The inclusion $k\colon \mathcal{A}_{\mathrm{qct}}(\mathcal{X}) \hookrightarrow \mathcal{A}_{\vec{\mathrm{c}}}(\mathcal{X})$ has the right adjoint $\mathit{\Gamma}'_{\mathcal{X}}$. ($\mathit{\Gamma}'_{\mathcal{X}}(\mathcal{A}_{\vec{\mathrm{c}}}(\mathcal{X})) \subset \mathcal{A}_{\mathrm{qct}}(\mathcal{X})$, by Lemma 5.1.4 and Corollary 3.1.5.) So for all $\mathcal{A}_{\mathrm{qct}}(\mathcal{X})$-complexes $\mathcal{G}'$ and $\mathcal{A}_{\vec{\mathrm{c}}}(\mathcal{X})$-complexes $\mathcal{F}'$ there is a natural isomorphism of abelian-group complexes

$$\mathrm{Hom}^\bullet_{\mathcal{A}_{\mathrm{qct}}}(\mathcal{G}', \mathit{\Gamma}'_{\mathcal{X}}\mathcal{F}') \xrightarrow{\sim} \mathrm{Hom}^\bullet_{\mathcal{A}_{\vec{\mathrm{c}}}}(k\mathcal{G}', \mathcal{F}').$$

Note that if $\mathcal{F}'$ is K-injective over $\mathcal{A}_{\vec{\mathrm{c}}}(\mathcal{X})$ then $\mathit{\Gamma}'_{\mathcal{X}}\mathcal{F}'$ is K-injective over $\mathcal{A}_{\mathrm{qct}}(\mathcal{X})$, because $\mathit{\Gamma}'_{\mathcal{X}}$ has an exact left adjoint. Combining this isomorphism with the isomorphism (4.1.4.1) in the proof of Theorem 4.1, we can conclude just as in part 4 at the end of that proof, with the functor $f_{\mathrm{t}}^\times$ defined to be the composition

$$\mathbf{D}(\mathcal{Y}) \xrightarrow{\rho} \mathbf{K_I}(\mathcal{Y}) \xrightarrow{\mathcal{C}_\bullet} \mathbf{K_I}(\mathcal{A}_{\vec{\mathrm{c}}}(\mathcal{X})) \xrightarrow{\mathit{\Gamma}'_{\mathcal{X}}} \mathbf{K_I}(\mathcal{A}_{\mathrm{qct}}(\mathcal{X})) \xrightarrow{\text{natural}} \mathbf{D}(\mathcal{A}_{\mathrm{qct}}(\mathcal{X})).$$

(We have in mind here simply that the natural functor $\mathbf{D}(\mathcal{A}_{\mathrm{qct}}(\mathcal{X})) \to \mathbf{D}(\mathcal{A}_{\vec{\mathrm{c}}}(\mathcal{X}))$ has a right adjoint. That is easily seen to be true once one knows the existence of K-injective resolutions in $\mathbf{D}(\mathcal{A}_{\vec{\mathrm{c}}}(\mathcal{X}))$; but we don't know how to prove the latter other than by quoting the generalization to arbitrary Grothendieck categories [**Fe**, Theorem 2], [**AJS**, Theorem 5.4]. The preceding argument avoids this issue. One could also apply Brown Representability directly, as in the proof of Theorem 1 described in the Introduction.)

Now suppose only that the map f is separated. If $\mathcal{Y}$ is separated then so is $\mathcal{X}$, and the preceding argument holds. For arbitrary noetherian $\mathcal{Y}$ the existence of a bounded-below right adjoint for $\mathbf{R}f_*\colon \mathbf{D}_{\mathrm{qct}}(\mathcal{X}) \to \mathbf{D}(\mathcal{Y})$ results then from the following Mayer-Vietoris pasting argument, by induction on the least number of separated open subsets needed to cover $\mathcal{Y}$. Finally, to dispose of the assertion about the $\mathbf{R}\mathrm{Hom}^\bullet$'s apply homology to reduce it to $f_{\mathrm{t}}^\times$ being a right adjoint.

To reduce clutter, we will abuse notation—but only in the rest of the proof of Theorem 6.1—by writing "$f^\times$" in place of "$f_{\mathrm{t}}^\times$."

LEMMA 6.1.1. *Let* $f\colon \mathcal{X} \to \mathcal{Y} = \mathcal{Y}_1 \cup \mathcal{Y}_2$ ($\mathcal{Y}_i$ *open in* $\mathcal{Y}$) *be a map of formal schemes, with* $\mathcal{X}$ *noetherian. Consider the commutative diagrams*

$$\begin{array}{ccccc} \mathcal{X}_{12} := \mathcal{X}_1 \cap \mathcal{X}_2 & \xrightarrow{\ q_i\ } & \mathcal{X}_i & \xrightarrow{\ x_i\ } & \mathcal{X} \\ {\scriptstyle f_{12}}\big\downarrow & & {\scriptstyle f_i}\big\downarrow & & \big\downarrow{\scriptstyle f} \\ \mathcal{Y}_{12} := \mathcal{Y}_1 \cap \mathcal{Y}_2 & \xrightarrow[\ p_i\ ]{} & \mathcal{Y}_i & \xrightarrow[\ y_i\ ]{} & \mathcal{Y} \end{array} \qquad (i = 1, 2)$$

where $\mathcal{X}_i := f^{-1}\mathcal{Y}_i$ *and all the horizontal arrows represent inclusions. Suppose that for* $i = 1, 2, 12$, *the functor* $\mathbf{R}f_{i*}\colon \mathbf{D}_{\mathrm{qct}}(\mathcal{X}_i) \to \mathbf{D}(\mathcal{Y}_i)$ *has a right adjoint* $f_i^{\times}$. *Then* $\mathbf{R}f_*\colon \mathbf{D}_{\mathrm{qct}}(\mathcal{X}) \to \mathbf{D}(\mathcal{Y})$ *has a right adjoint* $f^{\times}$; *and with the inclusions* $y_{12} := y_i \circ p_i$, $x_{12} := x_i \circ q_i$, *there is for each* $\mathcal{F} \in \mathbf{D}(\mathcal{Y})$ *a natural* $\mathbf{D}(\mathcal{X})$*-triangle*

$$f^{\times}\mathcal{F} \to \mathbf{R}x_{1*}f_1^{\times}y_1^*\mathcal{F} \oplus \mathbf{R}x_{2*}f_2^{\times}y_2^*\mathcal{F} \xrightarrow{\lambda_{\mathcal{F}}} \mathbf{R}x_{12*}f_{12}^{\times}y_{12}^*\mathcal{F} \to (f^{\times}\mathcal{F})[1].$$

Remark. If we expect $f^{\times}$ to exist, and the natural maps $x_i^*f^{\times} \to f_i^{\times}y_i^*$ to be isomorphisms, then there should be such a triangle—the Mayer-Vietoris triangle of $f^{\times}\mathcal{F}$. This suggests we first define $\lambda_{\mathcal{F}}$, then let $f^{\times}\mathcal{F}$ be the vertex of a triangle based on $\lambda_{\mathcal{F}}$, and verify ...

PROOF. There are natural maps

$$\tau_1\colon \mathbf{R}f_{1*}f_1^{\times} \to \mathbf{1}, \qquad \tau_2\colon \mathbf{R}f_{2*}f_2^{\times} \to \mathbf{1}, \qquad \tau_{12}\colon \mathbf{R}f_{12*}f_{12}^{\times} \to \mathbf{1}.$$

For $i = 1, 2$, define the "base-change" map $\beta_i\colon q_i^*f_i^{\times} \to f_{12}^{\times}p_i^*$ to be adjoint under Theorem 6.1 to the map of functors

$$\mathbf{R}f_{12*}q_i^*f_i^{\times} \underset{\text{natural}}{\xrightarrow{\ \sim\ }} p_i^*\mathbf{R}f_{i*}f_i^{\times} \xrightarrow{\ \tau_i\ } p_i^*.$$

This β_i corresponds to a functorial map $\beta_i'\colon f_i^{\times} \to \mathbf{R}q_{i*}f_{12}^{\times}p_i^*$, from which we obtain a functorial map

$$\mathbf{R}x_{i*}f_i^{\times}y_i^* \longrightarrow \mathbf{R}x_{i*}\mathbf{R}q_{i*}f_{12}^{\times}p_i^*y_i^* \xrightarrow{\ \sim\ } \mathbf{R}x_{12*}f_{12}^{\times}y_{12}^*,$$

and hence a natural map, for any $\mathcal{F} \in \mathbf{D}(\mathcal{Y})$:

$$\check{D}^0(\mathcal{F}) := \mathbf{R}x_{1*}f_1^{\times}y_1^*\mathcal{F} \oplus \mathbf{R}x_{2*}f_2^{\times}y_2^*\mathcal{F} \xrightarrow{\lambda_{\mathcal{F}}} \mathbf{R}x_{12*}f_{12}^{\times}y_{12}^*\mathcal{F} =: \check{D}^1(\mathcal{F}).$$

Embed this map in a triangle $\check{D}(\mathcal{F})$, and denote the third vertex by $f^{\times}(\mathcal{F})$:

$$\check{D}(\mathcal{F})\colon\ f^{\times}\mathcal{F} \to \check{D}^0(\mathcal{F}) \xrightarrow{\lambda_{\mathcal{F}}} \check{D}^1(\mathcal{F}) \to (f^{\times}\mathcal{F})[1].$$

Since $\check{D}^0(\mathcal{F})$ and $\check{D}^1(\mathcal{F})$ are in $\mathbf{D}_{\mathrm{qct}}(\mathcal{X})$ (see Proposition 5.2.6), therefore so is $f^{\times}\mathcal{F}$ (Corollary 5.1.3).

This is the triangle in Lemma 6.1.1. Of course we must still show that this $f^{\times}$ is functorial, and right-adjoint to $\mathbf{R}f_*$. (Then by uniqueness of adjoints such a triangle will exist no matter which right adjoint $f^{\times}$ is used.)

Let us next construct a map $\tau_{\mathcal{F}}\colon \mathbf{R}f_*f^\times\mathcal{F} \to \mathcal{F}$ ($\mathcal{F} \in \mathbf{D}(\mathcal{Y})$). Set

$$\check{C}^0(\mathcal{F}) := \mathbf{R}y_{1*}y_1^*\mathcal{F} \oplus \mathbf{R}y_{2*}y_2^*\mathcal{F}, \qquad \check{C}^1(\mathcal{F}) := \mathbf{R}y_{12*}y_{12}^*\mathcal{F}.$$

We have then the Mayer-Vietoris $\mathbf{D}(\mathcal{Y})$-triangle

$$\check{C}(\mathcal{F})\colon\ \mathcal{F} \to \check{C}^0(\mathcal{F}) \xrightarrow{\mu_{\mathcal{F}}} \check{C}^1(\mathcal{F}) \to \mathcal{F}[1],$$

arising from the usual exact sequence (Čech resolution)

$$0 \to \mathcal{F} \to y_{1*}y_1^*\mathcal{F} \oplus y_{2*}y_2^*\mathcal{F} \to y_{12*}y_{12}^*\mathcal{F} \to 0,$$

where $\mathcal{F}$ may be taken to be K-injective. Checking commutativity of the following natural diagram is a purely category-theoretic exercise (cf. [**L4**, Lemma (4.8.1.2)] :

$$\begin{array}{ccc}
\mathbf{R}f_*\check{D}^0(\mathcal{F}) & \xrightarrow{\quad\mathbf{R}f_*\lambda_{\mathcal{F}}\quad} & \mathbf{R}f_*\check{D}^1(\mathcal{F}) \\
\Vert & & \Vert \\
\mathbf{R}f_*(\mathbf{R}x_{1*}f_1^\times y_1^*\mathcal{F} \oplus \mathbf{R}x_{2*}f_2^\times y_2^*\mathcal{F}) & & \mathbf{R}f_*\mathbf{R}x_{12*}f_{12}^\times y_{12}^*\mathcal{F} \\
\simeq\downarrow & & \downarrow\simeq \\
\mathbf{R}y_{1*}\mathbf{R}f_{1*}f_1^\times y_1^*\mathcal{F} \oplus \mathbf{R}y_{2*}\mathbf{R}f_{2*}f_2^\times y_2^*\mathcal{F} & & \mathbf{R}y_{12*}\mathbf{R}f_{12*}f_{12}^\times y_{12}^*\mathcal{F} \\
\tau_1\oplus\tau_2\downarrow & & \downarrow\tau_{12} \\
\mathbf{R}y_{1*}y_1^*\mathcal{F} \oplus \mathbf{R}y_{2*}y_2^*\mathcal{F} & & \mathbf{R}y_{12*}y_{12}^*\mathcal{F} \\
\Vert & & \Vert \\
\check{C}^0(\mathcal{F}) & \xrightarrow[\quad\mu_{\mathcal{F}}\quad]{} & \check{C}^1(\mathcal{F})
\end{array}$$

This commutative diagram extends to a map $\check{\tau}_{\mathcal{F}}$ of triangles:

$$\begin{array}{ccccccc}
\mathbf{R}f_*f^\times\mathcal{F} & \longrightarrow & \mathbf{R}f_*\check{D}^0(\mathcal{F}) & \longrightarrow & \mathbf{R}f_*\check{D}^1(\mathcal{F}) & \longrightarrow & \mathbf{R}f_*f^\times\mathcal{F}[1] \\
\tau_{\mathcal{F}}\downarrow & & \downarrow & & \downarrow & & \downarrow\tau_{\mathcal{F}}[1] \\
\mathcal{F} & \longrightarrow & \check{C}^0(\mathcal{F}) & \longrightarrow & \check{C}^1(\mathcal{F}) & \longrightarrow & \mathcal{F}[1]
\end{array}$$

The map $\tau_{\mathcal{F}}$ is not necessarily unique. But the next Lemma will show, for fixed $\mathcal{F}$, that *the pair* $(f^\times\mathcal{F}, \tau_{\mathcal{F}})$ *represents the functor*

$$\mathrm{Hom}_{\mathbf{D}(\mathcal{Y})}(\mathbf{R}f_*\mathcal{E}, \mathcal{F}) \qquad (\mathcal{E} \in \mathbf{D}_{\mathrm{qct}}(\mathcal{X})).$$

It follows formally that one can make $f^\times$ into a functor and $\tau\colon \mathbf{R}f_*f^\times \to \mathbf{1}$ into a morphism of functors in such a way that the pair $(f^\times, \tau)$ is a right adjoint for $\mathbf{R}f_*\colon \mathbf{D}_{\mathrm{qct}}(\mathcal{X}) \to \mathbf{D}(\mathcal{Y})$ (cf. [**M1**, p. 83, Corollary 2]); and that there is a unique isomorphism of functors $\Theta\colon f^\times T_2 \xrightarrow{\ \sim\ } T_1 f^\times$ (where T_1 and T_2 are the respective translations on $\mathbf{D}_{\mathrm{qct}}(\mathcal{X})$ and $\mathbf{D}(\mathcal{Y})$) such that $(f^\times, \Theta)$ is a Δ-functor Δ-adjoint to $\mathbf{R}f_*$ (cf. [**L4**, Proposition (3.3.8)]). That will complete the proof of Lemma 6.1.1. $\square$

LEMMA 6.1.2. *For* $\mathcal{E} \in \mathbf{D}_{\mathrm{qct}}(\mathcal{X})$, *and with* $f^\times\mathcal{F}$, $\tau_{\mathcal{F}}$ *as above, the composition*

$$\mathrm{Hom}_{\mathbf{D}_{\mathrm{qct}}(\mathcal{X})}(\mathcal{E},\, f^\times\mathcal{F}) \xrightarrow{\text{natural}} \mathrm{Hom}_{\mathbf{D}(\mathcal{Y})}(\mathbf{R}f_*\mathcal{E}, \mathbf{R}f_*f^\times\mathcal{F}) \xrightarrow{\text{via } \tau_{\mathcal{F}}} \mathrm{Hom}_{\mathbf{D}(\mathcal{Y})}(\mathbf{R}f_*\mathcal{E}, \mathcal{F})$$

is an isomorphism.

PROOF. In the following diagram, to save space we write $H_{\mathcal{X}}$ for $\mathrm{Hom}_{\mathbf{D}_{\mathrm{qct}}(\mathcal{X})}$, $H_{\mathcal{Y}}$ for $\mathrm{Hom}_{\mathbf{D}(\mathcal{Y})}$, and f_* for $\mathbf{R}f_*$:

$$
\begin{array}{ccccc}
H_{\mathcal{X}}(\mathcal{E}, (\check{D}^0\mathcal{F})[-1]) & \longrightarrow & H_{\mathcal{Y}}(f_*\mathcal{E}, f_*((\check{D}^0\mathcal{F})[-1])) & \longrightarrow & H_{\mathcal{Y}}(f_*\mathcal{E}, (\check{C}^0\mathcal{F})[-1]) \\
\downarrow & & \downarrow & & \downarrow \\
H_{\mathcal{X}}(\mathcal{E}, (\check{D}^1\mathcal{F})[-1]) & \longrightarrow & H_{\mathcal{Y}}(f_*\mathcal{E}, f_*((\check{D}^1\mathcal{F})[-1])) & \longrightarrow & H_{\mathcal{Y}}(f_*\mathcal{E}, (\check{C}^1\mathcal{F})[-1]) \\
\downarrow & & \downarrow & & \downarrow \\
H_{\mathcal{X}}(\mathcal{E}, f^\times\mathcal{F}) & \longrightarrow & H_{\mathcal{Y}}(f_*\mathcal{E}, f_*f^\times\mathcal{F}) & \longrightarrow & H_{\mathcal{Y}}(f_*\mathcal{E}, \mathcal{F}) \\
\downarrow & & \downarrow & & \downarrow \\
H_{\mathcal{X}}(\mathcal{E}, \check{D}^0\mathcal{F}) & \longrightarrow & H_{\mathcal{Y}}(f_*\mathcal{E}, f_*\check{D}^0\mathcal{F}) & \longrightarrow & H_{\mathcal{Y}}(f_*\mathcal{E}, \check{C}^0\mathcal{F}) \\
\downarrow & & \downarrow & & \downarrow \\
H_{\mathcal{X}}(\mathcal{E}, \check{D}^1\mathcal{F}) & \longrightarrow & H_{\mathcal{Y}}(f_*\mathcal{E}, f_*\check{D}^1\mathcal{F}) & \longrightarrow & H_{\mathcal{Y}}(f_*\mathcal{E}, \check{C}^1\mathcal{F})
\end{array}
$$

The first column maps to the second via functoriality of f_*, and the second to the third via the above triangle map $\check{\tau}_{\mathcal{F}}$; so the diagram commutes. The columns are exact [**H1**, p. 23, Prop. 1.1 b)], and thus if each of the first two and last two rows is shown to compose to an isomorphism, then the same holds for the middle row, proving Lemma 6.1.2.

Let's look at the fourth row. With notation as in Lemma 6.1.1 (and again, with all the appropriate **R**'s omitted), we want the left column in the following natural diagram to compose to an isomorphism:

$$
\begin{array}{ccc}
H_{\mathcal{X}}(\mathcal{E}, x_{i*}f_i^\times y_i^*\mathcal{F}) & \xrightarrow{\ \simeq\ } & H_{\mathcal{X}_i}(x_i^*\mathcal{E}, f_i^\times y_i^*\mathcal{F}) \\
\downarrow & & \downarrow \\
H_{\mathcal{Y}}(f_*\mathcal{E}, f_*x_{i*}f_i^\times y_i^*\mathcal{F}) & & H_{\mathcal{Y}_i}(f_{i*}x_i^*\mathcal{E}, f_{i*}f_i^\times y_i^*\mathcal{F}) \\
{\scriptstyle\simeq}\downarrow & & \downarrow{\scriptstyle\simeq} \\
H_{\mathcal{Y}}(f_*\mathcal{E}, y_{i*}f_{i*}f_i^\times y_i^*\mathcal{F}) & \xrightarrow{\ \simeq\ } & H_{\mathcal{Y}_i}(y_i^*f_*\mathcal{E}, f_{i*}f_i^\times y_i^*\mathcal{F}) \\
{\scriptstyle\text{via } \tau_i}\downarrow & & \downarrow{\scriptstyle\text{via } \tau_i} \\
H_{\mathcal{Y}}(f_*\mathcal{E}, y_{i*}y_i^*\mathcal{F}) & \xrightarrow{\ \simeq\ } & H_{\mathcal{Y}_i}(y_i^*f_*\mathcal{E}, y_i^*\mathcal{F})
\end{array}
$$

Here the horizontal arrows represent adjunction isomorphisms. Checking that the diagram commutes is a straightforward category-theoretic exercise. By hypothesis, the right column composes to an isomorphism. Hence so does the left one.

The argument for the fifth row is similar. Using the (easily checked) fact that the morphism $f_*\check{D}^0 \to \check{C}^0$ is Δ-functorial, we find that the first row is, up to isomorphism, the same as the fourth row with $\mathcal{F}[-1]$ in place of $\mathcal{F}$, so it too composes to an isomorphism. Similarly, isomorphism for the second row follows from that for the fifth. $\square$

EXAMPLES 6.1.3. (1) Let $f\colon \mathcal{X} \to \mathcal{Y}$ be a map of quasi-compact formal schemes with $\mathcal{X}$ *properly algebraic,* and let $f^\times$ be the right adjoint given by Corollary 4.1.1. Using Proposition 5.2.1 we find then that $f_{\mathrm{t}}^\times := \mathbf{R}\varGamma'_{\mathcal{X}} \circ f^\times$ is a right adjoint for the restriction of $\mathbf{R}f_*$ to $\mathbf{D}_{\mathrm{qct}}(\mathcal{X})$.

(2) For a noetherian formal scheme $\mathcal{X}$, Theorem 6.1 gives a right adjoint $\mathbf{1}^! := \mathbf{1}_{\mathrm{t}}^\times$ to the inclusion $\mathbf{D}_{\mathrm{qct}}(\mathcal{X}) \hookrightarrow \mathbf{D}(\mathcal{X})$. If $\mathcal{G} \in \widetilde{\mathbf{D}}_{\mathrm{qc}}(\mathcal{X})$ (i.e., $\mathbf{R}\varGamma'_{\mathcal{X}}\mathcal{G} \in \mathbf{D}_{\mathrm{qct}}(\mathcal{X})$, see Definition 5.2.9), then the natural $\mathbf{D}_{\mathrm{qct}}(\mathcal{X})$-map $\mathbf{R}\varGamma'_{\mathcal{X}}\mathcal{G} \to \mathbf{1}^!\mathcal{G}$ (corresponding to the natural $\mathbf{D}(\mathcal{X})$-map $\mathbf{R}\varGamma'_{\mathcal{X}}\mathcal{G} \to \mathcal{G}$) is an *isomorphism,* see Proposition 5.2.1.

(3) If $\mathcal{X}$ is *separated* or if $\mathcal{X}$ is *finite-dimensional,* then we have the equivalence $\boldsymbol{j}_{\mathcal{X}}^{\mathrm{t}}\colon \mathbf{D}(\mathcal{A}_{\mathrm{qct}}(\mathcal{X})) \xrightarrow{\approx} \mathbf{D}_{\mathrm{qct}}(\mathcal{X})$ of Proposition 5.3.1, and we can take $\mathbf{1}^! := \boldsymbol{j}_{\mathcal{X}}^{\mathrm{t}} \circ \mathbf{R}Q_{\mathcal{X}}^{\mathrm{t}}$, see Corollary 5.1.5 and Lemma 5.2.2.

(4) Let $f\colon \mathcal{X} \to \mathcal{Y}$ be a closed immersion of noetherian formal schemes (see [**GD**, p. 442]). The functor $f_*\colon \mathcal{A}(\mathcal{X}) \to \mathcal{A}(\mathcal{Y})$ is exact, so $\mathbf{R}f_* = f_*$. Let $\mathcal{I}$ be the kernel of the surjective map $\mathcal{O}_{\mathcal{Y}} \twoheadrightarrow f_*\mathcal{O}_{\mathcal{X}}$ and let $\bar{\mathcal{Y}}$ be the ringed space $(\mathcal{Y}, \mathcal{O}_{\mathcal{Y}}/\mathcal{I})$, so that f factors naturally as $\mathcal{X} \xrightarrow{\bar f} \bar{\mathcal{Y}} \xrightarrow{i} \mathcal{Y}$, the map $\bar f$ being flat. The inverse isomorphisms $\mathcal{A}(\mathcal{X}) \underset{\bar f^*}{\overset{\bar f_*}{\rightleftarrows}} \mathcal{A}(\bar{\mathcal{Y}})$ extend to inverse isomorphisms $\mathbf{D}(\mathcal{X}) \underset{\bar f^*}{\overset{\bar f_*}{\rightleftarrows}} \mathbf{D}(\bar{\mathcal{Y}})$

The functor $\mathcal{H}_{\mathcal{I}}\colon \mathcal{A}(\mathcal{Y}) \to \mathcal{A}(\bar{\mathcal{Y}})$ defined by $\mathcal{H}_{\mathcal{I}}(F) := \mathcal{H}om(\mathcal{O}_{\mathcal{Y}}/\mathcal{I}, F)$ has an exact left adjoint, namely $i_*\colon \mathcal{A}(\bar{\mathcal{Y}}) \to \mathcal{A}(\mathcal{Y})$, so $\mathcal{H}_{\mathcal{I}}$ preserves K-injectivity and $\mathbf{R}\mathcal{H}_{\mathcal{I}}$ is right-adjoint to $i_*\colon \mathbf{D}(\bar{\mathcal{Y}}) \to \mathbf{D}(\mathcal{Y})$ (see proof of Lemma 5.2.2). Hence the functor $f^\natural\colon \mathbf{D}(\mathcal{Y}) \to \mathbf{D}(\bar{\mathcal{Y}})$ defined by

$$f^\natural(\mathcal{F}) := \bar f^*\mathbf{R}\mathcal{H}_{\mathcal{I}}(\mathcal{F}) = \bar f^*\mathbf{R}\mathcal{H}om^\bullet(\mathcal{O}_{\mathcal{Y}}/\mathcal{I}, \mathcal{F}) \qquad (\mathcal{F} \in \mathbf{D}(\mathcal{Y})) \tag{6.1.3.1}$$

is right-adjoint to $f_* = i_*\bar f_*$, and $f_*\colon \mathbf{D}_{\mathrm{qct}}(\mathcal{X}) \to \mathbf{D}(\mathcal{Y})$ has the right adjoint

$$f_{\mathrm{t}}^\times := f^! := \mathbf{1}^! \circ f^\natural.$$

We recall that $\mathcal{G} \in \mathcal{A}(\mathcal{X})$ is quasi-coherent iff $\bar f_*\mathcal{G} \in \mathcal{A}_{\mathrm{qc}}(\bar{\mathcal{Y}})$ iff $f_*\mathcal{G} \in \mathcal{A}_{\mathrm{qc}}(\mathcal{Y})$, see [**GD**, p. 115, (5.3.15), (5.3.13)]. Also, by looking at stalks (see §1.2.1) we find that $f_*\mathcal{G} \in \mathcal{A}_{\mathrm{t}}(\mathcal{Y}) \Rightarrow \mathcal{G} \in \mathcal{A}_{\mathrm{t}}(\mathcal{X})$. Hence Remark 5.2.10(4) together with the isomorphism $\mathbf{R}f_*\mathbf{R}\varGamma'_{\mathcal{X}} \cong \mathbf{R}\varGamma'_{\mathcal{Y}}\mathbf{R}f_*$ of Corollary 5.2.11(d) yields that $f^\natural\widetilde{\mathbf{D}}_{\mathrm{qc}}^+(\mathcal{Y}) \subset \widetilde{\mathbf{D}}_{\mathrm{qc}}^+(\mathcal{X})$; and given Corollary 5.1.3, Proposition 3.2.4 yields $f^\natural\mathbf{D}_{\mathrm{qct}}^+(\mathcal{Y}) \subset \mathbf{D}_{\mathrm{qct}}^+(\mathcal{X})$. Thus if $\mathcal{F} \in \widetilde{\mathbf{D}}_{\mathrm{qc}}^+(\mathcal{Y})$ then by (2) above, $f^!\mathcal{F} \cong \mathbf{R}\varGamma'_{\mathcal{X}}f^\natural\mathcal{F}$; and if $\mathcal{F} \in \mathbf{D}_{\mathrm{qct}}^+(\mathcal{Y})$ then $f^!\mathcal{F} \cong f^\natural\mathcal{F}$.

(5) Let $f\colon \mathcal{X} \to \mathcal{Y}$ be any map satisfying the hypotheses of Theorem 6.1. Let $\mathcal{J} \subset \mathcal{O}_{\mathcal{X}}$ and $\mathcal{I} \subset \mathcal{O}_{\mathcal{Y}}$ be ideals of definition such that $\mathcal{I}\mathcal{O}_{\mathcal{X}} \subset \mathcal{J}$, and let

$$X_n := (\mathcal{X}, \mathcal{O}_{\mathcal{X}}/\mathcal{J}^n) \xrightarrow{f_n} (\mathcal{Y}, \mathcal{O}_{\mathcal{Y}}/\mathcal{I}^n) =: Y_n \qquad (n > 0)$$

be the scheme-maps induced by f, so that each f_n also satisfies the hypotheses of Theorem 6.1. As the target of the functor $(f_n)_{\mathrm{t}}^\times$ is $\mathbf{D}_{\mathrm{qct}}(X_n) = \mathbf{D}_{\mathrm{qc}}(X_n)$, we write $f_n^\times$ for $(f_n)_{\mathrm{t}}^\times$ (see (1) above). If $j_n\colon X_n \hookrightarrow \mathcal{X}$ and $i_n\colon Y_n \hookrightarrow \mathcal{Y}$ are the canonical closed immersions then $fj_n = i_nf_n$, and so $j_n^!f_{\mathrm{t}}^\times = f_n^\times i_n^!$.

The functor $j_n^\natural\colon \mathbf{D}(\mathcal{X}) \to \mathbf{D}(X_n)$ being as in (6.1.3.1), we have, using (4),

$$\boldsymbol{h}_n\mathcal{G} := \mathbf{R}\mathcal{H}om^\bullet(\mathcal{O}_{\mathcal{X}}/\mathcal{J}^n, \mathcal{G}) = j_{n*}j_n^\natural\mathcal{G} \cong j_{n*}j_n^!\mathcal{G} \qquad (\mathcal{G} \in \widetilde{\mathbf{D}}_{\mathrm{qc}}^+(\mathcal{X})).$$

Hence for $\mathcal{G} := f_{\mathrm{t}}^\times\mathcal{F}$ ($\mathcal{F} \in \mathbf{D}^+(\mathcal{Y})$), Lemma 5.4.1 gives a "homotopy colimit" triangle

$$\oplus_{n\ge1}\, j_{n*}f_n^\times i_n^!\mathcal{F} \longrightarrow \oplus_{n\ge1}\, j_{n*}f_n^\times i_n^!\mathcal{F} \longrightarrow f_{\mathrm{t}}^\times\mathcal{F} \xrightarrow{+}$$

Once again, $\widetilde{\mathbf{D}}_{\mathrm{qc}}(\mathcal{X}) := (\mathbf{R}\varGamma'_{\mathcal{X}})^{-1}\mathbf{D}_{\mathrm{qct}}(\mathcal{X})$ (Definition 5.2.9).

COROLLARY 6.1.4. (a) *Let $f\colon \mathcal{X} \to \mathcal{Y}$ be a map of noetherian formal schemes. Suppose that f is separated or that $\mathcal{X}$ has finite Krull dimension, or else restrict to bounded-below complexes. Let $\mathbf{\Lambda}_{\mathcal{X}}\colon \mathbf{D}(\mathcal{X}) \to \mathbf{D}(\mathcal{X})$ be the bounded-below Δ-functor*

$$\mathbf{\Lambda}_{\mathcal{X}}(-) := \mathbf{R}\mathcal{H}om^{\bullet}(\mathbf{R}\varGamma'_{\mathcal{X}}\mathcal{O}_{\mathcal{X}}, -),$$

and let $f^{\#}\colon \mathbf{D}(\mathcal{Y}) \to \widetilde{\mathbf{D}}_{\mathrm{qc}}(\mathcal{X})$ be the Δ-functor $f^{\#} := \mathbf{\Lambda}_{\mathcal{X}} f_{\mathrm{t}}^{\times}$ (see Example 5.2.10(3)).

The functor $f^{\#}$ is bounded-below, and is right-adjoint to

$$\mathbf{R}f_*\mathbf{R}\varGamma'_{\mathcal{X}}\colon \widetilde{\mathbf{D}}_{\mathrm{qc}}(\mathcal{X}) \xrightarrow{5.2.6} \mathbf{D}_{\mathrm{qct}}(\mathcal{Y}) \hookrightarrow \mathbf{D}(\mathcal{Y}).$$

(*In particular with functor, the functor*

$$\mathbf{R}f_*\mathbf{R}\varGamma'_{\mathcal{X}}\boldsymbol{j}\colon \mathbf{D}(\mathcal{A}_{\vec{c}}(\mathcal{X})) \to \mathbf{D}(\mathcal{Y})$$

has the bounded-below right adjoint $\mathbf{R}Q_{\mathcal{X}}f^{\#}$—see Proposition 3.2.3.)

In fact there is a map of Δ-functors

$$\tau^{\#}\colon \mathbf{R}f_*\mathbf{R}\varGamma'_{\mathcal{X}}f^{\#} \to \mathbf{1}$$

such that for all $\mathcal{G} \in \widetilde{\mathbf{D}}_{\mathrm{qc}}(\mathcal{X})$ and $\mathcal{F} \in \mathbf{D}(\mathcal{Y})$, the composed map

$$\begin{aligned}\mathbf{R}\mathrm{Hom}^{\bullet}_{\mathcal{X}}(\mathcal{G}, f^{\#}\mathcal{F}) &\xrightarrow{\text{natural}} \mathbf{R}\mathrm{Hom}^{\bullet}_{\mathcal{Y}}(\mathbf{R}f_*\mathbf{R}\varGamma'_{\mathcal{X}}\mathcal{G}, \mathbf{R}f_*\mathbf{R}\varGamma'_{\mathcal{X}}f^{\#}\mathcal{F})\\ &\xrightarrow{\text{via } \tau^{\#}} \mathbf{R}\mathrm{Hom}^{\bullet}_{\mathcal{Y}}(\mathbf{R}f_*\mathbf{R}\varGamma'_{\mathcal{X}}\mathcal{G}, \mathcal{F})\end{aligned}$$

is an isomorphism.

(b) *If $g\colon \mathcal{Y} \to \mathcal{Z}$ is another such map then there is a natural isomorphism*

$$(gf)^{\#} \xrightarrow{\ \sim\ } f^{\#}g^{\#}.$$

PROOF. (a) The functor $\mathbf{\Lambda}_{\mathcal{X}}$ is bounded below because $\mathbf{R}\varGamma'_{\mathcal{X}}\mathcal{O}_{\mathcal{X}}$ is locally isomorphic to the bounded complex $\mathcal{K}^{\bullet}_{\infty}$ in the proof of Proposition 5.2.1(a), hence homologically bounded-above. Since $\mathbf{\Lambda}_{\mathcal{X}}$ is right-adjoint to $\mathbf{R}\varGamma'_{\mathcal{X}}$ (see (5.2.10.1)), (a) follows directly from Theorem 6.1.

(b) Propositions 5.2.6 and 5.2.1(a) show that for any $\mathcal{G} \in \mathbf{D}_{\mathrm{qct}}(\mathcal{X})$ we have $\mathbf{R}\varGamma'_{\mathcal{Y}}\mathbf{R}f_*\mathcal{G} \cong \mathbf{R}f_*\mathcal{G}$, and hence the functors $f_{\mathrm{t}}^{\times}\mathbf{\Lambda}_{\mathcal{Y}}$ and $f_{\mathrm{t}}^{\times}$ are both right-adjoint to $\mathbf{R}f_*|_{\mathbf{D}_{\mathrm{qct}}(\mathcal{X})}$, so they are isomorphic. Then Theorem 6.1(b) yields functorial isomorphisms

$$(gf)^{\#} = \mathbf{\Lambda}_{\mathcal{X}}(gf)_{\mathrm{t}}^{\times} \xrightarrow{\ \sim\ } \mathbf{\Lambda}_{\mathcal{X}}f_{\mathrm{t}}^{\times}g_{\mathrm{t}}^{\times} \xrightarrow{\ \sim\ } \mathbf{\Lambda}_{\mathcal{X}}f_{\mathrm{t}}^{\times}\mathbf{\Lambda}_{\mathcal{Y}}g_{\mathrm{t}}^{\times} = f^{\#}g^{\#}.$$

□

Here are some "identities" involving the dualizing functors $f^{\times}$ (Theorem 4.1), $f_{\mathrm{t}}^{\times}$ (Theorem 6.1), and $f^{\#} := \mathbf{\Lambda}_{\mathcal{X}}f_{\mathrm{t}}^{\times}$ (Corollary 6.1.4).

Note that $\mathbf{\Lambda}_{\mathcal{X}}$ is right-adjoint to $\mathbf{R}\varGamma'_{\mathcal{X}}$, see (5.2.10.1). Simple arguments show that the natural maps are isomorphisms $\mathbf{\Lambda}_{\mathcal{X}} \xrightarrow{\ \sim\ } \mathbf{\Lambda}_{\mathcal{X}}\mathbf{\Lambda}_{\mathcal{X}}$, $\mathbf{R}\varGamma'_{\mathcal{X}} \xrightarrow{\ \sim\ } \mathbf{R}\varGamma'_{\mathcal{X}}\mathbf{\Lambda}_{\mathcal{X}}$, see (b) and (d) in Remark 6.3.1(1).

COROLLARY 6.1.5. *With the notation of Corollary* 6.1.4,

(a) *There are natural isomorphisms*

$$\begin{aligned}\mathbf{R}\Gamma'_{\mathfrak{X}} f^{\#} &\xrightarrow{\sim} f_{\mathrm{t}}^{\times}, & f^{\#} &\xrightarrow{\sim} \mathbf{\Lambda}_{\mathfrak{X}} f_{\mathrm{t}}^{\times},\\ \mathbf{R}\Gamma'_{\mathfrak{X}} f_{\mathrm{t}}^{\times} &\xrightarrow{\sim} f_{\mathrm{t}}^{\times}, & f^{\#} &\xrightarrow{\sim} \mathbf{\Lambda}_{\mathfrak{X}} f^{\#}.\end{aligned}$$

(b) *The natural functorial maps* $\mathbf{R}\Gamma'_{\mathfrak{Y}} \to \mathbf{1} \to \mathbf{\Lambda}_{\mathfrak{Y}}$ *induce isomorphisms*

$$\begin{aligned}f_{\mathrm{t}}^{\times}\mathbf{R}\Gamma'_{\mathfrak{Y}} &\xrightarrow{\sim} f_{\mathrm{t}}^{\times} \xrightarrow{\sim} f_{\mathrm{t}}^{\times}\mathbf{\Lambda}_{\mathfrak{Y}},\\ f^{\#}\mathbf{R}\Gamma'_{\mathfrak{Y}} &\xrightarrow{\sim} f^{\#} \xrightarrow{\sim} f^{\#}\mathbf{\Lambda}_{\mathfrak{Y}}.\end{aligned}$$

(c) *There are natural pairs of maps*

$$\begin{aligned}f_{\mathrm{t}}^{\times} &\xrightarrow{\alpha_1} \mathbf{R}\Gamma'_{\mathfrak{X}} \boldsymbol{j} f^{\times} \xrightarrow{\alpha_2} f_{\mathrm{t}}^{\times},\\ f^{\#} &\xrightarrow{\beta_1} \mathbf{\Lambda}_{\mathfrak{X}} \boldsymbol{j} f^{\times} \xrightarrow{\beta_2} f^{\#},\end{aligned}$$

each of which composes to an identity map. If $\mathfrak{X}$ *is properly algebraic then all of these maps are isomorphisms.*

(d) *If* f *is* adic *then the isomorphism* $\mathbf{R}f_*\mathbf{R}\Gamma'_{\mathfrak{X}}\boldsymbol{j} \xleftarrow{\sim} \mathbf{R}\Gamma'_{\mathfrak{Y}}\mathbf{R}f_*\boldsymbol{j}$ *in* 5.2.11(d) *induces an isomorphism of the right adjoints* (*see Theorem* 4.1, *Proposition* 3.2.3)

$$f^{\times}\mathbf{\Lambda}_{\mathfrak{Y}} \xrightarrow{\sim} \mathbf{R}Q_{\mathfrak{X}} f^{\#}.$$

PROOF. (a) The second isomorphism (first row) is the identity map. Proposition 5.2.1 yields the third. The first is the composition

$$\mathbf{R}\Gamma'_{\mathfrak{X}} f^{\#} = \mathbf{R}\Gamma'_{\mathfrak{X}}\mathbf{\Lambda}_{\mathfrak{X}} f_{\mathrm{t}}^{\times} \xrightarrow{\sim} \mathbf{R}\Gamma'_{\mathfrak{X}} f_{\mathrm{t}}^{\times} \xrightarrow{\sim} f_{\mathrm{t}}^{\times}.$$

The fourth is the composition

$$f^{\#} = \mathbf{\Lambda}_{\mathfrak{X}} f_{\mathrm{t}}^{\times} \xrightarrow{\sim} \mathbf{\Lambda}_{\mathfrak{X}}\mathbf{\Lambda}_{\mathfrak{X}} f_{\mathrm{t}}^{\times} \xrightarrow{\sim} \mathbf{\Lambda}_{\mathfrak{X}} f^{\#}.$$

(b) The first isomorphism results from $\mathbf{R}\Gamma'_{\mathfrak{Y}}$ being right adjoint to the inclusion $\mathbf{D}_{\mathrm{t}}(\mathfrak{Y}) \hookrightarrow \mathbf{D}(\mathfrak{Y})$ (see Proposition 5.2.1(c)). For the second, check that $f_{\mathrm{t}}^{\times}$ and $f_{\mathrm{t}}^{\times}\mathbf{\Lambda}_{\mathfrak{Y}}$ are both right-adjoint to $\mathbf{R}f_*|_{\mathbf{D}_{\mathrm{qct}}(\mathfrak{X})}\cdots$ (Or, consider the composition $f_{\mathrm{t}}^{\times} \xrightarrow{\sim} f_{\mathrm{t}}^{\times}\mathbf{R}\Gamma'_{\mathfrak{Y}} \xrightarrow{\sim} f_{\mathrm{t}}^{\times}\mathbf{R}\Gamma'_{\mathfrak{Y}}\mathbf{\Lambda}_{\mathfrak{Y}} \xrightarrow{\sim} f_{\mathrm{t}}^{\times}\mathbf{\Lambda}_{\mathfrak{Y}}$.) Then apply $\mathbf{\Lambda}_{\mathfrak{X}}$ to the first row to get the second row.

(c) With $\boldsymbol{k}\colon \mathbf{D}(\mathcal{A}_{\mathrm{qct}}(\mathfrak{X})) \to \mathbf{D}(\mathcal{A}_{\vec{c}}(\mathfrak{X}))$ the natural functor, let

$$\alpha\colon \boldsymbol{k}\mathbf{R}Q^{\mathrm{t}}_{\mathfrak{X}} f_{\mathrm{t}}^{\times} \to f^{\times}$$

be adjoint to $\mathbf{R}f_*\boldsymbol{j}\boldsymbol{k}\mathbf{R}Q^{\mathrm{t}}_{\mathfrak{X}} f_{\mathrm{t}}^{\times} \overset{5.3.1}{=} \mathbf{R}f_* f_{\mathrm{t}}^{\times} \xrightarrow{\tau_{\mathrm{t}}} \mathbf{1}$. By Corollary 5.2.3, $\boldsymbol{j}(\alpha)\colon f_{\mathrm{t}}^{\times} \to \boldsymbol{j}f^{\times}$ factors naturally as

$$f_{\mathrm{t}}^{\times} \xrightarrow{\alpha_1} \mathbf{R}\Gamma'_{\mathfrak{X}}\boldsymbol{j}f^{\times} \to \boldsymbol{j}f^{\times}.$$

Let α_2 be the map adjoint to the natural composition $\mathbf{R}f_*\mathbf{R}\Gamma'_{\mathfrak{X}}\boldsymbol{j}f^{\times} \to \mathbf{R}f_*\boldsymbol{j}f^{\times} \to \mathbf{1}$. One checks that $\tau_{\mathrm{t}} \circ \mathbf{R}f_*(\alpha_2\alpha_1) = \tau_{\mathrm{t}}$ (τ_{t} as in Theorem 6.1), whence $\alpha_2\alpha_1 = $ identity.

The pair β_1, β_2 is obtained from α_1, α_2 by application of the functor $\mathbf{\Lambda}_{\mathfrak{X}}$—see Corollary 5.2.3. (Symmetrically, the pair α_1, α_2 is obtained from β_1, β_2 by application of the functor $\mathbf{R}\Gamma'_{\mathfrak{X}}$.)

When $\mathfrak{X}$ is properly algebraic, the functor $\boldsymbol{j}$ is fully faithful (Corollary 3.3.4); and it follows that $\mathbf{R}\Gamma'_{\mathfrak{X}}\boldsymbol{j}f^{\times}$ and $f_{\mathrm{t}}^{\times}$ are both right-adjoint to $\mathbf{R}f_*|_{\mathbf{D}_{\mathrm{qct}}(\mathfrak{X})}$.

(d) Straightforward. □

The next three corollaries deal with compatibilities between formal (local) and ordinary (global) Grothendieck duality.

COROLLARY 6.1.6. *Let $f_0\colon X \to Y$ be a map of noetherian ordinary schemes. Suppose either that f_0 is separated or that X is finite-dimensional, or else restrict to bounded-below complexes. Let $W \subset Y$ and $Z \subset f_0^{-1}W$ be closed subsets, $\kappa_{\mathcal{Y}}\colon \mathcal{Y} = Y_{/W} \to Y$ and $\kappa_{\mathcal{X}}\colon \mathcal{X} = X_{/Z} \to X$ the respective completion maps, and $f\colon \mathcal{X} \to \mathcal{Y}$ the map induced by f_0.*

$$\begin{array}{ccc} \mathcal{X}{:=}\, X_{/Z} & \xrightarrow{\ \kappa_{\mathcal{X}}\ } & X \\ {\scriptstyle f}\big\downarrow & & \big\downarrow{\scriptstyle f_0} \\ \mathcal{Y}{:=}\, Y_{/W} & \xrightarrow[\kappa_{\mathcal{Y}}]{} & Y \end{array}$$

With $f_0^{\times} := (f_0)_{\mathrm{t}}^{\times}$ right-adjoint to $\mathbf{R}f_{*}\colon \mathbf{D}_{\mathrm{qc}}(X) \to \mathbf{D}(Y)$, let τ_{t}' be the composition

$$\mathbf{R}f_*\kappa_{\mathcal{X}}^*\mathbf{R}\varGamma_Z f_0^{\times}\kappa_{\mathcal{Y}*} \underset{5.2.7}{\xrightarrow{\ \sim\ }} \kappa_{\mathcal{Y}}^*\mathbf{R}f_{0*}\mathbf{R}\varGamma_Z f_0^{\times}\kappa_{\mathcal{Y}*} \longrightarrow \kappa_{\mathcal{Y}}^*\mathbf{R}f_{0*}f_0^{\times}\kappa_{\mathcal{Y}*} \longrightarrow \kappa_{\mathcal{Y}}^*\kappa_{\mathcal{Y}*} \longrightarrow \mathbf{1}.$$

Then for all $\mathcal{E} \in \mathbf{D}_{\mathrm{qct}}(\mathcal{X})$ and $\mathcal{F} \in \mathbf{D}(\mathcal{Y})$, the composed map

$$\begin{aligned}\alpha(\mathcal{E},\mathcal{F})\colon \mathrm{Hom}_{\mathbf{D}(\mathcal{X})}(\mathcal{E}, \kappa_{\mathcal{X}}^*\mathbf{R}\varGamma_Z f_0^{\times}\kappa_{\mathcal{Y}*}\mathcal{F}) &\longrightarrow \mathrm{Hom}_{\mathbf{D}(\mathcal{Y})}(\mathbf{R}f_*\mathcal{E}, \mathbf{R}f_*\kappa_{\mathcal{X}}^*\mathbf{R}\varGamma_Z f_0^{\times}\kappa_{\mathcal{Y}*}\mathcal{F}) \\ &\underset{\text{via } \tau_{\mathrm{t}}'}{\longrightarrow} \mathrm{Hom}_{\mathbf{D}(\mathcal{Y})}(\mathbf{R}f_*\mathcal{E}, \mathcal{F})\end{aligned}$$

is an isomorphism. Hence the map adjoint to τ_{t}' is an isomorphism of functors

$$\kappa_{\mathcal{X}}^*\mathbf{R}\varGamma_Z f_0^{\times}\kappa_{\mathcal{Y}*} \xrightarrow{\ \sim\ } f_{\mathrm{t}}^{\times}.$$

PROOF. For any $\mathcal{E} \in \mathbf{D}_{\mathrm{qct}}(\mathcal{X})$, set $\mathcal{E}_0 := \kappa_{\mathcal{X}*}\mathcal{E} \in \mathbf{D}_{\mathrm{qc}Z}(X)$ (Proposition 5.2.4). Proposition 5.2.4 and [**AJL**, p. 7, Lemma (0.4.2)] give natural isomorphisms

$$\mathrm{Hom}_{\mathbf{D}(\mathcal{X})}(\mathcal{E}, \kappa_{\mathcal{X}}^*\mathbf{R}\varGamma_Z\mathcal{G}) \xrightarrow{\ \sim\ } \mathrm{Hom}_{\mathbf{D}(X)}(\mathcal{E}_0, \mathbf{R}\varGamma_Z\mathcal{G}) \xrightarrow{\ \sim\ } \mathrm{Hom}_{\mathbf{D}(X)}(\mathcal{E}_0, \mathcal{G}) \qquad (\mathcal{G} \in \mathbf{D}_{\mathrm{qc}}(X)).$$

(In other words, $\kappa_{\mathcal{X}}^*\mathbf{R}\varGamma_Z\mathcal{G} = (\kappa_{\mathcal{X}})_{\mathrm{t}}^{\times}\mathcal{G}$.) One checks then that the map $\alpha(\mathcal{E},\mathcal{F})$ factors as the sequence of natural isomorphisms

$$\begin{aligned}\mathrm{Hom}_{\mathbf{D}(\mathcal{X})}(\mathcal{E}, \kappa_{\mathcal{X}}^*\mathbf{R}\varGamma_Z f_0^{\times}\kappa_{\mathcal{Y}*}\mathcal{F}) &\xrightarrow{\ \sim\ } \mathrm{Hom}_{\mathbf{D}(X)}(\mathcal{E}_0, f_0^{\times}\kappa_{\mathcal{Y}*}\mathcal{F}) \\ &\xrightarrow{\ \sim\ } \mathrm{Hom}_{\mathbf{D}(Y)}(\mathbf{R}f_{0*}\mathcal{E}_0, \kappa_{\mathcal{Y}*}\mathcal{F}) \\ &\xrightarrow{\ \sim\ } \mathrm{Hom}_{\mathbf{D}(\mathcal{Y})}(\kappa_{\mathcal{Y}}^*\mathbf{R}f_{0*}\mathcal{E}_0, \mathcal{F}) \\ &\xrightarrow{\ \sim\ } \mathrm{Hom}_{\mathbf{D}(\mathcal{Y})}(\mathbf{R}f_*\kappa_{\mathcal{X}}^*\mathcal{E}_0, \mathcal{F}) \qquad \text{(Corollary 5.2.7)} \\ &\xrightarrow{\ \sim\ } \mathrm{Hom}_{\mathbf{D}(\mathcal{Y})}(\mathbf{R}f_*\mathcal{E}, \mathcal{F}).\end{aligned}$$ □

COROLLARY 6.1.7. *With hypotheses as in Corollary* 6.1.6:
(a) *There are natural isomorphisms*

$$\mathbf{R}\varGamma_{\mathcal{X}}'\kappa_{\mathcal{X}}^*f_0^{\times}\kappa_{\mathcal{Y}*} = (\kappa_{\mathcal{X}})_{\mathrm{t}}^{\times}f_0^{\times}\kappa_{\mathcal{Y}*} \xrightarrow{\ \sim\ } f_{\mathrm{t}}^{\times},$$
$$\mathbf{\Lambda}_{\mathcal{X}}\kappa_{\mathcal{X}}^*f_0^{\times}\kappa_{\mathcal{Y}*} = \kappa_{\mathcal{X}}^{\#}f_0^{\times}\kappa_{\mathcal{Y}*} \xrightarrow{\ \sim\ } f^{\#};$$

and if f_0 is proper, $Y = \mathrm{Spec}(A)$ (A adic), $Z = f_0^{-1}W$, then with $f^{\times}$ as in Corollary 4.1.1:

$$\kappa_{\mathcal{X}}^*f_0^{\times}\kappa_{\mathcal{Y}*} \xrightarrow{\ \sim\ } f^{\times}.$$

(b) *The functor $f_{0,Z}^{\times} := \mathbf{R}\varGamma_Z f_0^{\times}\colon \mathbf{D}(Y) \to \mathbf{D}_{\mathrm{qc}Z}(X)$ is right-adjoint to the functor $\mathbf{R}f_*|_{\mathbf{D}_{\mathrm{qc}Z}(X)}$; and there is an isomorphism*

$$\kappa_{\mathcal{X}}^*f_{0,Z}^{\times}\kappa_{\mathcal{Y}*} \xrightarrow{\ \sim\ } f_{\mathrm{t}}^{\times}.$$

(c) *If X is separated then, with notation as in Section* 3.3, *the functor*

$$f^{\#}_{0,Z} := \boldsymbol{j}_X \mathbf{R}Q_X \mathbf{R}\mathcal{H}om^{\bullet}_X(\mathbf{R}\varGamma_Z\mathcal{O}_X, f_0^{\times}-)\colon \mathbf{D}(Y) \to \mathbf{D}_{\mathrm{qc}}(X)$$

is right-adjoint to $\mathbf{R}f_{0*}\mathbf{R}\varGamma_Z|_{\mathbf{D}_{\mathrm{qc}}(X)}$; and if $\mathfrak{X}$ is properly algebraic, so that we have the equivalence $\boldsymbol{j}_{\mathfrak{X}}\colon \mathbf{D}(\mathcal{A}_{\vec{c}}(\mathfrak{X})) \to \mathbf{D}_{\vec{c}}(\mathfrak{X})$ (Corollary 3.3.4), *there is an isomorphism*

$$\kappa_{\mathfrak{X}}^* f^{\#}_{0,Z}\kappa_{\mathcal{Y}*} \xrightarrow{\sim} \boldsymbol{j}_{\mathfrak{X}}\mathbf{R}Q_{\mathfrak{X}} f^{\#}.$$

PROOF. (a) The first isomorphism combines Corollary 6.1.6 (in proving which we noted that $\kappa_{\mathfrak{X}}^*\mathbf{R}\varGamma_Z\mathcal{G} = (\kappa_{\mathfrak{X}})_{\mathrm{t}}^{\times}\mathcal{G}$ for $\mathcal{G} \in \mathbf{D}_{\mathrm{qc}}(X)$) and Proposition 5.2.4. The second follows from $f^{\#} = \mathbf{\Lambda}_{\mathfrak{X}} f_{\mathrm{t}}^{\times}$. The third is Corollary 4.1.2.

(b) The first assertion is easily checked; and the isomorphism is given by Corollary 6.1.6.

(c) When X is separated, $\boldsymbol{j}_X$ is an equivalence [**AJL**, p. 12, Proposition (1.3)], and then the first assertion is easily checked.

From Corollary 6.1.6 and Proposition 5.2.4 we get an isomorphism

$$\mathbf{R}\varGamma_Z f_0^{\times}\kappa_{\mathcal{Y}*} \xrightarrow{\sim} \kappa_{\mathfrak{X}*} f_{\mathrm{t}}^{\times}.$$

As in Corollary 5.2.3, the natural map is an isomorphism

$$\mathbf{R}\mathcal{H}om^{\bullet}_X(\mathbf{R}\varGamma_Z\mathcal{O}_X, \mathcal{G}) \xrightarrow{\sim} \mathbf{R}\mathcal{H}om^{\bullet}_X(\mathbf{R}\varGamma_Z\mathcal{O}_X, \mathbf{R}\varGamma_Z\mathcal{G}) \qquad (\mathcal{G} \in \mathbf{D}_{\mathrm{qc}}(X)).$$

When $\mathfrak{X}$ is properly algebraic, $\boldsymbol{j}_{\mathfrak{X}}\mathbf{R}Q_{\mathfrak{X}} \cong \kappa_{\mathfrak{X}}^*\boldsymbol{j}_X\mathbf{R}Q_X\kappa_{\mathfrak{X}*}$ (Proposition 3.2.3). So then we have a sequence of natural isomorphisms

$$\begin{aligned}
\kappa_{\mathfrak{X}}^* f^{\#}_{0,Z}\kappa_{\mathcal{Y}*} &=\!= \kappa_{\mathfrak{X}}^*\boldsymbol{j}_X\mathbf{R}Q_X\mathbf{R}\mathcal{H}om^{\bullet}_X(\mathbf{R}\varGamma_Z\mathcal{O}_X, f_0^{\times}\kappa_{\mathcal{Y}*}-)\\
&\xrightarrow{\sim} \kappa_{\mathfrak{X}}^*\boldsymbol{j}_X\mathbf{R}Q_X\mathbf{R}\mathcal{H}om^{\bullet}_X(\mathbf{R}\varGamma_Z\mathcal{O}_X, \mathbf{R}\varGamma_Z f_0^{\times}\kappa_{\mathcal{Y}*}-)\\
&\xrightarrow{\sim} \kappa_{\mathfrak{X}}^*\boldsymbol{j}_X\mathbf{R}Q_X\mathbf{R}\mathcal{H}om^{\bullet}_X(\mathbf{R}\varGamma_Z\mathcal{O}_X, \kappa_{\mathfrak{X}*}f_{\mathrm{t}}^{\times}-)\\
&\xrightarrow{\sim} \kappa_{\mathfrak{X}}^*\boldsymbol{j}_X\mathbf{R}Q_X\kappa_{\mathfrak{X}*}\mathbf{R}\mathcal{H}om^{\bullet}_{\mathfrak{X}}(\kappa_{\mathfrak{X}}^*\mathbf{R}\varGamma_Z\mathcal{O}_X, f_{\mathrm{t}}^{\times}-)\\
&\xrightarrow{\sim} \boldsymbol{j}_{\mathfrak{X}}\mathbf{R}Q_{\mathfrak{X}}\mathbf{R}\mathcal{H}om^{\bullet}_{\mathfrak{X}}(\mathbf{R}\varGamma'_{\mathfrak{X}}\mathcal{O}_{\mathfrak{X}}, f_{\mathrm{t}}^{\times}-)\\
&=\!= \boldsymbol{j}_{\mathfrak{X}}\mathbf{R}Q_{\mathfrak{X}} f^{\#}.
\end{aligned}$$

□

The following instance of "flat base change" will be needed in the proof of the general base-change Theorem 3.

COROLLARY 6.1.8. *In Corollary* 6.1.6, *assume further that $Z = f_0^{-1}W$. Then the natural map is an isomorphism*

$$\mathbf{R}\varGamma_Z f_0^{\times}\mathcal{F} \xrightarrow{\sim} \mathbf{R}\varGamma_Z f_0^{\times}\kappa_{\mathcal{Y}*}\kappa_{\mathcal{Y}}^*\mathcal{F} \qquad (\mathcal{F} \in \mathbf{D}(Y)),$$

and so there is a composed isomorphism

$$\zeta\colon \mathbf{R}\varGamma'_{\mathfrak{X}}\kappa_{\mathfrak{X}}^* f_0^{\times}\mathcal{F} \underset{5.2.4(\mathrm{c})}{\xrightarrow{\sim}} \kappa_{\mathfrak{X}}^*\mathbf{R}\varGamma_Z f_0^{\times}\mathcal{F} \xrightarrow{\sim} \kappa_{\mathfrak{X}}^*\mathbf{R}\varGamma_Z f_0^{\times}\kappa_{\mathcal{Y}*}\kappa_{\mathcal{Y}}^*\mathcal{F} \underset{6.1.7(\mathrm{b})}{\xrightarrow{\sim}} f_{\mathrm{t}}^{\times}\kappa_{\mathcal{Y}}^*\mathcal{F}.$$

PROOF. First, $\mathbf{R}f_{0*}(\mathbf{D}_{\mathrm{qc}Z}(X)) \subset \mathbf{D}_{\mathrm{qc}W}(Y)$. For, by [**L4**, Proposition (3.9.2)], $\mathbf{R}f_{0*}(\mathbf{D}_{\mathrm{qc}}(X)) \subset \mathbf{D}_{\mathrm{qc}}(Y)$; and then the assertion follows from the natural isomorphism of functors (from $\mathbf{D}_{\mathrm{qc}}(X)$ to $\mathbf{D}_{\mathrm{qc}}(Y)$) $\mathbf{R}\varGamma_W\mathbf{R}f_{0*} \cong \mathbf{R}f_{0*}\mathbf{R}\varGamma_{f^{-1}W}$, because $\mathcal{G} \in \mathbf{D}_{\mathrm{qc}Z}(X)$ (resp. $\mathcal{H} \in \mathbf{D}_{\mathrm{qc}W}(Y)$) iff $\mathbf{R}\varGamma_Z\mathcal{G} \cong \mathcal{G}$ (resp. $\mathbf{R}\varGamma_W\mathcal{H} \cong \mathcal{H}$), cf. Proposition 5.2.1(a) and its proof. (The said functorial isomorphism arises from the corresponding one without the $\mathbf{R}$'s, since $\mathbf{R}f_{0*}$ preserves K-flabbiness, see [**Sp**, 5.12, 5.15(b), 6.4, 6.7].

Now Corollary 5.2.5 gives that the natural map is an isomorphism

$$\mathrm{Hom}_{\mathbf{D}(Y)}(\mathbf{R}f_{0*}\mathcal{E},\mathcal{F}) \xrightarrow{\sim} \mathrm{Hom}_{\mathbf{D}(Y)}(\mathbf{R}f_{0*}\mathcal{E},\kappa_{\mathcal{Y}*}\kappa_{\mathcal{Y}}^{*}\mathcal{F}) \qquad (\mathcal{E}\in\mathbf{D}_{\mathrm{qc}Z}(X)),$$

and the conclusion follows from the adjunction in Corollary 6.1.7(b). □

6.2. The next Proposition is a special case of Greenlees-May Duality for formal schemes (see [**AJL**′, Proposition 0.3.1]). It is the key to many statements in this paper concerning complexes with coherent homology.

PROPOSITION 6.2.1. *Let $\mathfrak{X}$ be a locally noetherian formal scheme, $\mathcal{E}\in\mathbf{D}(\mathfrak{X})$. Then for all $\mathcal{F}\in\mathbf{D}_{\mathrm{c}}(\mathfrak{X})$ the natural map $\mathbf{R}\varGamma'_{\mathfrak{X}}\mathcal{E}\to\mathcal{E}$ induces an isomorphism*

$$\mathbf{R}\mathcal{H}om^{\bullet}(\mathcal{E},\mathcal{F}) \xrightarrow{\sim} \mathbf{R}\mathcal{H}om^{\bullet}(\mathbf{R}\varGamma'_{\mathfrak{X}}\mathcal{E},\mathcal{F}).$$

PROOF. The canonical isomorphism (cf. (5.2.10.1))

$$\mathbf{R}\mathcal{H}om^{\bullet}(\mathbf{R}\varGamma'_{\mathfrak{X}}\mathcal{E},\mathcal{F}) \xrightarrow{\sim} \mathbf{R}\mathcal{H}om^{\bullet}(\mathcal{E},\mathbf{R}\mathcal{H}om^{\bullet}(\mathbf{R}\varGamma'_{\mathfrak{X}}\mathcal{O}_{\mathfrak{X}},\mathcal{F}))$$

reduces the question to where $\mathcal{E}=\mathcal{O}_{\mathfrak{X}}$. It suffices then—as in the proof of Corollary 5.2.3—that for affine $\mathfrak{X}=\mathrm{Spf}(A)$, the natural map be an isomorphism

$$\mathrm{Hom}_{\mathbf{D}(\mathfrak{X})}(\mathcal{O}_{\mathfrak{X}},\mathcal{F}) \xrightarrow{\sim} \mathrm{Hom}_{\mathbf{D}(\mathfrak{X})}(\mathbf{R}\varGamma'_{\mathfrak{X}}\mathcal{O}_{\mathfrak{X}},\mathcal{F}) \qquad (\mathcal{F}\in\mathbf{D}_{\mathrm{c}}(\mathfrak{X})).$$

Let I be an ideal of definition of the adic ring A, set $Z:=\mathrm{Supp}(A/I)$, and let $\kappa\colon\mathfrak{X}\to X:=\mathrm{Spec}(A)$ be the completion map. The categorical equivalences in Proposition 3.3.1 and the isomorphism $\kappa^{*}\mathbf{R}\varGamma'_{Z}\mathcal{O}_{X}\xrightarrow{\sim}\mathbf{R}\varGamma'_{\mathfrak{X}}\mathcal{O}_{\mathfrak{X}}$ in Proposition 5.2.4 make the problem whether for all $F\in\mathbf{D}_{\mathrm{c}}(X)$ (e.g., $F=\mathbf{R}Q\kappa_{*}\mathcal{F}:=\boldsymbol{j}_{X}\mathbf{R}Q_{X}\kappa_{*}\mathcal{F}$) the natural map is an isomorphism

$$\mathrm{Hom}_{\mathbf{D}(X)}(\mathcal{O}_{X},F) \xrightarrow{\sim} \mathrm{Hom}_{\mathbf{D}(X)}(\mathbf{R}\varGamma_{Z}\mathcal{O}_{X},F).$$

Now, the canonical functor $\boldsymbol{j}_{X}:\mathbf{D}(\mathcal{A}_{\mathrm{qc}}(X))\to\mathbf{D}(X)$ induces an equivalence of categories $\mathbf{D}(\mathcal{A}_{\mathrm{qc}}(X))\xrightarrow{\approx}\mathbf{D}_{\mathrm{qc}}(X)$ (see beginning of §3.3), and so we may assume that F is a K-flat quasi-coherent complex. Lemma 5.2.2 shows that $\boldsymbol{j}_{X}\mathbf{R}Q_{X}$ is right-adjoint to the inclusion $\mathbf{D}_{\mathrm{qc}}(X)\hookrightarrow\mathbf{D}(X)$. The natural map

$$\mathbf{R}\mathcal{H}om^{\bullet}(\mathcal{O}_{X},F)\to\mathbf{R}\mathcal{H}om^{\bullet}(\mathbf{R}\varGamma_{Z}\mathcal{O}_{X},F)$$

factors then as

$$\begin{aligned} \mathbf{R}\mathcal{H}om^{\bullet}(\mathcal{O}_{X},F)=F &\underset{3.3.1}{\xrightarrow{\sim}} \boldsymbol{j}_{X}\mathbf{R}Q_{X}\kappa_{*}\kappa^{*}F \\ &\longrightarrow \kappa_{*}\kappa^{*}F \qquad\qquad (6.2.1.1)\\ &\underset{\lambda}{\xrightarrow{\sim}} \varprojlim_{n} F/(I\mathcal{O}_{X})^{n}F \underset{\Phi}{\xrightarrow{\sim}} \mathbf{R}\mathcal{H}om^{\bullet}(\mathbf{R}\varGamma_{Z}\mathcal{O}_{X},F), \end{aligned}$$

where the map λ, obtained by applying κ_{*} to the natural map from $\kappa^{*}F$ to the completion $F_{/Z}$, is a $\mathbf{D}(X)$-isomorphism by [**AJL**, p. 6, Proposition (0.4.1)]; and Φ is the isomorphism $\Phi(F,\mathcal{O}_{X})$ of [**AJL**, §2]. (The fact that Φ is an isomorphism is essentially the main result in [**AJL**].) Also, by adjointness, the natural map is an isomorphism

$$\mathrm{Hom}_{\mathbf{D}(X)}(\mathcal{O}_{X},\boldsymbol{j}_{X}\mathbf{R}Q_{X}\kappa_{*}\kappa^{*}F) \xrightarrow{\sim} \mathrm{Hom}_{\mathbf{D}(X)}(\mathcal{O}_{X},\kappa_{*}\kappa^{*}F).$$

Conclude now by applying the functor $\mathrm{H}^{0}\mathbf{R}\Gamma(X,-)$ to (6.2.1.1). □

COROLLARY 6.2.2. *Let $f\colon\mathfrak{X}\to\mathcal{Y}$ be as in Corollary 6.1.4, and assume further that f is adic. Then for all $\mathcal{F}\in\mathbf{D}_{\mathrm{c}}(\mathcal{Y})$ the map corresponding to the natural composition $\mathbf{R}f_{*}\mathbf{R}\varGamma'_{\mathfrak{X}}jf^{\times}\mathcal{F}\to\mathbf{R}f_{*}jf^{\times}\mathcal{F}\to\mathcal{F}$ (see Theorem 4.1) is an isomorphism*

$$f^{\times}\mathcal{F} \xrightarrow{\sim} \mathbf{R}Q_{\mathfrak{X}}f^{\#}\mathcal{F}.$$

PROOF. By Proposition 6.2.1, $\mathcal{F} \cong \mathbf{\Lambda}_{\mathcal{Y}}\mathcal{F} := \mathbf{R}\mathcal{H}om^{\bullet}(\mathbf{R}\varGamma'_{\mathcal{Y}}\mathcal{O}_{\mathcal{Y}}, \mathcal{F})$; so this Corollary is a special case of Corollary 6.1.5(d). $\square$

COROLLARY 6.2.3. *In Corollary* 6.1.6, *suppose* $Y = \operatorname{Spec}(A)$ *(A adic) and that the the map f_0 is* proper. *Then with the customary notation $f_0^!$ for $f_0^{\times}$ we have, for any $\mathcal{F} \in \mathbf{D}_{\mathrm{c}}^+(\mathcal{Y})$, a natural isomorphism*

$$\kappa_{\mathcal{X}}^* f_0^! \kappa_{\mathcal{Y}*}\mathcal{F} \xrightarrow{\sim} f^{\#}\mathcal{F} \in \mathbf{D}_{\mathrm{c}}^+(\mathcal{X}).$$

PROOF. The natural map $f_0^! \boldsymbol{j}_Y \mathbf{R}Q_Y \kappa_{\mathcal{Y}*} \to f_0^! \kappa_{\mathcal{Y}*}$ is an isomorphism of functors from $\mathbf{D}(\mathcal{Y})$ to $\mathbf{D}_{\mathrm{qc}}(X)$, both being right-adjoint to $\kappa_{\mathcal{Y}}^* \mathbf{R} f_{0*}$. Proposition 3.3.1 gives $\boldsymbol{j}_Y \mathbf{R}Q_Y \kappa_{\mathcal{Y}*}\mathcal{F} \in \mathbf{D}_{\mathrm{c}}^+(Y)$; so by [**V**, p. 396, Lemma 1], $f_0^! \kappa_{\mathcal{Y}*}\mathcal{F} \in \mathbf{D}_{\mathrm{c}}^+(X)$.[19] Hence Proposition 6.2.1 and Corollary 6.1.7(a) yield isomorphisms

$$\kappa_{\mathcal{X}}^* f_0^! \kappa_{\mathcal{Y}*}\mathcal{F} \xrightarrow{\sim} \mathbf{R}\mathcal{H}om^{\bullet}(\mathbf{R}\varGamma'_{\mathcal{X}}\mathcal{O}_{\mathcal{X}}, \kappa_{\mathcal{X}}^* f_0^! \kappa_{\mathcal{Y}*}\mathcal{F}) =: \mathbf{\Lambda}_{\mathcal{X}} \kappa_{\mathcal{X}}^* f_0^! \kappa_{\mathcal{Y}*}\mathcal{F} \xrightarrow{\sim} f^{\#}\mathcal{F}. \quad \square$$

6.3. More relations, involving the functors $\mathbf{R}\varGamma'_{\mathcal{X}}$ and $\mathbf{\Lambda}_{\mathcal{X}} := \mathbf{R}\mathcal{H}om^{\bullet}(\mathbf{R}\varGamma'_{\mathcal{X}}\mathcal{O}_{\mathcal{X}}, -)$ on a locally noetherian formal scheme $\mathcal{X}$, will now be summarized.

REMARKS 6.3.1. Let $\mathcal{X}$ be a locally noetherian formal scheme.

(1) The functor $\boldsymbol{\varGamma} := \mathbf{R}\varGamma'_{\mathcal{X}} \colon \mathbf{D}(\mathcal{X}) \to \mathbf{D}(\mathcal{X})$ admits a natural map $\boldsymbol{\varGamma} \xrightarrow{\gamma} \mathbf{1}$, which induces a functorial isomorphism

$$\text{(A)} \qquad \operatorname{Hom}(\boldsymbol{\varGamma}\mathcal{E}, \boldsymbol{\varGamma}\mathcal{F}) \xrightarrow{\sim} \operatorname{Hom}(\boldsymbol{\varGamma}\mathcal{E}, \mathcal{F}) \qquad (\mathcal{E}, \mathcal{F} \in \mathbf{D}(\mathcal{X})),$$

see Proposition 5.2.1(c). Moreover $\boldsymbol{\varGamma}$ has a right adjoint, viz. $\mathbf{\Lambda} := \mathbf{\Lambda}_{\mathcal{X}}$ (see (5.2.10.1)).

The rest of (1) consists of (well-known) formal consequences of these properties.

Since γ is functorial, it holds that $\gamma(\mathcal{F}) \circ \gamma(\boldsymbol{\varGamma}\mathcal{F}) = \gamma(\mathcal{F}) \circ \boldsymbol{\varGamma}(\gamma(\mathcal{F})) \colon \boldsymbol{\varGamma\varGamma}\mathcal{F} \to \mathcal{F}$, so injectivity of the map in (A) (with $\mathcal{E} = \boldsymbol{\varGamma}\mathcal{F}$) yields $\gamma(\boldsymbol{\varGamma}\mathcal{F}) = \boldsymbol{\varGamma}(\gamma(\mathcal{F})) \colon \boldsymbol{\varGamma\varGamma}\mathcal{F} \to \boldsymbol{\varGamma}\mathcal{F}$; and one finds after setting $\mathcal{F} = \boldsymbol{\varGamma}\mathcal{G}$ in (A) that this functorial map is an *isomorphism*

$$\text{(a)} \qquad \gamma(\boldsymbol{\varGamma}) = \boldsymbol{\varGamma}(\gamma) \colon \boldsymbol{\varGamma\varGamma} \xrightarrow{\sim} \boldsymbol{\varGamma}.$$

Conversely, given (a) one can deduce that the map in (A) is an isomorphism, whose inverse takes $\alpha \colon \boldsymbol{\varGamma}\mathcal{E} \to \mathcal{F}$ to the composition $\boldsymbol{\varGamma}\mathcal{E} \xrightarrow{\sim} \boldsymbol{\varGamma\varGamma}\mathcal{E} \xrightarrow{\boldsymbol{\varGamma}\alpha} \boldsymbol{\varGamma}\mathcal{F}$.[20] The composed functorial map $\lambda \colon \mathbf{1} \to \mathbf{\Lambda}\boldsymbol{\varGamma} \xrightarrow{\mathbf{\Lambda}(\gamma)} \mathbf{\Lambda}$ induces an isomorphism

$$\text{(B)} \qquad \operatorname{Hom}(\mathbf{\Lambda}\mathcal{E}, \mathbf{\Lambda}\mathcal{F}) \xrightarrow{\sim} \operatorname{Hom}(\mathcal{E}, \mathbf{\Lambda}\mathcal{F}) \qquad (\mathcal{E}, \mathcal{F} \in \mathbf{D}(\mathcal{X})),$$

or equivalently (as above), λ induces an isomorphism

$$\text{(b)} \qquad \lambda(\mathbf{\Lambda}) = \mathbf{\Lambda}(\lambda) \colon \mathbf{\Lambda} \xrightarrow{\sim} \mathbf{\Lambda\Lambda}.$$

Moreover, the isomorphism (A) transforms via adjointness to an isomorphism

$$\operatorname{Hom}(\mathcal{E}, \mathbf{\Lambda}\boldsymbol{\varGamma}\mathcal{F}) \xrightarrow{\sim} \operatorname{Hom}(\mathcal{E}, \mathbf{\Lambda}\mathcal{F}) \qquad (\mathcal{E}, \mathcal{F} \in \mathbf{D}(\mathcal{X})),$$

---

[19]For $\mathcal{G} \in \mathbf{D}_{\mathrm{c}}^+(Y)$ one has $f_0^!\mathcal{G} \in \mathbf{D}_{\mathrm{c}}^+(X)$: The question being local on X one reduces to where *either* X is a projective space $\mathbf{P}_Y^n$ and f_0 is projection, so that $f_0^!\mathcal{G} = f_0^*\mathcal{G} \otimes \Omega_{X/Y}^n[n] \in \mathbf{D}_{\mathrm{c}}^+(X)$, *or* f_0 is a closed immersion and $f_{0*}f_0^!\mathcal{G} = \mathbf{R}\mathcal{H}om_Y^{\bullet}(f_{0*}\mathcal{O}_X, \mathcal{F}) \in \mathbf{D}_{\mathrm{c}}^+(Y)$ [**H1**, p. 92, Proposition 3.3] whence, again, $f_0^!\mathcal{G} \in \mathbf{D}_{\mathrm{c}}^+(X)$ [**GD**, p. 115, (5.3.13)].

[20]The *idempotence* of $\boldsymbol{\varGamma}$, expressed by (a) or (A), can be interpreted as follows.

Set $\mathbf{D} := \mathbf{D}(\mathcal{X})$, $\mathbf{S} := \{\mathcal{E} \in \mathbf{D} \mid \boldsymbol{\varGamma}(\mathcal{E}) = 0\}$, so that $\boldsymbol{\varGamma}$ factors uniquely as $\mathbf{D} \xrightarrow{q} \mathbf{D}/\mathbf{S} \xrightarrow{\bar{\boldsymbol{\varGamma}}} \mathbf{D}$ where q is the "Verdier quotient" functor. Then $\bar{\boldsymbol{\varGamma}}$ *is left-adjoint to* q, so that $\mathbf{S} \subset \mathbf{D}$ admits a "Bousfield colocalization." It follows from (c) and (d) below that $\mathbf{S} = \{\mathcal{E} \in \mathbf{D} \mid \mathbf{\Lambda}(\mathcal{E}) = 0\}$, and (b) below means that *the functor* $\bar{\mathbf{\Lambda}} \colon \mathbf{D}/\mathbf{S} \to \mathbf{D}$ *defined by* $\mathbf{\Lambda} = \bar{\mathbf{\Lambda}} \circ q$ *is right-adjoint to* q; thus $\mathbf{S} \subset \mathbf{D}$ also admits a "Bousfield localization." And $\mathbf{D}/\mathbf{S}$ is equivalent, via $\bar{\boldsymbol{\varGamma}}$ and $\bar{\mathbf{\Lambda}}$ respectively, to the categories $\mathbf{D}_{\mathrm{t}} \subset \mathbf{D}$ and $\mathbf{D}\hat{} \subset \mathbf{D}$ introduced below—categories denoted by $\mathbf{S}^{\perp}$ and ${}^{\perp}\mathbf{S}$ in [**N2**, Chapter 8].

whose meaning is that γ induces an isomorphism

$$\text{(c)} \qquad \boldsymbol{\Lambda}\boldsymbol{\varGamma} \xrightarrow{\ \sim\ } \boldsymbol{\Lambda}.$$

Similarly, (B) means that λ induces the conjugate isomorphism

$$\text{(d)} \qquad \boldsymbol{\varGamma}\boldsymbol{\Lambda} \xleftarrow{\ \sim\ } \boldsymbol{\varGamma}.$$

Similarly, that $\boldsymbol{\Lambda}(\lambda(\mathcal{F}))$—or $\gamma(\boldsymbol{\varGamma}(\mathcal{E}))$—is an isomorphism (respectively that $\lambda(\boldsymbol{\Lambda}(\mathcal{F}))$—or $\boldsymbol{\varGamma}(\gamma(\mathcal{E}))$—is an isomorphism) is equivalent to the first (respectively the second) of the following maps (induced by λ and γ respectively) being an isomorphism:

$$\text{(AB)} \qquad \operatorname{Hom}(\boldsymbol{\varGamma}\mathcal{E},\mathcal{F}) \xrightarrow{\ \sim\ } \operatorname{Hom}(\boldsymbol{\varGamma}\mathcal{E},\boldsymbol{\Lambda}\mathcal{F}) \xleftarrow{\ \sim\ } \operatorname{Hom}(\mathcal{E},\boldsymbol{\Lambda}\mathcal{F}).$$

That (c) is an isomorphism also means that the functor $\boldsymbol{\Lambda}$ factors, via $\boldsymbol{\varGamma}$, through the essential image $\mathbf{D}_{\mathrm{t}}(\mathscr{X})$ of $\boldsymbol{\varGamma}$ (i.e., the full subcategory $\mathbf{D}_{\mathrm{t}}(\mathscr{X})$ whose objects are isomorphic to $\boldsymbol{\varGamma}\mathcal{E}$ for some $\mathcal{E}$); and similarly (d) being an isomorphism means that $\boldsymbol{\varGamma}$ factors, via $\boldsymbol{\Lambda}$, through the essential image $\mathbf{D}\hat{\ }(\mathscr{X})$ of $\boldsymbol{\Lambda}$; and the isomorphisms $\boldsymbol{\varGamma}\boldsymbol{\Lambda}\boldsymbol{\varGamma} \cong \boldsymbol{\varGamma}$ and $\boldsymbol{\Lambda}\boldsymbol{\varGamma}\boldsymbol{\Lambda} \cong \boldsymbol{\Lambda}$ deduced from (a)–(d) signify that $\boldsymbol{\Lambda}$ and $\boldsymbol{\varGamma}$ induce quasi-inverse equivalences between the categories $\mathbf{D}_{\mathrm{t}}(\mathscr{X})$ and $\mathbf{D}\hat{\ }(\mathscr{X})$.

(2) If $\mathscr{X}$ is properly algebraic, the natural functor $\boldsymbol{j}\colon \mathbf{D}(\mathcal{A}_{\vec{c}}(\mathscr{X})) \to \mathbf{D}_{\vec{c}}(\mathscr{X})$ is an *equivalence,* and the inclusion $\mathbf{D}_{\vec{c}}(\mathscr{X}) \hookrightarrow \mathbf{D}(\mathscr{X})$ has a right adjoint $\mathbf{Q} := \boldsymbol{j}\mathbf{R}Q_{\mathscr{X}}$ (Corollary 3.3.4.) Then (easily checked, given Corollary 3.1.5 and Proposition 5.2.1) all of (1) holds with $\mathbf{D}$, $\mathbf{D}_{\mathrm{t}}$, and $\boldsymbol{\Lambda}$ replaced by $\mathbf{D}_{\vec{c}}$, $\mathbf{D}_{\mathrm{qct}}$, and $\boldsymbol{\Lambda}^{\vec{c}} := \mathbf{Q}\boldsymbol{\Lambda}$, respectively.

(3) As in (1), $\boldsymbol{\Lambda}$ induces an equivalence from $\mathbf{D}_{\mathrm{qct}}(\mathscr{X})$ to $\mathbf{D}_{\hat{\mathrm{qc}}}(\mathscr{X})$, the essential image of $\boldsymbol{\Lambda}|_{\mathbf{D}_{\mathrm{qct}}(\mathscr{X})}$—or, since $\boldsymbol{\Lambda} \cong \boldsymbol{\Lambda}\boldsymbol{\varGamma}$, of $\boldsymbol{\Lambda}|_{\mathbf{D}_{\mathrm{qc}}(\mathscr{X})}$ (Proposition 5.2.1). So for any $f\colon \mathscr{X} \to \mathscr{Y}$ as in Corollary 6.1.5, the functor

$$\boldsymbol{\Lambda}_{\mathscr{Y}}\mathbf{R}f_{*}\mathbf{R}\varGamma'_{\mathscr{X}}\colon \mathbf{D}_{\hat{\mathrm{qc}}}(\mathscr{X}) \to \mathbf{D}_{\hat{\mathrm{qc}}}(\mathscr{Y})$$

has the right adjoint $\boldsymbol{\Lambda}_{\mathscr{X}}f_{\mathrm{t}}^{\times}\mathbf{R}\varGamma'_{\mathscr{Y}} = \boldsymbol{\Lambda}_{\mathscr{X}}f_{\mathrm{t}}^{\times} = f^{\#}$. There result two "parallel" adjoint pseudofunctors [**L4**, (3.6.7)(d)] (where "3.6.6" should be "3.6.2"):

$$(\mathbf{R}f_{*}, f_{\mathrm{t}}^{\times}) \text{ (on } \mathbf{D}_{\mathrm{qct}}) \quad \text{and} \quad (\boldsymbol{\Lambda}_{\mathscr{Y}}\mathbf{R}f_{*}\mathbf{R}\varGamma'_{\mathscr{X}}, f^{\#}) \text{ (on } \mathbf{D}_{\hat{\mathrm{qc}}}).$$

Both of these correspond to the same adjoint pseudofunctor on the quotient $\mathbf{D}_{\mathrm{qc}}/(\mathbf{S} \cap \mathbf{D}_{\mathrm{qc}})$, see footnote under (1).

If f is *adic* then $\mathbf{R}f_{*}\boldsymbol{\Lambda}_{\mathscr{X}} \cong \boldsymbol{\Lambda}_{\mathscr{Y}}\mathbf{R}f_{*}$ (Corollary 5.2.11(c)), and so Proposition 5.2.6 gives that $\mathbf{R}f_{*}(\mathbf{D}_{\hat{\mathrm{qc}}}(\mathscr{X})) \subset \mathbf{D}_{\hat{\mathrm{qc}}}(\mathscr{Y})$. Moreover, there are functorial isomorphisms

$$\boldsymbol{\Lambda}_{\mathscr{Y}}\mathbf{R}f_{*}\mathbf{R}\varGamma'_{\mathscr{X}}\boldsymbol{\Lambda}_{\mathscr{X}} \cong \mathbf{R}f_{*}\boldsymbol{\Lambda}_{\mathscr{X}}\mathbf{R}\varGamma'_{\mathscr{X}}\boldsymbol{\Lambda}_{\mathscr{X}} \cong \mathbf{R}f_{*}\boldsymbol{\Lambda}_{\mathscr{X}}.$$

Thus for adic f, $\boldsymbol{\Lambda}_{\mathscr{Y}}\mathbf{R}f_{*}\mathbf{R}\varGamma'_{\mathscr{X}}$ can be replaced above by $\mathbf{R}f_{*}$.

When f is *proper* more can be said, see Theorem 8.4.

7. Flat base change.

A *fiber square* of adic formal schemes is a commutative diagram

$$\begin{array}{ccc} \mathscr{V} & \xrightarrow{\ v\ } & \mathscr{X} \\ {\scriptstyle g}\big\downarrow & & \big\downarrow{\scriptstyle f} \\ \mathscr{U} & \xrightarrow[\ u\ ]{} & \mathscr{Y} \end{array}$$

such that the natural map is an *isomorphism* $\mathscr{V} \xrightarrow{\ \sim\ } \mathscr{X} \times_{\mathscr{Y}} \mathscr{U}$. If $\mathscr{I}$, $\mathscr{J}$, $\mathscr{K}$ are ideals of definition of $\mathscr{Y}$, $\mathscr{X}$, $\mathscr{U}$ respectively, then $\mathcal{L} := \mathscr{J}\mathcal{O}_{\mathscr{V}} + \mathscr{K}\mathcal{O}_{\mathscr{V}}$ is an ideal of definition of $\mathscr{V}$, and the scheme $V := (\mathscr{V}, \mathcal{O}_{\mathscr{V}}/\mathcal{L})$ is the fiber product of the $(\mathscr{Y}, \mathcal{O}_{\mathscr{Y}}/\mathscr{I})$-schemes $(\mathscr{X}, \mathcal{O}_{\mathscr{X}}/\mathscr{J})$ and $(\mathscr{U}, \mathcal{O}_{\mathscr{U}}/\mathscr{K})$, see [**GD**, p. 417, Proposition (10.7.3)]. By [**GD**, p. 414, Corollaire (10.6.4)], if V is locally noetherian and the $\mathcal{O}_V$-module $\mathcal{L}/\mathcal{L}^2$ is of finite type then $\mathcal{V}$ is locally noetherian. That happens whenever $\mathfrak{X}$, $\mathfrak{Y}$ and $\mathfrak{U}$ are locally noetherian and either u or f is of pseudo-finite type.

Our goal is to prove Theorem 7.4 (= Theorem 3 of the Introduction). That is, given a fiber square as above, with $\mathfrak{X}$, $\mathfrak{Y}$, $\mathfrak{U}$ and $\mathcal{V}$ noetherian, f *pseudo-proper,* and u *flat,* we want to establish a functorial isomorphism

$$\beta_{\mathcal{F}}\colon \mathbf{R}\varGamma'_{\mathcal{V}}v^*f_{\mathrm{t}}^{\times}\mathcal{F} \xrightarrow{\sim} g_{\mathrm{t}}^{\times}\mathbf{R}\varGamma'_{\mathfrak{U}}u^*\mathcal{F}\ (\cong g_{\mathrm{t}}^{\times}u^*\mathcal{F}) \qquad (\mathcal{F}\in\widetilde{\mathbf{D}}^{+}_{\mathrm{qc}}(\mathfrak{Y})).$$

Some consequences of this theorem will be given in Section 8.

In order to define $\beta_{\mathcal{F}}$ (Definition 7.3) we first need to set up a canonical isomorphism $\mathbf{R}\varGamma'_{\mathfrak{U}}u^*\mathbf{R}f_* \xrightarrow{\sim} \mathbf{R}\varGamma'_{\mathfrak{U}}\mathbf{R}g_*v^*$. This is done in Proposition 7.2. (When u is *adic* as well as flat, $\mathbf{R}\varGamma'_{\mathfrak{U}}$ can be omitted.)

Our proof of Theorem 7.4 has the weakness that it *assumes* the case when f is a proper map of noetherian ordinary schemes. As far as we know, the published proofs of this latter result make use of finite-dimensionality hypotheses on the schemes involved (see [**V**, p. 392, Thm. 2], [**H1**, p. 383, Cor. 3.4]), or projectivity hypotheses on f [**H1**, p. 191, 5]). There is however an outline of a proof for the general case, even without noetherian hypotheses, in [**L5**]—see Corollary 4.3 there.[21]

To begin with, here are several properties of formal-scheme maps (see §1.2.2) which propagate across fiber squares.

PROPOSITION 7.1. (a) *Let* $f\colon \mathfrak{X}\to\mathfrak{Y}$ *and* $u:\mathfrak{U}\to\mathfrak{Y}$ *be maps of locally noetherian formal schemes, such that the fiber product* $\mathfrak{X}\times_{\mathfrak{Y}}\mathfrak{U}$ *is locally noetherian (a condition which holds, e.g., if either* f *or* u *is of pseudo-finite type, see* [**GD**, p. 414, Corollaire (10.6.4)]). *If* f *is* separated (*resp.* affine, *resp.* pseudo-proper, *resp.* pseudo-finite, *resp.* of pseudo-finite type, *resp.* adic) *then so is the projection* $\mathfrak{X}\times_{\mathfrak{Y}}\mathfrak{U}\to\mathfrak{U}$.

(b) *With* $f\colon\mathfrak{X}\to\mathfrak{Y}$ *and* $u:\mathfrak{U}\to\mathfrak{Y}$ *as in* (a), *assume either that* u *is adic or that* f *is of pseudo-finite type. If* u *is flat then so is the projection* $\mathfrak{X}\times_{\mathfrak{Y}}\mathfrak{U}\to\mathfrak{X}$.

(c) *Let* $f\colon\mathfrak{X}\to\mathfrak{Y}$, $u:\mathfrak{U}\to\mathfrak{Y}$ *be maps of locally noetherian formal schemes, with* u *flat and locally over* $\mathfrak{Y}$ *the completion of a finite-type map of ordinary schemes. Then* $\mathfrak{X}\times_{\mathfrak{Y}}\mathfrak{U}$ *is locally noetherian, and the projection* $\mathfrak{X}\times_{\mathfrak{Y}}\mathfrak{U}\to\mathfrak{X}$ *is flat.*

PROOF. (a) The adicity assertion is obvious, and the rest follows from corresponding assertions for the ordinary schemes obtained by factoring out defining ideals.

(b) It's enough to treat the case when $\mathfrak{Y}$, $\mathfrak{X}$, and $\mathfrak{U}$ are the formal spectra, respectively, of noetherian adic rings (A,I), (B,J) and (C,K) such that B and C are A-algebras with $J\supset IB$ and $K\supset IC$, and such that $B\mathbin{\widehat{\otimes}}_A C$ is noetherian (since $\mathfrak{X}\times_{\mathfrak{Y}}\mathfrak{U}$ is locally noetherian, see [**GD**, p. 414, Corollaire (10.6.5)]). By the following Lemma 7.1.1, the problem is to show that if C is A-flat and *either* $K=IC$ (u adic), *or* B/J is a finitely-generated A-algebra (f of pseudo-finite type), then $B\mathbin{\widehat{\otimes}}_A C$ is B-flat.

The local criterion of flatness [**B**, p. 98, §5.2, Thm. 1 and p. 101, §5.4, Prop. 2] reduces the problem further to showing that for all $n>0$, $(B\mathbin{\widehat{\otimes}}_A C)/J^n(B\mathbin{\widehat{\otimes}}_A C)$ is (B/J^n)-flat, i.e., that $(B/J^n)\mathbin{\widehat{\otimes}}_A C$ is (B/J^n)-flat. But, C being A-flat, if $K=IC$ then $(B/J^n)\mathbin{\widehat{\otimes}}_A C=(B/J^n)\otimes_{A/I^n}(C/I^nC)$ is clearly B/J^n-flat; while if B/J is

[21]Details may eventually appear in [**L4**]. It is quite possible that the argument can be adapted to give a direct proof for formal schemes too.

a finitely-generated A-algebra, then $(B/J^n)\otimes_A C$ is noetherian and (B/J^n)-flat, whence so is its K-adic completion $(B/J^n)\,\widehat{\otimes}_A C$.

(c) Proceeding as in the proof of (b), we may assume C to be the K'-adic completion of a finite-type A-algebra C' (K' a C'-ideal). If C is A-flat then by [**B**, §5.4, Proposition 4], the localization $C'':=C'[(1+K')^{-1}]$ is A-flat, so the noetherian B-algebra $B\otimes_A C''$ is B-flat, as is its (noetherian) completion $B\,\widehat{\otimes}_A C$. □

LEMMA 7.1.1. *Let $\varphi : A\to C$ be a continuous homomorphism of noetherian adic rings. Then C is A-flat iff the corresponding map* $\mathrm{Spf}(\varphi)\colon \mathrm{Spf}(C)\to\mathrm{Spf}(A)$ *is flat, i.e., iff for each open prime $q\subset C$, $C_{\{q\}}$ is $A_{\{\varphi^{-1}q\}}$-flat.*

PROOF. Recall that if K is an ideal of definition of C and $q\supset K$ is an open prime ideal in C, then with $C\setminus q$ ordered by divisibility,

$$C_{\{q\}}:=\mathcal{O}_{\mathrm{Spf}(C),q}=\varinjlim_{f\in C\setminus q} C_{\{f\}}$$

where $C_{\{f\}}$ is the K-adic completion of the localization C_f.

Now for each $f\notin q$ and $n>0$ the canonical map $C_f/K^nC_f\to C_{\{f\}}/K^nC_{\{f\}}$ is bijective, so the $\varinjlim$ of these maps is an isomorphism $C_q/K^nC_q \xrightarrow{\ \sim\ } C_{\{q\}}/K^nC_{\{q\}}$, whence so is the K-adic completion $\widehat{C_q}\xrightarrow{\ \sim\ }\widehat{C_{\{q\}}}$ of the canonical map $C_q\to C_{\{q\}}$. We can therefore apply [**B**, §5.4, Proposition 4] twice to get that C_q is $A_{\varphi^{-1}q}$-flat iff $C_{\{q\}}$ is $A_{\{\varphi^{-1}q\}}$-flat. So if C is A-flat then $\mathrm{Spf}(\varphi)$ is flat; and the converse holds because C is A-flat iff C_m is $A_{\varphi^{-1}m}$-flat for every maximal ideal m in C, and every such m is open since C is complete. □

PROPOSITION 7.2. (a) *Consider a fiber square of noetherian formal schemes*

$$\begin{array}{ccc}\mathcal{V} & \xrightarrow{\ v\ } & \mathcal{X}\\ {\scriptstyle g}\big\downarrow & & \big\downarrow{\scriptstyle f}\\ \mathcal{U} & \xrightarrow[\ u\ ]{} & \mathcal{Y}\end{array}$$

with u and v flat. Let

$$\psi_{\mathcal{G}}\colon \mathbf{R}g_*\mathbf{R}\varGamma'_{\mathcal{V}}v^*\mathcal{G}\to\mathbf{R}\varGamma'_{\mathcal{U}}\mathbf{R}g_*v^*\mathcal{G}\qquad(\mathcal{G}\in\mathbf{D}_{\mathrm{qc}}(\mathcal{X}))$$

be the unique map whose composition with the natural map $\mathbf{R}\varGamma'_{\mathcal{U}}\mathbf{R}g_*v^*\mathcal{G}\to\mathbf{R}g_*v^*\mathcal{G}$ is the natural map $\mathbf{R}g_*\mathbf{R}\varGamma'_{\mathcal{V}}v^*\mathcal{E}\to\mathbf{R}g_*v^*\mathcal{E}$. (The existence of $\psi_{\mathcal{G}}$ is given by Propositions 5.2.1 and 5.2.6.) Then for all $\mathcal{E}\in\mathbf{D}_{\mathrm{qct}}(\mathcal{X})$, $\psi_{\mathcal{E}}$ is an isomorphism.

In particular, if u (hence v) is adic *then $\psi_{\mathcal{E}}$ can be identified with the identity map of $\mathbf{R}g_*v^*\mathcal{E}$.*

(b) *Let $\mathcal{X}$, $\mathcal{Y}$, $\mathcal{U}$ be noetherian formal schemes, let $f\colon\mathcal{X}\to\mathcal{Y}$ and $u\colon\mathcal{U}\to\mathcal{Y}$ be maps, with u flat, and assume further that one of the following holds:*

(i) *u is adic, and $\mathcal{V}:=\mathcal{X}\times_{\mathcal{Y}}\mathcal{U}$ is noetherian,*

(ii) *f is of pseudo-finite type,*

(iii) *u is locally the completion of a finite-type map of ordinary schemes;*

so that by Proposition 7.1 *we have a fiber square as in* (a). *Let*

$$\theta_{\mathcal{G}}\colon u^*\mathbf{R}f_*\mathcal{G}\to\mathbf{R}g_*v^*\mathcal{G}\qquad(\mathcal{G}\in\mathbf{D}(\mathcal{X}))$$

be adjoint to the canonical map $\mathbf{R}f_*\mathcal{G}\to\mathbf{R}f_*\mathbf{R}v_*v^*\mathcal{G}=\mathbf{R}u_*\mathbf{R}g_*v^*\mathcal{G}$.

Then for all $\mathcal{E} \in \mathbf{D}_{\mathrm{qct}}(\mathfrak{X})$, *the map* $\theta'_{\mathcal{E}} := \mathbf{R}\varGamma'_{\mathfrak{U}}(\theta_{\mathcal{E}})$ *is an* isomorphism

$$\theta'_{\mathcal{E}} \colon \mathbf{R}\varGamma'_{\mathfrak{U}} u^* \mathbf{R} f_* \mathcal{E} \xrightarrow{\sim} \mathbf{R}\varGamma'_{\mathfrak{U}} \mathbf{R} g_* v^* \mathcal{E}.$$

In particular, if u *(hence* v*) is* adic *then* $\theta_{\mathcal{E}}$ *itself is an isomorphism.*

(c) *Under the hypotheses of* (a) *resp.* (b), *if* f *(hence* g*) is adic then* $\psi_{\mathcal{E}}$ *resp.* $\theta'_{\mathcal{E}}$ *is an isomorphism for all* $\mathcal{E} \in \widetilde{\mathbf{D}}_{\mathrm{qc}}(\mathfrak{X})$ *(see Definition* 5.2.9).

PROOF. (a) Let $\mathcal{J}$ be an ideal of definition of $\mathfrak{X}$, and $\mathcal{K}$ of $\mathfrak{U}$, so that $\mathcal{J}\mathcal{O}_{\mathfrak{V}} + \mathcal{K}\mathcal{O}_{\mathfrak{V}}$ is an ideal of definition of $\mathfrak{V}$. The obvious equality $\varGamma_{\mathcal{J}\mathcal{O}_{\mathfrak{V}}+\mathcal{K}\mathcal{O}_{\mathfrak{V}}} = \varGamma_{\mathcal{K}\mathcal{O}_{\mathfrak{V}}}\varGamma_{\mathcal{J}\mathcal{O}_{\mathfrak{V}}}$, applied to K-injective $\mathcal{O}_{\mathfrak{V}}$-complexes, leads to a natural functorial map

$$\mathbf{R}\varGamma'_{\mathfrak{V}} \underset{1.2.1}{\overset{\mathrm{def}}{=}} \mathbf{R}\varGamma_{\mathcal{J}\mathcal{O}_{\mathfrak{V}}+\mathcal{K}\mathcal{O}_{\mathfrak{V}}} \longrightarrow \mathbf{R}\varGamma_{\mathcal{K}\mathcal{O}_{\mathfrak{V}}}\mathbf{R}\varGamma_{\mathcal{J}\mathcal{O}_{\mathfrak{V}}}$$

which is an *isomorphism,* as one checks locally via [**AJL**, p. 20, Corollary (3.1.3)]. Also, there are natural isomorphisms

$$\mathbf{R}\varGamma_{\mathcal{J}\mathcal{O}_{\mathfrak{V}}} v^*\mathcal{E} \underset{5.2.8(\mathrm{b})}{\xrightarrow{\sim}} v^*\mathbf{R}\varGamma'_{\mathfrak{X}}\mathcal{E} = v^*\mathbf{R}\varGamma_{\mathcal{J}}\mathcal{E} \underset{5.2.1(\mathrm{a})}{\xrightarrow{\sim}} v^*\mathcal{E} \qquad (\mathcal{E} \in \mathbf{D}_{\mathrm{qct}}(\mathfrak{X})).$$

Thus the natural map $\mathbf{R}\varGamma'_{\mathfrak{V}} \to \mathbf{R}\varGamma_{\mathcal{K}\mathcal{O}_{\mathfrak{V}}}$ induces an *isomorphism*—the composition

$$\mathbf{R}g_*\mathbf{R}\varGamma'_{\mathfrak{V}}v^*\mathcal{E} \xrightarrow{\sim} \mathbf{R}g_*\mathbf{R}\varGamma_{\mathcal{K}\mathcal{O}_{\mathfrak{V}}}\mathbf{R}\varGamma_{\mathcal{J}\mathcal{O}_{\mathfrak{V}}}v^*\mathcal{E} \xrightarrow{\sim} \mathbf{R}g_*\mathbf{R}\varGamma_{\mathcal{K}\mathcal{O}_{\mathfrak{V}}}v^*\mathcal{E}.$$

Since $(*)$: $\mathbf{R}g_*\mathbf{R}\varGamma_{\mathcal{K}\mathcal{O}_{\mathfrak{V}}}v^*\mathcal{E} \cong \mathbf{R}g_*\mathbf{R}\varGamma'_{\mathfrak{V}}v^*\mathcal{E} \in \mathbf{D}_{\mathrm{t}}(\mathfrak{U})$ (Propositions 5.2.1 and 5.2.6) therefore we can imitate the proof of Proposition 5.2.8(d)—*without* the boundedness imposed there on $\mathcal{G}$, since that would be needed only to get $(*)$—to see that the map $\mathbf{R}g_*\mathbf{R}\varGamma_{\mathcal{K}\mathcal{O}_{\mathfrak{V}}}v^*\mathcal{E} \to \mathbf{R}g_*v^*\mathcal{E}$ induced by $\mathbf{R}\varGamma_{\mathcal{K}\mathcal{O}_{\mathfrak{V}}} \to \mathbf{1}$ factors uniquely as

$$\mathbf{R}g_*\mathbf{R}\varGamma_{\mathcal{K}\mathcal{O}_{\mathfrak{V}}}v^*\mathcal{E} \xrightarrow{\sim} \mathbf{R}\varGamma'_{\mathfrak{U}}\mathbf{R}g_*v^*\mathcal{E} \longrightarrow \mathbf{R}g_*v^*\mathcal{E},$$

with the first map an isomorphism. It follows that $\psi_{\mathcal{E}}$ is the composed isomorphism

$$\mathbf{R}g_*\mathbf{R}\varGamma'_{\mathfrak{V}}v^*\mathcal{E} \xrightarrow{\sim} \mathbf{R}g_*\mathbf{R}\varGamma_{\mathcal{K}\mathcal{O}_{\mathfrak{V}}}\mathbf{R}\varGamma_{\mathcal{J}\mathcal{O}_{\mathfrak{V}}}v^*\mathcal{E} \xrightarrow{\sim} \mathbf{R}g_*\mathbf{R}\varGamma_{\mathcal{K}\mathcal{O}_{\mathfrak{V}}}v^*\mathcal{E} \xrightarrow{\sim} \mathbf{R}\varGamma'_{\mathfrak{U}}\mathbf{R}g_*v^*\mathcal{E}.$$

The last statement in (a) (for adic u) results then from Corollary 5.2.11(b) and Propositions 5.2.6 and 5.2.1(a).

(b) Once $\theta'_{\mathcal{E}}$ is shown to be an isomorphism, the last statement in (b) (for adic u) follows from Corollary 5.2.11(b), and Propositions 5.2.6 and 5.2.1(a).

To show that $\theta'_{\mathcal{E}}$ is an isomorphism, it suffices to show that the composition

$$\psi_{\mathcal{E}}^{-1}\theta'_{\mathcal{E}} \colon \mathbf{R}\varGamma'_{\mathfrak{U}}u^*\mathbf{R}f_*\mathcal{E} \to \mathbf{R}g_*\mathbf{R}\varGamma'_{\mathfrak{V}}v^*\mathcal{E} \qquad (\mathcal{E} \in \mathbf{D}_{\mathrm{qct}}(\mathfrak{X})).$$

is an isomorphism. We use Lemma 5.4.1 to reduce the problem, as follows.

First, the functors u^*, v^*, $\mathbf{R}\varGamma'_{\mathfrak{U}}$ and $\mathbf{R}\varGamma'_{\mathfrak{V}}$ are bounded, and commute with direct sums: for u^* and v^* that is clear, and for $\mathbf{R}\varGamma'_{\mathfrak{U}}$ and $\mathbf{R}\varGamma'_{\mathfrak{V}}$ it holds because they can be realized locally by tensoring with a bounded flat complex (see proof of Proposition 5.2.1). Furthermore, Lemma 5.1.4, Proposition 5.2.1, and Proposition 3.3.5 show that $\mathbf{R}\varGamma'_{\mathfrak{V}}v^*\mathbf{D}_{\mathrm{qct}}(\mathfrak{X}) \subset \mathbf{D}_{\mathrm{qct}}(\mathfrak{V})$; and the functor $\mathbf{R}g_*$ (resp. $\mathbf{R}f_*$) is bounded on, and commutes with direct sums in, $\mathbf{D}_{\mathrm{qct}}(\mathfrak{V})$ (resp. $\mathbf{D}_{\mathrm{qct}}(\mathfrak{X})$), see Propositions 5.1.4, 3.4.3 and 3.5.2. Hence, standard way-out reasoning allows us to assume that $\mathcal{E} \in \mathbf{D}^{+}_{\mathrm{qct}}(\mathfrak{X})$.

Next, let $\mathcal{J}$ be an ideal of definition of $\mathfrak{X}$, X_n $(n > 0)$ the scheme $(\mathfrak{X}, \mathcal{O}_{\mathfrak{X}}/\mathcal{J}^n)$, and $j_n \colon X_n \hookrightarrow \mathfrak{X}$ the associated closed immersion. The functor $j_{n*} \colon \mathcal{A}(X_n) \to \mathcal{A}(\mathfrak{X})$ is exact, so it extends to a functor $\mathbf{D}(X_n) \to \mathbf{D}(\mathfrak{X})$. The functor $j_n^{\natural} \colon \mathbf{D}(\mathfrak{X}) \to \mathbf{D}(X_n)$ being defined as in (6.1.3.1), we have

$$h_n(\mathcal{G}) := \mathbf{R}\mathcal{H}om^{\bullet}(\mathcal{O}_{\mathfrak{X}}/\mathcal{J}^n, \mathcal{G}) = j_{n*}j_n^{\natural}\mathcal{G} \qquad (\mathcal{G} \in \mathbf{D}(\mathfrak{X})).$$

If $\mathcal E\in\mathbf D^+_{\mathrm{qct}}(\mathcal X)$ then $\mathcal E=\mathbf R\varGamma_{\mathcal J}\mathcal E$ (Proposition 5.2.1(a)), and, as noted just after (6.1.3.1), $j_n^\natural\mathcal E\in\mathbf D_{\mathrm{qc}}(X_n)$. Hence, from the triangle in Lemma 5.4.1 (with $\mathcal G$ replaced by an $\mathcal E\in\mathbf D^+_{\mathrm{qct}}(\mathcal X)$) we derive a diagram of triangles

$$\begin{array}{ccccccc}
\mathbf R\varGamma'_{\mathcal U}u^*\mathbf Rf_*(\oplus_{n\ge1}\boldsymbol h_n\mathcal E) & \longrightarrow & \mathbf R\varGamma'_{\mathcal U}u^*\mathbf Rf_*(\oplus_{n\ge1}\boldsymbol h_n\mathcal E) & \longrightarrow & \mathbf R\varGamma'_{\mathcal U}u^*\mathbf Rf_*(\mathbf R\varGamma_{\mathcal J}\mathcal E) & \xrightarrow{+} \\
\simeq\big\downarrow & & \simeq\big\downarrow & & \big\downarrow\simeq & \\
\oplus_{n\ge1}\mathbf R\varGamma'_{\mathcal U}u^*\mathbf Rf_*\boldsymbol h_n\mathcal E & \longrightarrow & \oplus_{n\ge1}\mathbf R\varGamma'_{\mathcal U}u^*\mathbf Rf_*\boldsymbol h_n\mathcal E & \longrightarrow & \mathbf R\varGamma'_{\mathcal U}u^*\mathbf Rf_*\mathcal E & \xrightarrow{+} \\
\oplus\big\downarrow\psi^{-1}_{\boldsymbol h_n\mathcal E}\theta'_{\boldsymbol h_n\mathcal E} & & \oplus\big\downarrow\psi^{-1}_{\boldsymbol h_n\mathcal E}\theta'_{\boldsymbol h_n\mathcal E} & & \big\downarrow\psi_{\mathcal E}^{-1}\theta'_{\mathcal E} & \\
\oplus_{n\ge1}\mathbf Rg_*\mathbf R\varGamma'_{\mathcal V}v^*\boldsymbol h_n\mathcal E & \longrightarrow & \oplus_{n\ge1}\mathbf Rg_*\mathbf R\varGamma'_{\mathcal V}v^*\boldsymbol h_n\mathcal E & \longrightarrow & \mathbf Rg_*\mathbf R\varGamma'_{\mathcal V}v^*\mathcal E & \xrightarrow{+} \\
\simeq\big\downarrow & & \simeq\big\downarrow & & \big\downarrow\simeq & \\
\mathbf Rg_*\mathbf R\varGamma'_{\mathcal V}v^*(\oplus_{n\ge1}\boldsymbol h_n\mathcal E) & \longrightarrow & \mathbf Rg_*\mathbf R\varGamma'_{\mathcal V}v^*(\oplus_{n\ge1}\boldsymbol h_n\mathcal E) & \longrightarrow & \mathbf Rg_*\mathbf R\varGamma'_{\mathcal V}v^*(\mathbf R\varGamma_{\mathcal J}\mathcal E) & \xrightarrow{+}
\end{array}$$

From this diagram we see that if each $\psi^{-1}_{\boldsymbol h_n\mathcal E}\theta'_{\boldsymbol h_n\mathcal E}$ is an isomorphism, then so is $\psi_{\mathcal E}^{-1}\theta'_{\mathcal E}$. So we need only prove (b) when $\mathcal E=j_{n*}\mathcal F$ with $\mathcal F:=j_n^\natural\mathcal E\in\mathbf D_{\mathrm{qc}}(X_n)$. Let us show that in fact *for any $n>0$ and any $\mathcal F\in\mathbf D_{\mathrm{qc}}(X_n)$, $\theta'_{j_{n*}\mathcal F}$ is an isomorphism.*

The assertion (b) is local both on $\mathcal Y$ and on $\mathcal U$. Indeed, for (b) to hold it suffices, for every diagram of fiber squares

$$\begin{array}{ccccccc}
\mathcal V & \xleftarrow{\;j'\;} & \mathcal V' & \xrightarrow{\;v'\;} & \mathcal X' & \xrightarrow{\;j\;} & \mathcal X \\
{\scriptstyle g}\big\downarrow & & {\scriptstyle g'}\big\downarrow & & \big\downarrow{\scriptstyle f'} & & \big\downarrow{\scriptstyle f} \\
\mathcal U & \xleftarrow[\;i'\;]{} & \mathcal U' & \xrightarrow[\;u'\;]{} & \mathcal Y' & \xrightarrow[\;i\;]{} & \mathcal Y
\end{array}$$

where $\mathcal Y'$ ranges over a base of open subsets of $\mathcal Y$, $\mathcal U'$ ranges over a base of open subsets of $u^{-1}\mathcal Y'$, u' is induced by u, and i, i' are the inclusions, that $i'^*\theta'_{\mathcal E}$ $(=\theta'_{\mathcal E}|_{\mathcal U'})$ be an isomorphism. Now when u is an open immersion, $\theta_{\mathcal G}$ is an isomorphism for all $\mathcal G\in\mathbf D(\mathcal X)$. (One may assume $\mathcal G$ to be K-injective and note that v^*, having the exact left adjoint "extension by zero," preserves K-injectivity, so that $\theta_{\mathcal G}$ becomes the usual isomorphism $u^*f_*\mathcal G\xrightarrow{\sim}g_*v^*\mathcal G$). Thus there are functorial isomorphisms $i'^*\mathbf Rg_*\xrightarrow{\sim}\mathbf Rg'_*j'^*$ and $i^*\mathbf Rf_*\xrightarrow{\sim}\mathbf Rf'_*j^*$; and similarly there is an isomorphism $i'^*\mathbf R\varGamma'_{\mathcal U}\xrightarrow{\sim}\mathbf R\varGamma'_{\mathcal U'}i'^*$. So it suffices that the composition

$$i'^*\mathbf R\varGamma'_{\mathcal U}u^*\mathbf Rf_*\mathcal E\xrightarrow{i'^*\theta'_{\mathcal E}}i'^*\mathbf R\varGamma'_{\mathcal U}\mathbf Rg_*v^*\mathcal E\xrightarrow{\sim}\mathbf R\varGamma'_{\mathcal U'}i'^*\mathbf Rg_*v^*\mathcal E\xrightarrow{\sim}\mathbf R\varGamma'_{\mathcal U'}\mathbf Rg'_*j'^*v^*\mathcal E$$

be an isomorphism; and with a bit of patience one identifies this composition with

$$\mathbf R\varGamma'_{\mathcal U'}u'^*i^*\mathbf Rf_*\mathcal E\xrightarrow{\sim}\mathbf R\varGamma'_{\mathcal U'}u'^*\mathbf Rf'_*j^*\mathcal E\xrightarrow{\theta'_{j^*\mathcal E}}\mathbf R\varGamma'_{\mathcal U'}\mathbf Rg'_*v'^*j^*\mathcal E,$$

thereby reducing to showing that $\theta'_{j^*\mathcal E}$ is an isomorphism. Thus one may assume that both $\mathcal Y$ and $\mathcal U$ are affine, say $\mathcal Y=\mathrm{Spf}(A)$ and $\mathcal U=\mathrm{Spf}(C)$ with C a flat A-algebra (Lemma 7.1.1).

Suppose next that $\mathcal X$ and $\mathcal Y$ are ordinary schemes, so that $\mathcal Y=\mathrm{Spec}(A)$. In cases (i) and (ii) of (b), set $C'':=C$, and in case (iii) let C'' be as in the proof of part (c) of Proposition 7.1. In any case, C'' is A-flat, C is the K-adic completion of C'' for some C''-ideal K, $\mathcal X\times_{\mathcal Y}\mathrm{Spf}(C)$ is the K-adic completion of $\mathcal X\times_{\mathcal Y}\mathrm{Spec}(C'')$, and we have a natural commutative diagram

$$
\begin{array}{ccccc}
\mathcal{X}\times_{\mathcal{Y}}\mathrm{Spf}(C) & \xrightarrow{v_2} & \mathcal{X}\times_{\mathcal{Y}}\mathrm{Spec}(C'') & \xrightarrow{v_1} & \mathcal{X} \\
{\scriptstyle g}\big\downarrow & & {\scriptstyle g_1}\big\downarrow & & \big\downarrow{\scriptstyle f} \\
\mathrm{Spf}(C) & \xrightarrow[u_2]{} & \mathrm{Spec}(C'') & \xrightarrow[u_1]{} & \mathcal{Y}
\end{array}
$$

With $\varGamma'$ denoting $\varGamma'_{\mathrm{Spf}(C)}$, $\theta'_{\mathcal{E}} =: \theta'(\mathcal{E}, f, u)$ factors naturally as the composition

$$\mathbf{R}\varGamma' u_2^* u_1^* \mathbf{R}f_*\mathcal{E} \xrightarrow{\mathbf{R}\varGamma' u_2^*(\theta(\mathcal{E},f,u_1))} \mathbf{R}\varGamma' u_2^* \mathbf{R}g_{1*} v_1^*\mathcal{E} \xrightarrow{\theta'(v_1^*\mathcal{E},g_1,u_2)} \mathbf{R}\varGamma' v_2^* \mathbf{R}g_* v_2^* v_1^*\mathcal{E}.$$

Here $\theta(\mathcal{E}, f, u_1)$ is an isomorphism because all the schemes involved are ordinary schemes. (One argues as in [**H1**, p. 111, Prop. 5.12], using [**AHK**, p. 35, (6.7)]; for a fussier treatment see [**L4**, Prop. (3.9.5)].) Also, $\theta'(v_1^*\mathcal{E}, g_1, u_2)$ is an isomorphism, in case (i) of (b) since then u_2 and v_2 are identity maps, and in cases (ii) and (iii) by Corollary 5.2.12 since then $\mathcal{X}\times_{\mathcal{Y}}\mathrm{Spec}(C'')$ is noetherian. Thus:

LEMMA 7.2.1. *Proposition* 7.2 *holds when $\mathcal{X}$ and $\mathcal{Y}$ are both ordinary schemes.*

We will also need the following special case of Proposition 7.2:

LEMMA 7.2.2. *Let $\mathcal{I}$ be an ideal of definition of $\mathcal{Y}$, Y_n the scheme $(\mathcal{Y}, \mathcal{O}_{\mathcal{Y}}/\mathcal{I}^n)$, and $i_n\colon Y_n \hookrightarrow \mathcal{Y}$ the canonical closed immersion. Let $u_n\colon Y_n\times_{\mathcal{Y}}\mathcal{U} \to Y_n$ and $p_n\colon Y_n\times_{\mathcal{Y}}\mathcal{U}\to\mathcal{U}$ be the projections (so that u_n is flat and p_n is a closed immersion, see* [**GD**, p. 442, (10.14.5)(ii)]). *Then the natural map is an isomorphism*

$$u^* i_{n*}\mathcal{G} \xrightarrow{\sim} p_{n*}u_n^*\mathcal{G} \qquad (\mathcal{G}\in\mathbf{D}_{\mathrm{qc}}(Y_n)).$$

PROOF. Since the functors u^*, i_{n*}, p_{n*}, and u_n^* are all exact, we may assume that $\mathcal{G}$ is a quasi-coherent $\mathcal{O}_{Y_n}$-module; and since those functors commute with $\varinjlim$ we may further assume $\mathcal{G}$ coherent, and then refer to [**GD**, p. 443, (10.14.6)]. □

Finally, for general noetherian formal schemes $\mathcal{X}$ and $\mathcal{Y}$, and $\mathcal{I}$ and Y_n as above, let $\mathcal{J}\supset\mathcal{I}\mathcal{O}_{\mathcal{X}}$ be an ideal of definition of $\mathcal{X}$, let X_n be the scheme $(\mathcal{X}, \mathcal{O}_{\mathcal{X}}/\mathcal{J}^n)$, and let $f_n\colon X_n\to Y_n$ be the map induced by f. Then for any $\mathcal{F}\in\mathbf{D}_{\mathrm{qc}}(X_n)$, it holds that $\mathbf{R}f_{n*}\mathcal{F}\in\mathbf{D}_{\mathrm{qc}}(Y_n)$. (See Proposition 5.2.6—though the simpler case $\mathcal{F}\in\mathbf{D}^+_{\mathrm{qc}}(X_n)$ would do for proving Proposition 7.2.) Associated to the natural diagram

(7.2.3)

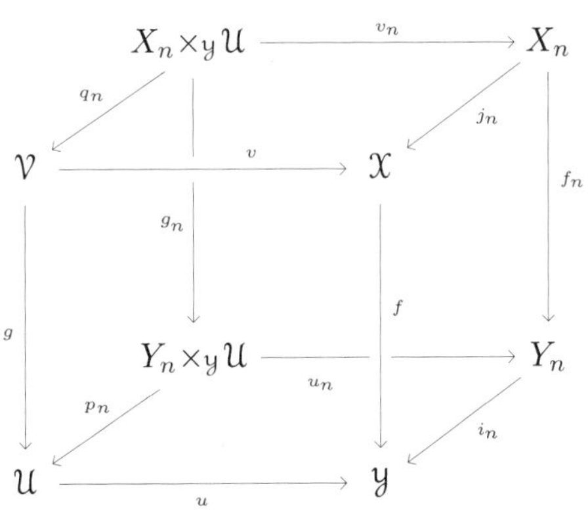

there is a composed isomorphism

$$\begin{aligned}
\mathbf{R}\varGamma'_{\mathfrak{U}}u^*\mathbf{R}f_*j_{n*}\mathcal{F} &\xrightarrow{\sim} \mathbf{R}\varGamma'_{\mathfrak{U}}u^*i_{n*}\mathbf{R}f_{n*}\mathcal{F} && (\mathcal{F}\in\mathbf{D}_{\mathrm{qc}}(X_n))\\
&\xrightarrow{\sim} \mathbf{R}\varGamma'_{\mathfrak{U}}p_{n*}u_n^*\mathbf{R}f_{n*}\mathcal{F} && (\text{Lemma 7.2.2})\\
&\xrightarrow{\sim} \mathbf{R}\varGamma'_{\mathfrak{U}}p_{n*}\mathbf{R}g_{n*}v_n^*\mathcal{F} && (\text{Lemma 7.2.1})\\
&\xrightarrow{\sim} \mathbf{R}\varGamma'_{\mathfrak{U}}\mathbf{R}g_*q_{n*}v_n^*\mathcal{F}\\
&\xrightarrow{\sim} \mathbf{R}\varGamma'_{\mathfrak{U}}\mathbf{R}g_*v^*j_{n*}\mathcal{F} && (\text{Lemma 7.2.2}),
\end{aligned}$$

which—the conscientious reader will verify—is just $\theta'_{j_{n*}\mathcal{F}}$.

Thus $\theta'_{j_{n*}\mathcal{F}}$ is indeed an isomorphism.

(c) By definition $\mathbf{R}\varGamma'_{\mathfrak{X}}(\widetilde{\mathbf{D}}_{\mathrm{qc}}(\mathfrak{X}))\subset\mathbf{D}_{\mathrm{qct}}(\mathfrak{X})$, and so by (a) and (b) it's enough to see, as follows, that the natural map $\mathbf{R}\varGamma'_{\mathfrak{X}}\mathcal{E}\to\mathcal{E}$ induces isomorphisms of the source and target of both $\psi_{\mathcal{E}}$ and $\theta'_{\mathcal{E}}$.

Proposition 5.2.8(c) gives the isomorphism $\mathbf{R}g_*\mathbf{R}\varGamma'_{\mathfrak{V}}v^*\mathbf{R}\varGamma'_{\mathfrak{X}}\mathcal{E}\xrightarrow{\sim}\mathbf{R}g_*\mathbf{R}\varGamma'_{\mathfrak{V}}v^*\mathcal{E}$, as well as the second of the following isomorphisms, the first and third of which follow from Corollary 5.2.11(d):

$$\mathbf{R}\varGamma'_{\mathfrak{U}}\mathbf{R}g_*v^*\mathbf{R}\varGamma'_{\mathfrak{X}}\mathcal{E}\cong\mathbf{R}g_*\mathbf{R}\varGamma'_{\mathfrak{V}}v^*\mathbf{R}\varGamma'_{\mathfrak{X}}\mathcal{E}\cong\mathbf{R}g_*\mathbf{R}\varGamma'_{\mathfrak{V}}v^*\mathcal{E}\cong\mathbf{R}\varGamma'_{\mathfrak{U}}\mathbf{R}g_*v^*\mathcal{E}.$$

Likewise, there are natural isomorphisms

$$\mathbf{R}\varGamma'_{\mathfrak{U}}u^*\mathbf{R}f_*\mathbf{R}\varGamma'_{\mathfrak{X}}\mathcal{E}\cong\mathbf{R}\varGamma'_{\mathfrak{U}}u^*\mathbf{R}\varGamma'_{\mathfrak{Y}}\mathbf{R}f_*\mathcal{E}\cong\mathbf{R}\varGamma'_{\mathfrak{U}}u^*\mathbf{R}f_*\mathcal{E}.$$

□

Notation and assumptions stay as in Proposition 7.2(a). Assume that f and g satisfy the hypotheses of Theorem 6.1, so that the functor $\mathbf{R}f_*\colon\mathbf{D}_{\mathrm{qct}}(\mathfrak{X})\to\mathbf{D}(\mathfrak{Y})$ has a right adjoint $f_{\mathrm{t}}^{\times}$, and similarly for g. Recall from Corollary 6.1.5(b) that there is a natural isomorphism $g_{\mathrm{t}}^{\times}\mathbf{R}\varGamma'_{\mathfrak{U}}\xrightarrow{\sim}g_{\mathrm{t}}^{\times}$.

DEFINITION 7.3. With conditions as in Proposition 7.2(b), the *base-change map*

$$\beta_{\mathcal{F}}\colon\mathbf{R}\varGamma'_{\mathfrak{V}}v^*f_{\mathrm{t}}^{\times}\mathcal{F}\to g_{\mathrm{t}}^{\times}\mathbf{R}\varGamma'_{\mathfrak{U}}u^*\mathcal{F}\qquad(\mathcal{F}\in\mathbf{D}(\mathfrak{Y}))$$

is defined to be the map adjoint to the natural composition

$$\mathbf{R}g_*\mathbf{R}\varGamma'_{\mathfrak{V}}v^*f_{\mathrm{t}}^{\times}\mathcal{F}\underset{\psi}{\xrightarrow{\sim}}\mathbf{R}\varGamma'_{\mathfrak{U}}\mathbf{R}g_*v^*f_{\mathrm{t}}^{\times}\mathcal{F}\underset{\theta'^{-1}}{\xrightarrow{\sim}}\mathbf{R}\varGamma'_{\mathfrak{U}}u^*\mathbf{R}f_*f_{\mathrm{t}}^{\times}\mathcal{F}\to\mathbf{R}\varGamma'_{\mathfrak{U}}u^*\mathcal{F}$$

where $\psi:=\psi_{f_{\mathrm{t}}^{\times}\mathcal{F}}$ and $\theta':=\theta'_{f_{\mathrm{t}}^{\times}\mathcal{F}}$. In particular, if u (hence v) is *adic* then

$$\beta_{\mathcal{F}}\colon v^*f_{\mathrm{t}}^{\times}\mathcal{F}\to g_{\mathrm{t}}^{\times}u^*\mathcal{F}$$

is the map adjoint to the natural composition

$$\mathbf{R}g_*v^*f_{\mathrm{t}}^{\times}\mathcal{F}\underset{\theta^{-1}}{\xrightarrow{\sim}}u^*\mathbf{R}f_*f_{\mathrm{t}}^{\times}\mathcal{F}\to u^*\mathcal{F}$$

where $\theta:=\theta_{f_{\mathrm{t}}^{\times}\mathcal{F}}$.

Notation. For a pseudo-proper (hence separated) map f (see §1.2.2), we write $f^!$ instead of $f_{\mathrm{t}}^{\times}$.

THEOREM 7.4. *Let $\mathcal{X}$, $\mathcal{Y}$ and $\mathcal{U}$ be noetherian formal schemes, let $f\colon \mathcal{X} \to \mathcal{Y}$ be a pseudo-proper map, and let $u\colon \mathcal{U} \to \mathcal{Y}$ be flat, so that in any fiber square*

$$\begin{CD}
\mathcal{V} @>{v}>> \mathcal{X}\\
@V{g}VV @VV{f}V\\
\mathcal{U} @>>{u}> \mathcal{Y}
\end{CD}$$

the formal scheme $\mathcal{V}$ is noetherian, g is pseudo-proper, and v is flat (Proposition 7.1). Then for all $\mathcal{F} \in \widetilde{\mathbf{D}}^{+}_{\mathrm{qc}}(\mathcal{Y}) := \widetilde{\mathbf{D}}_{\mathrm{qc}}(\mathcal{Y}) \cap \mathbf{D}^{+}(\mathcal{Y})$ the base-change map $\beta_{\mathcal{F}}$ is an isomorphism

$$\beta_{\mathcal{F}}\colon \mathbf{R}\varGamma'_{\mathcal{V}} v^{*} f^{!}\mathcal{F} \xrightarrow{\sim} g^{!}\,\mathbf{R}\varGamma'_{\mathcal{U}} u^{*}\mathcal{F}\ (\cong g^{!}u^{*}\mathcal{F}).$$

Remark. In [**N1**, p. 233, Example 6.5] Neeman gives an example where f is a finite map of ordinary schemes, u is an open immersion, $\mathcal{F} \in \mathbf{D}^{-}_{\mathrm{c}}(\mathcal{Y})$, and $\beta_{\mathcal{F}}$ is *not* an isomorphism.

PROOF. Recall diagram (7.2.3), in which, $\mathcal{I}$ and $\mathcal{J} \supset \mathcal{I}\mathcal{O}_{\mathcal{X}}$ being defining ideals of $\mathcal{Y}$ and $\mathcal{X}$ respectively, Y_n is the scheme $(\mathcal{Y}, \mathcal{O}_{\mathcal{Y}}/\mathcal{I}^n)$ and X_n is the scheme $(\mathcal{X}, \mathcal{O}_{\mathcal{X}}/\mathcal{J}^n)$. Let $\mathcal{K} \supset \mathcal{I}\mathcal{O}_{\mathcal{U}}$ be a defining ideal of $\mathcal{U}$, let $\mathcal{L} := \mathcal{J}\mathcal{O}_{\mathcal{V}} + \mathcal{K}\mathcal{O}_{\mathcal{V}}$, a defining ideal of $\mathcal{V}$, let V_n $(n > 0)$ be the scheme $(\mathcal{V}, \mathcal{O}_{\mathcal{V}}/\mathcal{L}^n)$, and let $l_n\colon V_n \hookrightarrow \mathcal{V}$ be the canonical closed immersion. Then by Example 6.1.3(4),

$$l_{n*}l_n^{!}\mathcal{G} = l_{n*}l_n^{\natural}\mathcal{G} = \mathbf{R}\mathcal{H}om(\mathcal{O}_{\mathcal{V}}/\mathcal{L}^n, \mathcal{G}) =: \boldsymbol{h}_n(\mathcal{G}) \qquad (\mathcal{G} \in \mathbf{D}^{+}_{\mathrm{qct}}(\mathcal{V})).$$

So in view of the natural isomorphism $\mathbf{R}\varGamma'_{\mathcal{V}} g^{!}u^{*}\mathcal{F} \xrightarrow{\sim} g^{!}u^{*}\mathcal{F}$ (Proposition 5.2.1(a)), Lemma 5.4.1 shows it sufficient to prove that the maps

$$\boldsymbol{h}_n(\beta_{\mathcal{F}})\colon l_{n*}l_n^{!}\mathbf{R}\varGamma'_{\mathcal{V}} v^{*} f^{!}\mathcal{F} \to l_{n*}l_n^{!}g^{!}u^{*}\mathcal{F} \qquad (n > 0)$$

are all isomorphisms.

Moreover, the closed immersion l_n factors uniquely as

$$V_n \xrightarrow{r_n} X_n \times_{\mathcal{Y}} \mathcal{U} \xrightarrow{q_n} \mathcal{V},$$

so we can replace $l_n^{!}$ by $r_n^{!}q_n^{!}$ (Theorem 6.1(b)). Thus *it will suffice to prove that the maps*

$$q_n^{!}(\beta_{\mathcal{F}})\colon q_n^{!}\mathbf{R}\varGamma'_{\mathcal{V}} v^{*} f^{!}\mathcal{F} \to q_n^{!}g^{!}u^{*}\mathcal{F} \qquad (n > 0)$$

are all isomorphisms.

In the cube (7.2.3), the front, top, rear, and bottom faces are fiber squares, denoted, respectively, by $\square$, $\square_{\mathrm{t}}$, $\square_{\mathrm{r}}$ and $\square_{\mathrm{b}}$; and we have the "composed" fiber square $\square_{\mathrm{c}}$:

$$\begin{CD}
X_n \times_{\mathcal{Y}} \mathcal{U} @>{v_n}>> X_n\\
@V{p_n g_n}V{= g q_n}V @V{i_n f_n}V{= f j_n}V\\
\mathcal{U} @>>{u}> \mathcal{Y}
\end{CD}$$

The proper map f_n and the closed immersions i_n and j_n are all of pseudo-finite type. Also, it follows from Proposition 7.1(b) that in addition to u, the maps u, u_n, v and v_n are all flat. So corresponding to the fibre squares $\square_{\bullet}$ we have base-change maps $\beta_{\bullet}$.

Consider the following diagram of functorial maps where, to save space, we set $\blacktriangle := X_n \times_{\mathcal{Y}} \mathcal{U}$ and $\blacktriangledown := Y_n \times_{\mathcal{Y}} \mathcal{U}$.

$$
\begin{array}{ccccccccc}
q_n^! \mathbf{R}\varGamma'_{\mathcal{V}} v^* f^! & \xleftarrow{\ \beta_{\mathrm{t}}\ } & \mathbf{R}\varGamma'_{\blacktriangle} v_n^* j_n^! f^! & \xleftarrow{\ \sim\ } & \mathbf{R}\varGamma'_{\blacktriangle} v_n^* (f j_n)^! & = & \mathbf{R}\varGamma'_{\blacktriangle} v_n^* (i_n f_n)^! & \xrightarrow{\ \sim\ } & \mathbf{R}\varGamma'_{\blacktriangle} v_n^* f_n^! i_n^! \\
{\scriptstyle q_n^!(\beta)}\downarrow & & & & \downarrow{\scriptstyle \beta_{\mathrm{c}}} & & & & \downarrow{\scriptstyle \beta_{\mathrm{r}}} \\
q_n^! g^! u^* & \xrightarrow{\ \sim\ } & (g q_n)^! u^* & = & (p_n g_n)^! u^* & \xleftarrow{\ \sim\ } & g_n^! p_n^! u^* & \xleftarrow[g_n^!(\beta_{\mathrm{b}})]{} & g_n^! \mathbf{R}\varGamma'_{\blacktriangledown} u_n^* i_n^!
\end{array}
$$

As above, we want to see that $q_n^!(\beta)$ is an isomorphism (in the category of functors from $\widetilde{\mathbf{D}}^+_{\mathrm{qc}}(\mathcal{Y})$ to $\mathbf{D}(X_n \times_{\mathcal{Y}} \mathcal{U})$). For that the following assertions clearly suffice:

(a) The preceding diagram commutes.
(b) The base-change maps β_{t} and β_{b} are isomorphisms.
(c) The base-change map β_{r} is an isomorphism.

Assertion (a) results from part (b) of the transitivity lemma 7.5.2 below. Since i_n and j_n are closed immersions, assertion (b) results from Lemma 7.6.1, which is just Theorem 7.4 for the case when f is a closed immersion. Since f is pseudo-proper therefore f_n is proper, and assertion (c) is essentially the case of Theorem 7.4—established in Lemma 7.7.1—when $\mathcal{X}$ and $\mathcal{Y}$ are ordinary schemes.

Thus these three Lemmas will complete the proof of Theorem 7.4. □

7.5. We will need some "transitivity" properties of the maps $\theta'_{\mathcal{E}}$ and $\beta_{\mathcal{F}}$ relative to horizontal and vertical composition of fiber squares of noetherian formal schemes, i.e., diagrams of the form

$$
\text{(7.5.0a)} \qquad
\begin{array}{ccccc}
\mathcal{V} & \xrightarrow{\ v_2\ } & \mathcal{V}_1 & \xrightarrow{\ v_1\ } & \mathcal{X} \\
{\scriptstyle g}\downarrow & & {\scriptstyle g_1}\downarrow & & \downarrow{\scriptstyle f} \\
\mathcal{U} & \xrightarrow[u_2]{} & \mathcal{U}_1 & \xrightarrow[u_1]{} & \mathcal{Y}
\end{array}
$$

$$
\text{(7.5.0b)} \qquad
\begin{array}{ccc}
\mathcal{V} & \xrightarrow{\ v\ } & \mathcal{X} \\
{\scriptstyle g_2}\downarrow & & \downarrow{\scriptstyle f_2} \\
\mathcal{W} & \xrightarrow{\ w\ } & \mathcal{Z} \\
{\scriptstyle g_1}\downarrow & & \downarrow{\scriptstyle f_1} \\
\mathcal{U} & \xrightarrow[u]{} & \mathcal{Y}
\end{array}
$$

where all squares are fiber squares, and the maps u, u_i, v, v_i, and w are all flat.

As we will be dealing with several fiber squares simultaneously we will indicate the square with which, for instance, the map $\theta_{\mathcal{G}}$ in Proposition 7.2 is associated, by writing $\theta_{f,u}(\mathcal{G})$ instead.

The transitivity properties begin with:

LEMMA 7.5.1. *Coming out of the fiber square diagrams* (7.5.0a) *and* (7.5.0b), *the following natural diagrams commute for all* $\mathcal{G} \in \mathbf{D}(\mathcal{X})$:

$$\begin{array}{ccccc}
(u_1u_2)^*\mathbf{R}f_*\mathcal{G} & & \xrightarrow{\quad\theta_{f,u_1u_2}(\mathcal{G})\quad} & & \mathbf{R}g_*(v_1v_2)^*\mathcal{G} \\
\simeq\big\downarrow & & & & \big\downarrow\simeq \\
u_2^*u_1^*\mathbf{R}f_*\mathcal{G} & \xrightarrow[u_2^*(\theta_{f,u_1}(\mathcal{G}))]{} & u_2^*\mathbf{R}g_{1*}v_1^*\mathcal{G} & \xrightarrow[\theta_{g_1,u_2}(v_1^*\mathcal{G})]{} & \mathbf{R}g_*v_2^*v_1^*\mathcal{G}
\end{array}$$

$$\begin{array}{ccccc}
u^*\mathbf{R}(f_1f_2)_*\mathcal{G} & & \xrightarrow{\quad\theta_{f_1f_2,u}(\mathcal{G})\quad} & & \mathbf{R}(g_1g_2)_*v^*\mathcal{G} \\
\simeq\big\downarrow & & & & \big\downarrow\simeq \\
u^*\mathbf{R}f_{1*}\mathbf{R}f_{2*}\mathcal{G} & \xrightarrow[\theta_{f_1,u}(\mathbf{R}f_{2*}\mathcal{G})]{} & \mathbf{R}g_{1*}w^*\mathbf{R}f_{2*}\mathcal{G} & \xrightarrow[\mathbf{R}g_{1*}(\theta_{f_2,w}(\mathcal{G}))]{} & \mathbf{R}g_{1*}\mathbf{R}g_{2*}v^*\mathcal{G}
\end{array}$$

PROOF. This is a formal exercise, based on adjointness of u^* and $\mathbf{R}u_*$, etc. Details are left to the reader. □

LEMMA 7.5.2. (a) *In the fiber square diagram* (7.5.0a) (*with* u_1, v_1, u_2 *and* v_2 *flat*), *let* $\mathcal{F} \in \mathbf{D}(\mathcal{Y})$ *be such that the maps* $\theta'_1 := \theta'_{f,u_1}(f_{\mathrm{t}}^{\times}\mathcal{F})$, $\theta'_2 := \theta'_{g_1,u_2}((g_1)_{\mathrm{t}}^{\times}u_1^*\mathcal{F})$ *and* $\theta''_2 := \theta'_{g_1,u_2}(\mathbf{R}\varGamma'_{\mathcal{V}_1}v_1^*f_{\mathrm{t}}^{\times}\mathcal{F})$ *of Proposition* 7.2 *are isomorphisms. Then the map* $\theta' := \theta'_{f,u_1u_2}(f_{\mathrm{t}}^{\times}\mathcal{F})$ *is an isomorphism, so the base-change maps* $\beta_1 := \beta_{f,u_1}(\mathcal{F})$, $\beta_2 := \beta_{g_1,u_2}(u_1^*\mathcal{F})$ *and* $\beta := \beta_{f,u_1u_2}(\mathcal{F})$ *can all be defined as in Definition* 7.3; *and the following natural diagram, all of whose uparrows are isomorphisms, commutes*:

$$\begin{array}{ccccc}
\mathbf{R}\varGamma'_{\mathcal{V}}(v_1v_2)^*f_{\mathrm{t}}^{\times}\mathcal{F} & & \xrightarrow{\quad\beta\quad} & & g_{\mathrm{t}}^{\times}\mathbf{R}\varGamma'_{\mathcal{U}}(u_1u_2)^*\mathcal{F} \\
\simeq\big\uparrow & & & & \big\uparrow\simeq \\
\mathbf{R}\varGamma'_{\mathcal{V}}v_2^*v_1^*f_{\mathrm{t}}^{\times}\mathcal{F} & & \mathbf{R}\varGamma'_{\mathcal{V}}v_2^*(g_1)_{\mathrm{t}}^{\times}u_1^*\mathcal{F} & \xrightarrow{\beta_2} & g_{\mathrm{t}}^{\times}\mathbf{R}\varGamma'_{\mathcal{U}}u_2^*u_1^*\mathcal{F} \\
\simeq\big\uparrow \text{5.2.8(c)} & & \simeq\big\uparrow \text{6.1.5(b)} & & \text{5.2.8(c)}\big\uparrow\simeq \\
\mathbf{R}\varGamma'_{\mathcal{V}}v_2^*\mathbf{R}\varGamma'_{\mathcal{V}_1}v_1^*f_{\mathrm{t}}^{\times}\mathcal{F} & \xrightarrow[\mathbf{R}\varGamma'_{\mathcal{V}}v_2^*(\beta_1)]{} & \mathbf{R}\varGamma'_{\mathcal{V}}v_2^*(g_1)_{\mathrm{t}}^{\times}\mathbf{R}\varGamma'_{\mathcal{U}_1}u_1^*\mathcal{F} & \xrightarrow[\beta_2]{} & g_{\mathrm{t}}^{\times}\mathbf{R}\varGamma'_{\mathcal{U}}u_2^*\mathbf{R}\varGamma'_{\mathcal{U}_1}u_1^*\mathcal{F}
\end{array}$$

(b) *In the fiber square diagram* (7.5.0b)—*where* u, v *and* w *are assumed flat—set* $f := f_1f_2$ *and* $g := g_1g_2$. *Let* $\mathcal{F} \in \mathbf{D}(\mathcal{Y})$ *be such that the maps* $\theta'_1 := \theta'_{f_1,u}((f_1)_{\mathrm{t}}^{\times}\mathcal{F})$, $\theta'_2 := \theta'_{f_2,w}(f_{\mathrm{t}}^{\times}\mathcal{F})$ *and* $\theta' := \theta'_{f,u}(f_{\mathrm{t}}^{\times}\mathcal{F})$ *of Proposition* 7.2 *are isomorphisms, so that the base-change maps* $\beta_1 := \beta_{f_1,u}(\mathcal{F})$, $\beta_2 := \beta_{f_2,w}((f_1)_{\mathrm{t}}^{\times}\mathcal{F})$ *and* $\beta := \beta_{f,u}(\mathcal{F})$ *are all defined. Then the following diagram, whose two uparrows are isomorphisms, commutes*:

$$\begin{array}{ccccc}
\mathbf{R}\varGamma'_{\mathcal{V}}v^*f_{\mathrm{t}}^{\times}\mathcal{F} & & \xrightarrow{\quad\beta\quad} & & g_{\mathrm{t}}^{\times}\mathbf{R}\varGamma'_{\mathcal{U}}u^*\mathcal{F} \\
\simeq\big\uparrow & & & & \big\uparrow\simeq \\
\mathbf{R}\varGamma'_{\mathcal{V}}v^*(f_2)_{\mathrm{t}}^{\times}(f_1)_{\mathrm{t}}^{\times}\mathcal{F} & \xrightarrow[\beta_2]{} & (g_2)_{\mathrm{t}}^{\times}\mathbf{R}\varGamma'_{\mathcal{W}}w^*(f_1)_{\mathrm{t}}^{\times}\mathcal{F} & \xrightarrow[(g_2)_{\mathrm{t}}^{\times}(\beta_1)]{} & (g_2)_{\mathrm{t}}^{\times}(g_1)_{\mathrm{t}}^{\times}\mathbf{R}\varGamma'_{\mathcal{U}}u^*\mathcal{F}
\end{array}$$

PROOF. (a) The map

$$\gamma := \mathbf{R}\varGamma'_{\mathcal U}u_2^*(\theta_{f,u_1}(f_{\mathrm t}^{\times}\mathcal F))\colon \mathbf{R}\varGamma'_{\mathcal U}u_2^*u_1^*\mathbf{R}f_*f_{\mathrm t}^{\times}\mathcal F \longrightarrow \mathbf{R}\varGamma'_{\mathcal U}u_2^*\mathbf{R}g_{1*}v_1^*f_{\mathrm t}^{\times}\mathcal F$$

is isomorphic, by Proposition 5.2.8(c), to

$$\mathbf{R}\varGamma'_{\mathcal U}u_2^*(\theta'_1)\colon \mathbf{R}\varGamma'_{\mathcal U}u_2^*\mathbf{R}\varGamma'_{\mathcal U_1}u_1^*\mathbf{R}f_*f_{\mathrm t}^{\times}\mathcal F \longrightarrow \mathbf{R}\varGamma'_{\mathcal U}u_2^*\mathbf{R}\varGamma'_{\mathcal U_1}\mathbf{R}g_{1*}v_1^*f_{\mathrm t}^{\times}\mathcal F,$$

and so is an isomorphism (since θ'_1 is).

The map

$$\theta'_{g_1,u_2}(v_1^*f_{\mathrm t}^{\times}\mathcal F)\colon \mathbf{R}\varGamma'_{\mathcal U}u_2^*\mathbf{R}g_{1*}v_1^*f_{\mathrm t}^{\times}\mathcal F \to \mathbf{R}\varGamma'_{\mathcal U}\mathbf{R}g_*v_2^*v_1^*f_{\mathrm t}^{\times}\mathcal F$$

is also an isomorphism, as it is isomorphic to

$$\theta''_2\colon \mathbf{R}\varGamma'_{\mathcal U}u_2^*\mathbf{R}g_{1*}\mathbf{R}\varGamma'_{\mathcal V_1}v_1^*f_{\mathrm t}^{\times}\mathcal F \to \mathbf{R}\varGamma'_{\mathcal U}\mathbf{R}g_*v_2^*\mathbf{R}\varGamma'_{\mathcal V_1}v_1^*f_{\mathrm t}^{\times}\mathcal F,$$

because the natural map $\mathbf{R}\varGamma'_{\mathcal U}u_2^*\mathbf{R}g_{1*}\mathbf{R}\varGamma'_{\mathcal V_1}v_1^*f_{\mathrm t}^{\times}\mathcal F \to \mathbf{R}\varGamma'_{\mathcal U}u_2^*\mathbf{R}g_{1*}v_1^*f_{\mathrm t}^{\times}\mathcal F$ is the composed isomorphism

$$\begin{aligned}\mathbf{R}\varGamma'_{\mathcal U}u_2^*\mathbf{R}g_{1*}\mathbf{R}\varGamma'_{\mathcal V_1}v_1^*f_{\mathrm t}^{\times}\mathcal F \xrightarrow[\mathbf{R}\varGamma'_{\mathcal U}u_2^*\psi_{f_{\mathrm t}^{\times}\mathcal F}]{\ \sim\ } {}&\mathbf{R}\varGamma'_{\mathcal U}u_2^*\mathbf{R}\varGamma'_{\mathcal U_1}\mathbf{R}g_{1*}v_1^*f_{\mathrm t}^{\times}\mathcal F\\ &\xrightarrow[5.2.8(c)]{\ \sim\ } \mathbf{R}\varGamma'_{\mathcal U}u_2^*\mathbf{R}g_{1*}v_1^*f_{\mathrm t}^{\times}\mathcal F\end{aligned}$$

(see Proposition 7.2(a)); and because $\mathbf{R}\varGamma'_{\mathcal U}\mathbf{R}g_*v_2^*\mathbf{R}\varGamma'_{\mathcal V_1}v_1^*f_{\mathrm t}^{\times}\mathcal F \to \mathbf{R}\varGamma'_{\mathcal U}\mathbf{R}g_*v_2^*v_1^*f_{\mathrm t}^{\times}\mathcal F$ is one of the maps in the commutative diagram (B) below, all of whose other maps are isomorphisms.

Thus in the next diagram, whose commutativity results easily from that of the first diagram in Lemma 7.5.1, all the maps other than θ' are isomorphisms, whence so is θ'.

$$\begin{array}{ccc}\mathbf{R}\varGamma'_{\mathcal U}\mathbf{R}g_*(v_1v_2)^*f_{\mathrm t}^{\times}\mathcal F & \xleftarrow{\quad\theta'\quad} & \mathbf{R}\varGamma'_{\mathcal U}u_2^*u_1^*\mathbf{R}f_*f_{\mathrm t}^{\times}\mathcal F\\ \simeq\big\downarrow & \text{(A)} & \simeq\big\downarrow\gamma\\ \mathbf{R}\varGamma'_{\mathcal U}\mathbf{R}g_*v_2^*v_1^*f_{\mathrm t}^{\times}\mathcal F & \xleftarrow[\theta'_{g_1,u_2}(v_1^*f_{\mathrm t}^{\times}\mathcal F)]{\sim} & \mathbf{R}\varGamma'_{\mathcal U}u_2^*\mathbf{R}g_{1*}v_1^*f_{\mathrm t}^{\times}\mathcal F\end{array}$$

Now it suffices to show that the diagram which is *adjoint* to the diagram in (a) without its southeast (bottom right) corner, commutes. That adjoint diagram is the outer border of the following one, where, to reduce clutter, we omit all occurrences of the symbols $\mathbf{R}$ and $\mathcal F$, write $f^{\times}$ for $f_{\mathrm t}^{\times}$, etc., and leave some obvious maps unlabeled:

$$\begin{array}{ccccccc}g_*\varGamma'_{\mathcal V}(v_1v_2)^*f^{\times} & \xrightarrow{\psi} & \varGamma'_{\mathcal U}g_*(v_1v_2)^*f^{\times} & \xrightarrow{\theta'^{-1}} & \varGamma'_{\mathcal U}u_2^*u_1^*f_*f^{\times} & \longrightarrow & \varGamma'_{\mathcal U}u_2^*u_1^*\\ \big\uparrow & & \big\uparrow & \text{(A)} & \big\uparrow\gamma^{-1} & & \Big\|\\ g_*\varGamma'_{\mathcal V}v_2^*v_1^*f^{\times} & \text{(B)} & \varGamma'_{\mathcal U}g_*v_2^*v_1^*f^{\times} & \xleftarrow[\theta'_{g_1,u_2}(v_1^*f^{\times})]{} & \varGamma'_{\mathcal U}u_2^*g_{1*}v_1^*f^{\times} & & \Big\|\\ \simeq\big\uparrow 5.2.8(c) & & \big\uparrow & & \big\uparrow & \text{(C)} & \Big\|\\ g_*\varGamma'_{\mathcal V}v_2^*\varGamma'_{\mathcal V_1}v_1^*f^{\times} & \xrightarrow[\psi]{} & \varGamma'_{\mathcal U}g_*v_2^*\varGamma'_{\mathcal V_1}v_1^*f^{\times} & \xrightarrow[\theta_2''^{-1}]{} & \varGamma'_{\mathcal U}u_2^*g_{1*}\varGamma'_{\mathcal V_1}v_1^*f^{\times} & & \Big\|\\ \beta_1\big\downarrow & & \beta_1\big\downarrow & & \big\downarrow\beta_1 & & \Big\|\\ g_*\varGamma'_{\mathcal V}v_2^*g_1^{\times}u_1^* & \xrightarrow[\psi]{} & \varGamma'_{\mathcal U}g_*v_2^*g_1^{\times}u_1^* & \xrightarrow[\theta_2'^{-1}]{} & \varGamma'_{\mathcal U}u_2^*g_{1*}g_1^{\times}u_1^* & \longrightarrow & \varGamma'_{\mathcal U}u_2^*u_1^*\end{array}$$

It suffices then that each one of the subrectangles commute.

For the three unlabeled subrectangles commutativity is clear.

As before, commutativity of subrectangle (A) follows from that of the first diagram in Lemma 7.5.1.

Commutativity of (B) is easily checked after composition with the natural map $\varGamma'_{\mathcal{U}}g_*(v_1v_2)^*f^\times \to g_*(v_1v_2)^*f^\times$. (See the characterization of ψ in Proposition 7.2(a).)

Commutativity of (C) results from that of the following diagram:

$$
\begin{array}{ccccccc}
g_{1*}v_1^*f^\times & = & g_{1*}v_1^*f^\times & \xleftarrow{\theta_{f,u_1}} & u_1^*f_*f^\times & \longrightarrow & u_1^* \\
\uparrow & \text{(D)} & \uparrow & & \uparrow & & \uparrow \\
g_{1*}\varGamma'_{\mathcal{V}_1}v_1^*f^\times & \xrightarrow[\psi]{} & \varGamma'_{\mathcal{U}_1}g_{1*}v_1^*f^\times & \xrightarrow[\theta_1'^{-1}]{} & \varGamma'_{\mathcal{U}_1}u_1^*f_*f^\times & \longrightarrow & \varGamma'_{\mathcal{U}_1}u_1^* \\
{\scriptstyle\beta_1}\downarrow & & & \text{(E)} & & & \downarrow \\
g_{1*}g_1^\times u_1^* & & & \xrightarrow{\qquad\qquad\qquad} & & & u_1^*
\end{array}
$$

Here subrectangle (D) commutes by the characterization of ψ in Proposition 7.2(a); and (E) commutes by the very definition of the base-change map β_1.

(b) As in (a), we consider the *adjoint* diagram, essentially the outer border of the following diagram (7.5.2.1).

(Note: The map $\psi\colon g_{1*}\varGamma'_{\mathcal{W}}w^*f_{2*}f_2^\times f_1^\times \to \varGamma'_{\mathcal{U}}g_{1*}w^*f_{2*}f_2^\times f_1^\times$ in the middle of diagram 7.5.2.1 is defined because $f_{2*}f_2^\times f_1^\times := \mathbf{R}f_{2*}(f_2)_{\mathrm{t}}^\times(f_1)_{\mathrm{t}}^\times\mathcal{F} \in \mathbf{D}_{\mathrm{qc}}(\mathcal{Z})$, by Proposition 5.2.6.)

For diagram 7.5.2.1, commutativity of subrectangle (B) (resp. (D)) is given by the definition of β_2 (resp. β_1.) Commutativity of (C) follows from that of the second diagram in Lemma 7.5.1. Commutativity of (A) is left as an exercise. (It is helpful to compose with the natural map $\varGamma'_{\mathcal{U}}g_{1*}g_{2*}v^*f_2^\times f_1^\times \to g_{1*}g_{2*}v^*f_2^\times f_1^\times$ and to use the characterization of ψ in Proposition 7.2(a).) The rest is straightforward. □

$$
\begin{array}{ccccccc}
g_*\varGamma'_{\mathcal{V}}v^*f_2^\times f_1^\times & \xrightarrow{\ \sim\ } & g_{1*}g_{2*}\varGamma'_{\mathcal{V}}v^*f_2^\times f_1^\times & & \xrightarrow{\text{via }\beta_2} & & g_{1*}g_{2*}g_2^\times\varGamma'_{\mathcal{W}}w^*f_1^\times \\
\simeq\downarrow & & \downarrow{\scriptstyle\psi} & & & & \downarrow \\
g_*\varGamma'_{\mathcal{V}}v^*f^\times & & g_{1*}\varGamma'_{\mathcal{W}}g_{2*}v^*f_2^\times f_1^\times & & \text{(B)} & & \\
\psi\downarrow & \text{(A)} & \simeq\downarrow{\scriptstyle g_{1*}(\theta_2'^{-1})} & & & & \downarrow \\
\varGamma'_{\mathcal{U}}g_*v^*f^\times & & g_{1*}\varGamma'_{\mathcal{W}}w^*f_{2*}f_2^\times f_1^\times & & \longrightarrow & & g_{1*}\varGamma'_{\mathcal{W}}w^*f_1^\times \\
\simeq\downarrow & & \downarrow{\scriptstyle\psi} & & & & \| \\
\varGamma'_{\mathcal{U}}g_{1*}g_{2*}v^*f_2^\times f_1^\times & \xleftarrow{\text{via }\theta_{f_2,w}(f^\times)} & \varGamma'_{\mathcal{U}}g_{1*}w^*f_{2*}f_2^\times f_1^\times & \longrightarrow & \varGamma'_{\mathcal{U}}g_{1*}w^*f_1^\times & \xleftarrow{\psi} & g_{1*}\varGamma'_{\mathcal{W}}w^*f_1^\times \\
\downarrow & & \downarrow{\scriptstyle\theta'_{f_1,u}(f_{2*}f^\times)^{-1}} & & \downarrow{\scriptstyle\theta_1'^{-1}} & & \downarrow \\
\simeq & \text{(C)} & \varGamma'_{\mathcal{U}}u^*f_{1*}f_{2*}f_2^\times f_1^\times & \longrightarrow & \varGamma'_{\mathcal{U}}u^*f_{1*}f_1^\times & \text{(D)} & {\scriptstyle g_{1*}(\beta_1)} \\
\downarrow & & \downarrow\simeq & & \downarrow & & \downarrow \\
\varGamma'_{\mathcal{U}}g_*v^*f^\times & \xrightarrow[\theta'^{-1}]{} & \varGamma'_{\mathcal{U}}u^*f_*f^\times & \longrightarrow & \varGamma'_{\mathcal{U}}u^* & \longleftarrow & g_{1*}g_1^\times\varGamma'_{\mathcal{U}}u^*
\end{array}
$$

(7.5.2.1)

7.6. This subsection, proving Lemma 7.6.1, is independent of the preceding one.

LEMMA 7.6.1. *Theorem 7.4 holds when f is a closed immersion.*

PROOF. The natural isomorphisms $\mathbf{R}\varGamma'_{\mathscr{V}}v^*f^!\mathbf{R}\varGamma'_{\mathscr{Y}}\mathcal{F} \xrightarrow{\sim} \mathbf{R}\varGamma'_{\mathscr{V}}v^*f^!\mathcal{F}$ and

$$g^!u^*\mathbf{R}\varGamma'_{\mathscr{Y}}\mathcal{F} \xrightarrow{\sim} g^!\mathbf{R}\varGamma'_{\mathscr{U}}u^*\mathbf{R}\varGamma'_{\mathscr{Y}}\mathcal{F} \underset{5.2.8(\mathrm{c})}{\xrightarrow{\sim}} g^!\mathbf{R}\varGamma'_{\mathscr{U}}u^*\mathcal{F} \xrightarrow{\sim} g^!u^*\mathcal{F}$$

(see Corollary 6.1.5(b)) let us replace $\mathcal{F}$ by $\mathbf{R}\varGamma'_{\mathscr{Y}}\mathcal{F}$, i.e., we may assume $\mathcal{F} \in \mathbf{D}^+_{\mathrm{qct}}(\mathscr{Y})$.

Recall from Example 6.1.3(4) that $\mathbf{R}f_* = f_*\colon \mathbf{D}(\mathscr{X}) \to \mathbf{D}(\mathscr{Y})$ has a right adjoint $f^\natural$ such that $f^\natural(\mathbf{D}^+_{\mathrm{qct}}(\mathscr{Y})) \subset \mathbf{D}^+_{\mathrm{qct}}(\mathscr{X})$; and that there is a natural isomorphism

$$j^{\mathscr{X}}_{\mathcal{G}}\colon \mathbf{R}\varGamma'_{\mathscr{X}}f^\natural\mathcal{G} \xrightarrow{\sim} \mathbf{1}^!f^\natural\mathcal{G} \cong f^!\mathcal{G} \qquad (\mathcal{G} \in \mathbf{D}^+_{\mathrm{qc}}(\mathscr{Y})).$$

The canonical map $f_*f^! \to \mathbf{1}$ is the natural composition

$$f_*f^! \underset{(j^{\mathscr{X}})^{-1}}{\xrightarrow{\sim}} f_*\mathbf{R}\varGamma'_{\mathscr{X}}f^\natural \to f_*f^\natural \to \mathbf{1}.$$

Similar remarks hold for g—also a closed immersion [**GD**, p. 442, (10.14.5)(ii)].

As in the proof of Lemma 7.2.2, the map $\theta_{\mathcal{E}}\colon u^*f_*\mathcal{E} \xrightarrow{\sim} g_*v^*\mathcal{E}$ of Proposition 7.2 is an isomorphism for all $\mathcal{E} \in \mathbf{D}_{\mathrm{qct}}(\mathscr{X})$. (Recall Lemma 3.1.5.) This being so, the base-change map $\beta_{\mathcal{F}}$ is easily seen to factor naturally as

$$\mathbf{R}\varGamma'_{\mathscr{V}}v^*f^!\mathcal{F} \to g^!g_*\mathbf{R}\varGamma'_{\mathscr{V}}v^*f^!\mathcal{F} \to g^!g_*v^*f^!\mathcal{F} \underset{\theta^{-1}}{\xrightarrow{\sim}} g^!u^*f_*f^!\mathcal{F} \to g^!u^*\mathcal{F}.$$

Also, we can define the functorial map $\beta^\natural_{\mathcal{C}}$ to be the natural composition

$$v^*f^\natural\mathcal{C} \to g^\natural g_*v^*f^\natural\mathcal{C} \underset{\theta^{-1}}{\xrightarrow{\sim}} g^\natural u^*f_*f^\natural\mathcal{C} \to g^\natural u^*\mathcal{C} \qquad (\mathcal{C} \in \mathbf{D}_{\mathrm{qc}}(\mathscr{Y})).$$

The maps $\beta^\natural_{\mathcal{F}}$ and $\beta_{\mathcal{F}}$ are related by commutativity of the following diagram, in which $\mathcal{J}$ is an ideal of definition of $\mathscr{Y}$ (so that $\mathcal{J}\mathcal{O}_{\mathscr{X}}$ is an ideal of definition of $\mathscr{X}$):

$$\begin{array}{ccccccc}
\mathbf{R}\varGamma'_{\mathscr{V}}v^*\mathbf{R}\varGamma'_{\mathscr{X}}f^\natural & \underset{5.2.8(\mathrm{b})}{\xrightarrow{\sim}} & \mathbf{R}\varGamma'_{\mathscr{V}}\mathbf{R}\varGamma_{\mathcal{J}\mathcal{O}_{\mathscr{V}}}v^*f^\natural & \xrightarrow{\sim} & \mathbf{R}\varGamma'_{\mathscr{V}}v^*f^\natural & \xrightarrow{\mathbf{R}\varGamma'_{\mathscr{V}}(\beta^\natural)} & \mathbf{R}\varGamma'_{\mathscr{V}}g^\natural u^* \\
{\scriptstyle \mathbf{R}\varGamma'_{\mathscr{V}}v^*(j^{\mathscr{X}})}\big\downarrow{\scriptstyle \simeq} & & & & & & {\scriptstyle \simeq}\big\downarrow{\scriptstyle j^{\mathscr{V}}} \\
\mathbf{R}\varGamma'_{\mathscr{V}}v^*f^! & & & \underset{\beta}{\xrightarrow{\qquad\qquad\qquad}} & & & g^!u^*
\end{array}$$

(For the unlabeled isomorphism, see the beginning of the proof of Proposition 7.2.) Since $\mathbf{R}\varGamma'_{\mathscr{V}}$ is right-adjoint to the inclusion $\mathbf{D}_{\mathrm{qct}}(\mathscr{V}) \hookrightarrow \mathbf{D}(\mathscr{V})$ (Proposition 5.2.1), we can verify this commutativity after composing with $g^!u^* \xrightarrow{\sim} \mathbf{R}\varGamma'_{\mathscr{V}}g^\natural u^* \to g^\natural u^*$, at which point the verification is straightforward.

Thus to prove Lemma 7.6.1 we need only show that $\beta^\natural_{\mathcal{F}}$ is an isomorphism, i.e. (since g is a closed immersion), that $g_*(\beta^\natural_{\mathcal{F}})$ is an isomorphism.

For that purpose, consider the unique functorial map

$$\sigma = \sigma(\mathcal{E},\mathcal{G})\colon u^*\mathbf{R}\mathcal{H}om^\bullet_{\mathscr{Y}}(\mathcal{E},\mathcal{G}) \to \mathbf{R}\mathcal{H}om^\bullet_{\mathscr{U}}(u^*\mathcal{E},u^*\mathcal{G}) \qquad (\mathcal{E} \in \mathbf{D}^-_{\mathrm{c}}(\mathscr{Y}),\ \mathcal{G} \in \mathbf{D}^+(\mathscr{Y}))$$

which for bounded-below injective complexes $\mathcal{G}$ is the natural composition

$$u^*\mathbf{R}\mathcal{H}om^\bullet_{\mathscr{Y}}(\mathcal{E},\mathcal{G}) \cong u^*\mathcal{H}om^\bullet_{\mathscr{Y}}(\mathcal{E},\mathcal{G}) \to \mathcal{H}om^\bullet_{\mathscr{U}}(u^*\mathcal{E},u^*\mathcal{G}) \to \mathbf{R}\mathcal{H}om^\bullet_{\mathscr{U}}(u^*\mathcal{E},u^*\mathcal{G}).$$

This map is an *isomorphism.* Indeed, it commutes with localization, so we need only check for affine $\mathscr{Y}$, and then, since every coherent $\mathcal{O}_{\mathscr{Y}}$-module is a homomorphic image of a finite-rank free one ([**GD**, p. 427, (10.10.2)]), a standard way-out argument reduces the problem to the trivial case $\mathcal{E} = \mathcal{O}_{\mathscr{Y}}$.

Take $\mathcal{E} := f_*\mathcal{O}_{\mathfrak{X}} =$ (say) $\mathcal{O}_{\mathfrak{Y}}/\mathcal{J}$. The source and target of $\sigma(\mathcal{O}_{\mathfrak{Y}}/\mathcal{J}, \mathcal{F})$ are

$$u^*\mathbf{R}\mathcal{H}om^\bullet(\mathcal{O}_{\mathfrak{Y}}/\mathcal{J}, \mathcal{F}) = u^*f_*f^\natural\mathcal{F} \cong g_*v^*f^\natural\mathcal{F},$$
$$\mathbf{R}\mathcal{H}om^\bullet(u^*(\mathcal{O}_{\mathfrak{Y}}/\mathcal{J}), u^*\mathcal{F}) = g_*g^\natural u^*\mathcal{F}.$$

Let $\mathcal{K}$ be a K-injective $\mathcal{O}_{\mathfrak{U}}$-complex quasi-isomorphic to $u^*\mathcal{F}$. Since the complexes $u^*\mathcal{H}om^\bullet_{\mathfrak{Y}}(\mathcal{O}_{\mathfrak{Y}}/\mathcal{J}, \mathcal{F})$ and $\mathcal{H}om^\bullet_{\mathfrak{U}}(u^*\mathcal{O}_{\mathfrak{Y}}/\mathcal{J}, \mathcal{K}) \cong \mathbf{R}\mathcal{H}om^\bullet_{\mathfrak{U}}(u^*\mathcal{O}_{\mathfrak{Y}}/\mathcal{J}, u^*\mathcal{F})$ are both annihilated by $\mathcal{J}\mathcal{O}_{\mathfrak{U}}$, we see that the isomorphism $\sigma(\mathcal{O}_{\mathfrak{Y}}/\mathcal{J}, \mathcal{F})$ is isomorphic to a map of the form $g_*(\varsigma)$ where $\varsigma\colon v^*f^\natural\mathcal{F} \to g^\natural u^*\mathcal{F}$ is a map in $\mathbf{D}(\mathfrak{V})$. It suffices then to show that $\varsigma = \beta^\natural_{\mathcal{F}}$, i.e. (by definition of $\beta^\natural_{\mathcal{F}}$), that the natural composition

$$u^*f_*f^\natural\mathcal{F} \xrightarrow{\ \sim\ } g_*v^*f^\natural\mathcal{F} \xrightarrow{g_*(\varsigma)} g_*g^\natural u^*\mathcal{F} \xrightarrow{\tau^\natural_{u^*\mathcal{F}}} u^*\mathcal{F}$$

is induced by the natural map

$$\tau^\natural_{\mathcal{F}}\colon f_*f^\natural\mathcal{F} = \mathbf{R}\mathcal{H}om^\bullet(\mathcal{O}_{\mathfrak{Y}}/\mathcal{J}, \mathcal{F}) \to \mathbf{R}\mathcal{H}om^\bullet(\mathcal{O}_{\mathfrak{Y}}, \mathcal{F}) = \mathcal{F}.$$

From Example 6.1.3(4) one sees, for injective $\mathcal{F}$, that $\tau^\natural_{\mathcal{F}}$ takes any homomorphism $\varphi\colon \mathcal{O}_{\mathfrak{Y}}/\mathcal{J} \to \mathcal{F}$ over an open subset of $\mathfrak{Y}$ to $\varphi(1)$; and similarly for $\tau^\natural_{u^*\mathcal{F}}$. The conclusion follows from the above definition of $\sigma(\mathcal{O}_{\mathfrak{Y}}/\mathcal{J}, \mathcal{F}) = g_*(\varsigma)$. □

7.7. In this subsection we prove Theorem 7.4 in case $f\colon \mathfrak{X} \to \mathfrak{Y}$ is a proper map of ordinary noetherian schemes, by reduction to the case where $\mathfrak{X}$, $\mathfrak{Y}$, $\mathfrak{U}$ and $\mathfrak{V}$ are *all* ordinary schemes—a case which we take for granted (see the introductory remarks for section 7). Of course when u is *adic* then $\mathfrak{U}$ is already an ordinary scheme, and no reduction is needed at all.

LEMMA 7.7.1. *Let $f\colon \mathfrak{X} \to \mathfrak{Y}$ be a proper map of ordinary noetherian schemes. For Theorem 7.4 to hold with this f it suffices that it hold whenever $\mathfrak{U}$ and $\mathfrak{V}$ are ordinary schemes as well.*

PROOF. Without yet assuming that $\mathfrak{X}$ and $\mathfrak{Y}$ are ordinary schemes, we can reduce Theorem 7.4 to the special case where the formal scheme $\mathfrak{U}$ is *affine* and $u(\mathfrak{U})$ is contained in an affine open subset of $\mathfrak{Y}$. Indeed, for the base-change map $\beta_{\mathcal{F}} = \beta_{f,u}(\mathcal{F})$ of Theorem 7.4 to be an isomorphism, it clearly suffices that for any composition of fiber squares

$$\begin{array}{ccccc}
\mathfrak{V}_0 & \xrightarrow{v_0} & \mathfrak{V} & \xrightarrow{v} & \mathfrak{X} \\
{\scriptstyle g_0}\big\downarrow & & {\scriptstyle g}\big\downarrow & & \big\downarrow{\scriptstyle f} \\
\mathfrak{U}_0 & \xrightarrow[u_0]{} & \mathfrak{U} & \xrightarrow[u]{} & \mathfrak{Y}
\end{array}$$

with u_0 the inclusion of an affine open $\mathfrak{U}_0 \subset \mathfrak{U}$ such that $u(\mathfrak{U}_0)$ is contained in an affine open subset of $\mathfrak{Y}$, the map

$$v_0^*(\beta_{\mathcal{F}})\colon v_0^*\mathbf{R}\varGamma'_{\mathfrak{V}}v^*f^!\mathcal{F} \to v_0^*g^!u^*\mathcal{F}$$

be an isomorphism. Remark 5.2.10(6) yields that $\mathcal{F} \in \widetilde{\mathbf{D}}^+_{\mathrm{qc}}(\mathfrak{Y}) \Rightarrow u^*\mathcal{F} \in \widetilde{\mathbf{D}}^+_{\mathrm{qc}}(\mathfrak{U})$. So if we assume the above-specified special case, then $\beta_{f,uu_0}(\mathcal{F})$ and $\beta_{g,u_0}(u^*\mathcal{F})$ are both isomorphisms. From Proposition 5.2.1(a) we have a natural isomorphism

$$v_0^*(\beta_{\mathcal{F}}) \cong \mathbf{R}\varGamma'_{\mathfrak{V}_0}v_0^*(\beta_{f,u}(\mathcal{F})),$$

so Lemma 7.5.2(a) shows that $v_0^*(\beta_{\mathcal{F}})$ is in fact an isomorphism.

With reference to the remarks just preceding Section 7.5, (a) and (b) having already been proved, only (c) remains, i.e., we need only prove Theorem 7.4 for the rear face of diagram (7.2.3).

In other words, with the notation of diagram (7.2.3), we may assume in proving Theorem 7.4 that $f = f_n$ (a proper map of ordinary schemes), and that $u = u_n$. Moreover Y_n is a closed subscheme of $\mathcal{Y}$, and so if $\mathcal{U}$ is affine and $u(\mathcal{U})$ is contained in an affine open subset of $\mathcal{Y}$, then $Y_n \times_{\mathcal{Y}} \mathcal{U}$ is affine and $u_n(Y_n \times_{\mathcal{Y}} \mathcal{U})$ is contained in an affine open subset of Y_n. It follows that $Y_n \times_{\mathcal{Y}} \mathcal{U}$ is the completion of an ordinary affine Y_n-scheme. (That can be seen via the one-one correspondence from maps between affine formal schemes to continuous homomorphisms between their associated rings [**GD**, p. 407, (10.4.6)]). Theorem 7.4 is thus reduced to the case depicted in the following diagram, where $f\colon \mathcal{X} \to \mathcal{Y}$ is now a proper map of ordinary noetherian schemes, U is an ordinary affine $\mathcal{Y}$-scheme, $\kappa\colon \mathcal{U} \to U$ is a completion map, and $u\colon \mathcal{U} \to \mathcal{Y}$ factors as shown.

$$
\begin{array}{ccccc}
\mathcal{X} \times_{\mathcal{Y}} \mathcal{U} & \longrightarrow & \mathcal{X} \times_{\mathcal{Y}} U & \longrightarrow & \mathcal{X} \\
{\scriptstyle g}\big\downarrow & (1) & \big\downarrow & (2) & \big\downarrow{\scriptstyle f} \\
\mathcal{U} & \underset{\kappa}{\longrightarrow} & U & \longrightarrow & \mathcal{Y}
\end{array}
$$

We will show that Theorem 7.4 holds for subdiagram (1) by identifying the base-change map associated to κ with the *isomorphism* ζ in Corollary 6.1.8. As subdiagram (2) is a fiber square of ordinary schemes, Lemma 7.7.1 will then result from the preceding reduction and the transitivity Lemma 7.5.2(a).

It is convenient to re-represent subdiagram (1) in the notation of Corollary 6.1.8. Consider then a diagram

$$
\begin{array}{ccc}
\mathcal{X} := X_{/Z} & \xrightarrow{\ \kappa_{\mathcal{X}}\ } & X \\
{\scriptstyle f}\big\downarrow & & \big\downarrow{\scriptstyle f_0} \\
\mathcal{Y} := Y_{/W} & \underset{\kappa_{\mathcal{Y}}}{\longrightarrow} & Y
\end{array}
$$

as in Corollary 6.1.6, with $Z = f_0^{-1}W$. That ζ *is* the base-change map means that ζ is adjoint to the natural composition

$$
\mathbf{R}f_*\mathbf{R}\varGamma'_{\mathcal{X}}\kappa_{\mathcal{X}}^*f_0^{\times} \underset{\psi}{\xrightarrow{\ \sim\ }} \mathbf{R}\varGamma'_{\mathcal{Y}}\mathbf{R}f_*\kappa_{\mathcal{X}}^*f_0^{\times} \underset{\theta'^{-1}}{\xrightarrow{\ \sim\ }} \mathbf{R}\varGamma'_{\mathcal{Y}}\kappa_{\mathcal{Y}}^*\mathbf{R}f_{0*}f_0^{\times} \longrightarrow \mathbf{R}\varGamma'_{\mathcal{Y}}\kappa_{\mathcal{Y}}^* \longrightarrow \kappa_{\mathcal{Y}}^*.
$$

But by definition, ζ is adjoint to the natural composition

$$
\mathbf{R}f_*\mathbf{R}\varGamma'_{\mathcal{X}}\kappa_{\mathcal{X}}^*f_0^{\times} \underset{5.2.4(\mathrm{c})}{\xrightarrow{\ \sim\ }} \mathbf{R}f_*\kappa_{\mathcal{X}}^*\mathbf{R}\varGamma_Z f_0^{\times} \longrightarrow \mathbf{R}f_*\kappa_{\mathcal{X}}^*\mathbf{R}\varGamma_Z f_0^{\times}\kappa_{\mathcal{Y}*}\kappa_{\mathcal{Y}}^* \underset{\tau'_{\mathrm{t}}(\kappa_{\mathcal{Y}}^*)}{\longrightarrow} \kappa_{\mathcal{Y}}^*
$$

with τ'_{t} as in Corollary 6.1.6—so that $\tau'_{\mathrm{t}}(\kappa_{\mathcal{Y}}^*)$ factors naturally as

$$
\begin{aligned}
\mathbf{R}f_*\kappa_{\mathcal{X}}^*\mathbf{R}\varGamma_Z f_0^{\times}\kappa_{\mathcal{Y}*}\kappa_{\mathcal{Y}}^* &\underset{5.2.7}{\xrightarrow{\ \sim\ }} \kappa_{\mathcal{Y}}^*\mathbf{R}f_{0*}\mathbf{R}\varGamma_Z f_0^{\times}\kappa_{\mathcal{Y}*}\kappa_{\mathcal{Y}}^* \\
&\longrightarrow \kappa_{\mathcal{Y}}^*\mathbf{R}f_{0*}f_0^{\times}\kappa_{\mathcal{Y}*}\kappa_{\mathcal{Y}}^* \\
&\longrightarrow \kappa_{\mathcal{Y}}^*\kappa_{\mathcal{Y}*}\kappa_{\mathcal{Y}}^* \\
&\xrightarrow{\ \pi\ } \kappa_{\mathcal{Y}}^*.
\end{aligned}
$$

It will suffice then to verify that the following natural diagram commutes (where, again, we omit all occurrences of $\mathbf{R}$):

$$\begin{array}{ccccccc}
f_*\Gamma'_{\mathfrak X}\kappa^*_{\mathfrak X} f_0^\times & \xrightarrow{\ \psi\ } & \Gamma'_{\mathcal Y} f_*\kappa^*_{\mathfrak X} f_0^\times & \xrightarrow{\ \theta'^{-1}\ } & \Gamma'_{\mathcal Y}\kappa^*_{\mathcal Y} f_{0*} f_0^\times & \longrightarrow & \Gamma'_{\mathcal Y}\kappa^*_{\mathcal Y} \\
{\scriptstyle 5.2.4(c)}\downarrow & & \text{(A)} & & \downarrow & & \downarrow \\
f_*\kappa^*_{\mathfrak X}\Gamma_Z f_0^\times & \underset{5.2.7}{\xrightarrow{\ \sim\ }} & \kappa^*_{\mathcal Y} f_{0*}\Gamma_Z f_0^\times & \longrightarrow & \kappa^*_{\mathcal Y} f_{0*} f_0^\times & \longrightarrow & \kappa^*_{\mathcal Y} \\
\downarrow & & \downarrow & & \downarrow & & {\scriptstyle \pi}\uparrow\downarrow{\scriptstyle \iota} \\
f_*\kappa^*_{\mathfrak X}\Gamma_Z f_0^\times \kappa_{\mathcal Y*}\kappa^*_{\mathcal Y} & \underset{5.2.7}{\xrightarrow{\ \sim\ }} & \kappa^*_{\mathcal Y} f_{0*}\Gamma_Z f_0^\times \kappa_{\mathcal Y*}\kappa^*_{\mathcal Y} & \longrightarrow & \kappa^*_{\mathcal Y} f_{0*} f_0^\times \kappa_{\mathcal Y*}\kappa^*_{\mathcal Y} & \longrightarrow & \kappa^*_{\mathcal Y}\kappa_{\mathcal Y*}\kappa^*_{\mathcal Y}
\end{array}$$

Given that $\pi\iota = 1$, the verification of commutativity is straightforward, except for subrectangle (A).

Now there is a functorial isomorphism $\alpha\colon \mathbf{R}f_{0*}\mathbf{R}\Gamma_Z \xrightarrow{\ \sim\ } \mathbf{R}\Gamma_W\mathbf{R}f_{0*}$ which arises in the obvious way, via "K-flabby" resolutions, from the equality $f_{0*}\Gamma_Z = \Gamma_W f_{0*}$ (see the last paragraph in the Remark following (3.2.5) in [**AJL**, p. 25]), and whose composition with the natural map $\mathbf{R}\Gamma_W\mathbf{R}f_{0*} \to \mathbf{R}f_{0*}$ is the natural map $\mathbf{R}f_{0*}\mathbf{R}\Gamma_Z \to \mathbf{R}f_{0*}$. And, again, we have the isomorphism $\mathbf{R}\Gamma'_{\mathcal Y}\kappa^*_{\mathcal Y} \xrightarrow{\ \sim\ } \kappa^*_{\mathcal Y}\mathbf{R}\Gamma_W$ of Proposition 5.2.4(c), whose composition with the natural map $\kappa^*_{\mathcal Y}\mathbf{R}\Gamma_W \to \kappa^*_{\mathcal Y}$ is the natural map $\mathbf{R}\Gamma'_{\mathcal Y}\kappa^*_{\mathcal Y} \to \kappa^*_{\mathcal Y}$. Hence commutativity of (A) follows from that of the outer border—consisting entirely of isomorphisms—of the following diagram:

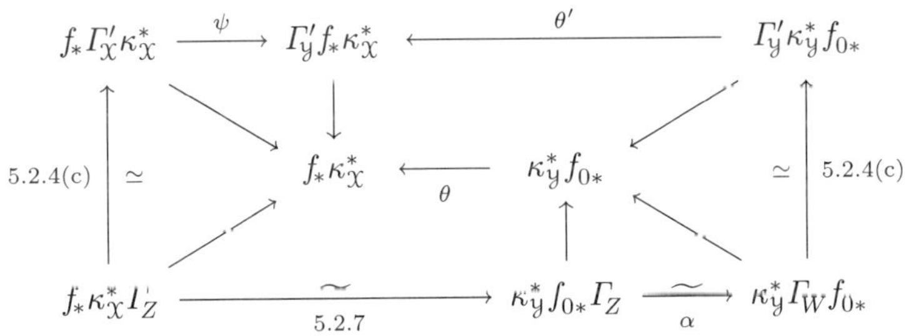

Since $\mathbf{R}\Gamma'_{\mathcal Y}$ is right-adjoint to the inclusion $\mathbf{D}_{\mathrm t}(\mathcal Y) \hookrightarrow \mathbf{D}(\mathcal Y)$ (Proposition 5.2.1), we can check commutativity *after* composing the outer border with the natural map $\mathbf{R}\Gamma'_{\mathcal Y} f_*\kappa^*_{\mathfrak X} \to f_*\kappa^*_{\mathfrak X}$, so that it suffices to check commutativity of all the subdiagrams of the preceding one. This is easy to do, as, with $\mathcal E := f_0^\times\mathcal F$, the maps denoted by $\theta_{\mathcal E}$ $(= \theta_{f_0,\kappa_{\mathcal Y}}(\mathcal E))$ in Corollary 5.2.7 and in Proposition 7.2 are the same.

This completes the proof of Lemma 7.7.1, and of Theorem 7.4. □

8. Consequences of the flat base change isomorphism.

We begin with a flat-base-change theorem for the functor $f^\# = \boldsymbol{\Lambda}_{\mathfrak X} f^!$ associated to a pseudo-proper map $f\colon \mathfrak X \to Y$ of noetherian formal schemes. (As before, $f^! := f_{\mathrm t}^\times$, and $f^\#$ is right-adjoint to the functor $\mathbf{R}f_*\mathbf{R}\Gamma'_{\mathfrak X}\colon \widetilde{\mathbf{D}}_{\mathrm{qc}}(\mathfrak X) \to \mathbf{D}(\mathcal Y)$, where $\widetilde{\mathbf{D}}_{\mathrm{qc}}(\mathfrak X)$ is the (full) Δ-subcategory of $\mathbf{D}(\mathfrak X)$ such that

$$\mathcal F \in \widetilde{\mathbf{D}}_{\mathrm{qc}}(\mathfrak X) \Leftrightarrow \mathbf{R}\Gamma'_{\mathfrak X}\mathcal F \in \mathbf{D}_{\mathrm{qct}}(\mathfrak X),$$

see Corollary 6.1.4.)

We deduce a sheafified version Theorem 8.2 of Theorem 2 of the Introduction (= Theorem 6.1 + Corollary 6.1.4). This is readily seen equivalent to the case of flat base change where $u\colon \mathcal{U} \to \mathcal{Y}$ is an open immersion; in other words, it expresses the local nature, over $\mathcal{Y}$, of $f^!$ and $f^{\#}$.

Section 8.3 establishes the local nature of $f^!$ and $f^{\#}$ over $\mathcal{X}$. From this we obtain that $f^{\#}(\mathbf{D}_{\mathrm{c}}^{+}(\mathcal{Y})) \subset \mathbf{D}_{\mathrm{c}}^{+}(\mathcal{X})$ (Proposition 8.3.2). This leads further to an improved base-change theorem for bounded-below complexes with coherent homology, and to Theorem 8.4, a duality theorem for such complexes under proper maps.

We consider as in Theorem 7.4 a fiber square of noetherian formal schemes

$$\begin{array}{ccc} \mathcal{V} & \xrightarrow{\ \ v\ \ } & \mathcal{X} \\ {\scriptstyle g}\big\downarrow & & \big\downarrow{\scriptstyle f} \\ \mathcal{U} & \xrightarrow[\ \ u\ \ ]{} & \mathcal{Y} \end{array}$$

with f and g pseudo-proper, u and v flat.

For any $\mathcal{F} \in \widetilde{\mathbf{D}}_{\mathrm{qc}}^{+}(\mathcal{Y})$ we have the composed isomorphism

$$\vartheta\colon \mathbf{R}\varGamma'_{\mathcal{V}}v^*f^{\#}\mathcal{F} \underset{5.2.8(\mathrm{c})}{\xrightarrow{\ \sim\ }} \mathbf{R}\varGamma'_{\mathcal{V}}v^*\mathbf{R}\varGamma'_{\mathcal{X}}f^{\#}\mathcal{F} \underset{6.1.5(\mathrm{a})}{\xrightarrow{\ \sim\ }} \mathbf{R}\varGamma'_{\mathcal{V}}v^*f^{!}\mathcal{F} \underset{7.4}{\xrightarrow{\ \sim\ }} g^!u^*\mathcal{F}.$$

In particular, $v^*f^{\#}\mathcal{F} \in \widetilde{\mathbf{D}}_{\mathrm{qc}}(\mathcal{V})$.

THEOREM 8.1. *Under the preceding conditions, let*

$$\beta^{\#}_{\mathcal{F}}\colon v^*f^{\#}\mathcal{F} \to g^{\#}u^*\mathcal{F} \qquad (\mathcal{F} \in \widetilde{\mathbf{D}}_{\mathrm{qc}}^{+}(\mathcal{Y}))$$

be the map adjoint to the natural composition

$$\mathbf{R}g_*\mathbf{R}\varGamma'_{\mathcal{V}}v^*f^{\#}\mathcal{F} \underset{\mathbf{R}g_*\vartheta}{\xrightarrow{\ \sim\ }} \mathbf{R}g_*g^!u^*\mathcal{F} \to u^*\mathcal{F}. \tag{8.1.1}$$

Then the map $\mathbf{\Lambda}_{\mathcal{V}}(\beta^{\#}_{\mathcal{F}})$ is an isomorphism

$$\mathbf{\Lambda}_{\mathcal{V}}(\beta^{\#}_{\mathcal{F}})\colon \mathbf{\Lambda}_{\mathcal{V}}v^*f^{\#}\mathcal{F} \xrightarrow{\ \sim\ } \mathbf{\Lambda}_{\mathcal{V}}g^{\#}u^*\mathcal{F} \underset{6.1.5(\mathrm{a})}{\cong} g^{\#}u^*\mathcal{F}.$$

Moreover, if u is an open immersion then $\beta^{\#}_{\mathcal{F}}$ itself is an isomorphism.

PROOF. The map $\beta^{\#}$ factors naturally as

$$v^*f^{\#} \to \mathbf{\Lambda}_{\mathcal{V}}v^*f^{\#} \underset{6.3.1(\mathrm{c})}{\xrightarrow{\ \sim\ }} \mathbf{\Lambda}_{\mathcal{V}}\mathbf{R}\varGamma'_{\mathcal{V}}v^*f^{\#} \underset{\mathbf{\Lambda}_{\mathcal{V}}\vartheta}{\xrightarrow{\ \sim\ }} \mathbf{\Lambda}_{\mathcal{V}}g^!u^* = g^{\#}u^*. \tag{8.1.2}$$

To see this, one needs to check that (8.1.2) is adjoint to (8.1.1). The natural map $\mathbf{1} \to \mathbf{\Lambda}_{\mathcal{V}}$ factors naturally as $\mathbf{1} \to \mathbf{\Lambda}_{\mathcal{V}}\mathbf{R}\varGamma'_{\mathcal{V}} \to \mathbf{\Lambda}_{\mathcal{V}}$ (easy check), and hence the adjointness in question amounts to the readily-verified commutativity of the outer border of the following diagram (with all occurrences of $\mathbf{R}$ left out):

$$\begin{array}{ccccc} g_*\varGamma'_{\mathcal{V}}\mathbf{\Lambda}_{\mathcal{V}}\varGamma'_{\mathcal{V}}v^*f^{\#} & \longleftarrow & g_*\varGamma'_{\mathcal{V}}v^*f^{\#} & \underset{\text{via } \vartheta}{\xrightarrow{\ \sim\ }} & g_*g^!u^* \\ {\scriptstyle \simeq}\big\downarrow & & \big\uparrow & & \big\uparrow \\ g_*\varGamma'_{\mathcal{V}}\mathbf{\Lambda}_{\mathcal{V}}v^*f^{\#} & \xrightarrow{\ \sim\ } & g_*\varGamma'_{\mathcal{V}}\mathbf{\Lambda}_{\mathcal{V}}\varGamma'_{\mathcal{V}}v^*f^{\#} & \underset{\text{via } \vartheta}{\xrightarrow{\ \sim\ }} & g_*\varGamma'_{\mathcal{V}}\mathbf{\Lambda}_{\mathcal{V}}g^!u^* \end{array}$$

That $\mathbf{\Lambda}_{\mathcal{V}}(\beta^{\#}_{\mathcal{F}})$ is an isomorphism results then from the idempotence of $\mathbf{\Lambda}_{\mathcal{V}}$ (Remark 6.3.1(b)).

When u—hence v—is an open immersion, we have isomorphisms (the first of which is obvious):

$$\mathbf{\Lambda}_{\mathcal{V}}v^*f^{\#} \xrightarrow{\ \sim\ } v^*\mathbf{\Lambda}_{\mathcal{X}}f^{\#} \underset{6.1.5(\mathrm{a})}{\xrightarrow{\ \sim\ }} v^*f^{\#},$$

and the last assertion follows. □

Next comes the sheafification of Theorem 2. Let $f\colon \mathcal{X}\to\mathcal{Y}$ be a map of locally noetherian formal schemes. For $\mathcal{G}$ and $\mathcal{E}\in\mathbf{D}(\mathcal{X})$ we have natural compositions

$$\mathbf{R}f_*\mathbf{R}\mathcal{H}om^\bullet_{\mathcal{X}}(\mathcal{G},\mathcal{E})\to\mathbf{R}f_*\mathbf{R}\mathcal{H}om^\bullet_{\mathcal{X}}(\mathbf{L}f^*\mathbf{R}f_*\mathcal{G},\mathcal{E})\xrightarrow[\text{[}\mathbf{Sp}\text{, p. 147, 6.7]}]{\sim}\mathbf{R}\mathcal{H}om^\bullet_{\mathcal{Y}}(\mathbf{R}f_*\mathcal{G},\mathbf{R}f_*\mathcal{E})$$

and

$$\mathbf{R}\mathcal{H}om^\bullet_{\mathcal{X}}(\mathcal{G},\mathcal{E})\to\mathbf{R}\mathcal{H}om^\bullet_{\mathcal{X}}(\mathbf{R}\varGamma'_{\mathcal{X}}\mathcal{G},\mathcal{E})\xrightarrow[5.2.3]{\sim}\mathbf{R}\mathcal{H}om^\bullet_{\mathcal{X}}(\mathbf{R}\varGamma'_{\mathcal{X}}\mathcal{G},\mathbf{R}\varGamma'_{\mathcal{X}}\mathcal{E}).$$

THEOREM 8.2. *Let $\mathcal{X}$ and $\mathcal{Y}$ be noetherian formal schemes and let $f\colon\mathcal{X}\to\mathcal{Y}$ be a pseudo-proper map. Then the following natural compositions are* isomorphisms:

$$\delta^\#\colon \mathbf{R}f_*\mathbf{R}\mathcal{H}om^\bullet_{\mathcal{X}}(\mathcal{G},f^\#\mathcal{F})\to\mathbf{R}\mathcal{H}om^\bullet_{\mathcal{Y}}(\mathbf{R}f_*\mathbf{R}\varGamma'_{\mathcal{X}}\mathcal{G},\mathbf{R}f_*\mathbf{R}\varGamma'_{\mathcal{X}}f^\#\mathcal{F})$$
$$\to\mathbf{R}\mathcal{H}om^\bullet_{\mathcal{Y}}(\mathbf{R}f_*\mathbf{R}\varGamma'_{\mathcal{X}}\mathcal{G},\mathcal{F})\qquad(\mathcal{G}\in\widetilde{\mathbf{D}}_{\mathrm{qc}}(\mathcal{X}),\ \mathcal{F}\in\widetilde{\mathbf{D}}^+_{\mathrm{qc}}(\mathcal{Y}));$$

$$\delta^!\colon \mathbf{R}f_*\mathbf{R}\mathcal{H}om^\bullet_{\mathcal{X}}(\mathcal{G},f^!\mathcal{F})\to\mathbf{R}\mathcal{H}om^\bullet_{\mathcal{Y}}(\mathbf{R}f_*\mathcal{G},\mathbf{R}f_*f^!\mathcal{F})\to\mathbf{R}\mathcal{H}om^\bullet_{\mathcal{Y}}(\mathbf{R}f_*\mathcal{G},\mathcal{F})$$
$$(\mathcal{G}\in\mathbf{D}_{\mathrm{qct}}(\mathcal{X}),\ \mathcal{F}\in\widetilde{\mathbf{D}}^+_{\mathrm{qc}}(\mathcal{Y})).$$

PROOF. The map $\delta^\#$ is an isomorphism iff the same is true of $\mathbf{R}\Gamma(\mathcal{U},\delta^\#)$ for all open $\mathcal{U}\subset\mathcal{Y}$. (For if $\mathcal{E}$—which may be assumed K-injective—is the vertex of a triangle based on $\delta^\#$, then $\delta^\#$ is an isomorphism $\Leftrightarrow\mathcal{E}\cong 0\Leftrightarrow H^i(\mathcal{E})=0$ for all $i\in\mathbb{Z}\Leftrightarrow$ the sheaf associated to the presheaf $\mathcal{U}\mapsto \mathrm{H}^i\Gamma(\mathcal{U},\mathcal{E})=\mathrm{H}^i\mathbf{R}\Gamma(\mathcal{U},\mathcal{E})$ vanishes for all i.) Set $\mathcal{V}:=f^{-1}\mathcal{U}$, and let $u\colon\mathcal{U}\hookrightarrow\mathcal{Y}$ and $v\colon\mathcal{V}\hookrightarrow\mathcal{X}$ be the respective inclusions. We have then the fiber square

$$\begin{array}{ccc}\mathcal{V} & \xrightarrow{v} & \mathcal{X}\\ {\scriptstyle g}\downarrow & & \downarrow{\scriptstyle f}\\ \mathcal{U} & \xrightarrow[u]{} & \mathcal{Y},\end{array}$$

and need only verify that $\mathbf{R}\Gamma(\mathcal{U},\delta^\#)$ is the composition of the following sequence of isomorphisms:

$$\begin{aligned}\mathbf{R}\Gamma(\mathcal{U},\mathbf{R}f_*\mathbf{R}\mathcal{H}om^\bullet_{\mathcal{X}}(\mathcal{G},f^\#\mathcal{F}))&\xrightarrow{\sim}\mathbf{R}\Gamma(\mathcal{V},\mathbf{R}\mathcal{H}om^\bullet_{\mathcal{X}}(\mathcal{G},f^\#\mathcal{F})) &&[\mathbf{Sp}, 6.4, 6.7, 5.15]\\ &\xrightarrow{\sim}\mathbf{R}\mathrm{Hom}^\bullet_{\mathcal{V}}(v^*\mathcal{G},v^*f^\#\mathcal{F}) &&[\mathbf{Sp}, 5.14, 5.12, 6.4]\\ &\xrightarrow{\sim}\mathbf{R}\mathrm{Hom}^\bullet_{\mathcal{V}}(v^*\mathcal{G},g^\#u^*\mathcal{F}) &&(\text{Theorem 8.1})\\ &\xrightarrow{\sim}\mathbf{R}\mathrm{Hom}^\bullet_{\mathcal{U}}(\mathbf{R}g_*\mathbf{R}\varGamma'_{\mathcal{V}}v^*\mathcal{G},u^*\mathcal{F}) &&(6.1.4, 5.2.10(6))\\ &\xrightarrow{\sim}\mathbf{R}\mathrm{Hom}^\bullet_{\mathcal{U}}(\mathbf{R}g_*v^*\mathbf{R}\varGamma'_{\mathcal{X}}\mathcal{G},u^*\mathcal{F}) &&(\text{elementary})\\ &\xrightarrow{\sim}\mathbf{R}\mathrm{Hom}^\bullet_{\mathcal{U}}(u^*\mathbf{R}f_*\mathbf{R}\varGamma'_{\mathcal{X}}\mathcal{G},u^*\mathcal{F}) &&(\text{elementary})\\ &\xrightarrow{\sim}\mathbf{R}\Gamma(\mathcal{U},\mathbf{R}\mathcal{H}om^\bullet_{\mathcal{Y}}(\mathbf{R}f_*\mathbf{R}\varGamma'_{\mathcal{X}}\mathcal{G},\mathcal{F})) &&(\text{as above}).\end{aligned}$$

This somewhat tedious verification is left to the reader (who may e.g., refer to the proof of $(4.3)^{\mathrm{o}}\Rightarrow(4.2)$ near the end of [**L5**]).

That $\delta^!$ is an isomorphism can be shown similarly—or be deduced via the natural map $f^!\cong\mathbf{R}\varGamma'_{\mathcal{X}}f^\#\to f^\#$ (Corollary 6.1.5), which for $\mathcal{G}\in\mathbf{D}_{\mathrm{qct}}(\mathcal{X})$ induces an isomorphism $\mathbf{R}\mathcal{H}om^\bullet_{\mathcal{X}}(\mathcal{G},f^!\mathcal{F})\xrightarrow{\sim}\mathbf{R}\mathcal{H}om^\bullet_{\mathcal{X}}(\mathcal{G},f^\#\mathcal{F})$ (Corollary 5.2.3). $\square$

8.3. For pseudo-proper $f\colon \mathcal{X}\to\mathcal{Y}$, the functors $f^!:=f_t^\times$ and $f^\#$ are *local on* $\mathcal{X}$, in the following sense.

PROPOSITION 8.3.1. *Let there be given a commutative diagram*

$$\begin{array}{ccc} \mathcal{U} & \xrightarrow{\;i_1\;} & \mathcal{X}_1 \\ {\scriptstyle i_2}\big\downarrow & & \big\downarrow{\scriptstyle f_1} \\ \mathcal{X}_2 & \xrightarrow[f_2]{} & \mathcal{Y} \end{array}$$

of noetherian formal schemes, with f_1 and f_2 pseudo-proper and i_1 and i_2 open immersions. Then there are functorial isomorphisms

$$i_1^*f_1^! \xrightarrow{\;\sim\;} i_2^*f_2^!, \qquad i_1^*f_1^\# \xrightarrow{\;\sim\;} i_2^*f_2^\#.$$

PROOF. The second isomorphism results from the first, since for any $\mathcal{F}\in\mathbf{D}(\mathcal{Y})$ and for $j=1,2$,

$$\begin{aligned} i_j^*f_j^\#\mathcal{F} &\overset{6.1.4}{=} i_j^*\mathbf{R}\mathcal{H}om^\bullet_{\mathcal{X}_j}(\mathbf{R}\varGamma'_{\mathcal{X}_j}\mathcal{O}_{\mathcal{X}_j}, f_j^!\mathcal{F}) \cong \mathbf{R}\mathcal{H}om^\bullet_{\mathcal{U}}(i_j^*\mathbf{R}\varGamma'_{\mathcal{X}_j}\mathcal{O}_{\mathcal{X}_j}, i_j^*f_j^!\mathcal{F})\\ &\cong \mathbf{R}\mathcal{H}om^\bullet_{\mathcal{U}}(\mathbf{R}\varGamma'_{\mathcal{U}}\mathcal{O}_{\mathcal{U}}, i_j^*f_j^!\mathcal{F}). \end{aligned}$$

For the first isomorphism, Verdier's proof of [**V**, p. 395, Corollary 1]—a special case of Proposition 8.3.1—applies verbatim, modulo the following extensions (a), (b) and (c) of some elementary properties of schemes to formal schemes.

(a) Since pseudo-proper maps are separated, the graph of i_j is a *closed immersion* $\gamma\colon\mathcal{U}\hookrightarrow\mathcal{X}_j\times_{\mathcal{Y}}\mathcal{U}$ (see [**GD**, p. 445, (10.15.4)], where the "finite-type" hypothesis is used only to ensure that $\mathcal{X}_j\times_{\mathcal{Y}}\mathcal{U}$ is locally noetherian, a condition which holds here by the first paragraph in Section 7. And if $\mathcal{U}\to\mathcal{Y}$ is an open immersion, then so is γ (since then both $\pi_j\colon\mathcal{X}_j\times_{\mathcal{Y}}\mathcal{U}\to\mathcal{X}_j$ and $i_j=\pi_j\gamma$ are open immersions).

(b) If $s\colon\mathcal{U}\to\mathcal{V}$ is an open and closed immersion, then the exact functors s_* and s^* are adjoint, and by Example 6.1.3(4) there is a functorial isomorphism

$$s^!\mathcal{F}\cong s^\natural\mathcal{F}\cong s^*\mathcal{F} \qquad (\mathcal{F}\in\mathbf{D}_{\mathrm{qct}}(\mathcal{V})).$$

(c) (Formal extension of [**GD**, p. 325, (6.10.6)].) Let $\mathcal{U}\overset{\gamma}{\hookrightarrow}\mathcal{W}\overset{w}{\hookrightarrow}\mathcal{Z}$ be maps of locally noetherian formal schemes such that γ is a closed immersion and w is an open immersion. (We are interested specifically in the case $\mathcal{W}:=\mathcal{X}_2\times_{\mathcal{Y}}\mathcal{U}$ and $\mathcal{Z}:=\mathcal{X}_2\times_{\mathcal{Y}}\mathcal{X}_1$, see (a).) Set $u:=w\gamma$. Then *the closure $\overline{\mathcal{U}}$ of $u(\mathcal{U})$ is a formal subscheme of $\mathcal{Z}$, and the map $\mathcal{U}\to\overline{\mathcal{U}}$ induced by u is an open immersion.*

Indeed, $\overline{\mathcal{U}}$ is the support of $\mathcal{O}_{\mathcal{Z}}/\mathcal{J}$ where $\mathcal{J}$ is the kernel of the natural map $\mathcal{O}_{\mathcal{Z}}\to u_*\mathcal{O}_{\mathcal{U}}$; and it follows from [**GD**, p. 441, (10.14.1)] that we need only show that $\mathcal{J}$ is *coherent*. The question being local, we may assume that $\mathcal{Z}$ is affine, say $\mathcal{Z}=\mathrm{Spf}(A)$. Cover $\mathcal{U}$ by a finite number of affine open subschemes $\mathcal{U}_i$ $(1\le i\le n)$, with inclusions $u_i\colon\mathcal{U}_i\hookrightarrow\mathcal{U}$. Then there is a natural injection

$$u_*\mathcal{O}_{\mathcal{U}}\hookrightarrow u_*(\oplus_{i=1}^n u_{i*}\mathcal{O}_{\mathcal{U}_i})\cong\oplus_{i=1}^n(uu_i)_*\mathcal{O}_{\mathcal{U}_i},$$

so that $\mathcal{J}$ is the intersection of the kernels of the natural maps $\mathcal{O}_{\mathcal{Z}}\to(uu_i)_*\mathcal{O}_{\mathcal{U}_i}$, giving us a reduction to the case where $\mathcal{U}$ itself is affine, say $\mathcal{U}=\mathrm{Spf}(B)$. Now if I is the kernel of the ring-homomorphism $\rho\colon A\to B$ corresponding to u, then for any $f\in A$ the kernel of the induced map $\rho_{\{f\}}\colon A_{\{f\}}\to B_{\{f\}}$ is $I_{\{f\}}$; and one deduces that $\mathcal{J}$ is the coherent $\mathcal{O}_{\mathcal{Z}}$-module denoted by $I^\triangle$ in [**GD**, p. 427, (10.10.2)]. □

PROPOSITION 8.3.2. *If $f\colon \mathcal{X} \to \mathcal{Y}$ is a pseudo-proper map of noetherian formal schemes then*

$$f^{\#}(\mathbf{D}_{\mathrm{c}}^{+}(\mathcal{Y})) \subset \mathbf{D}_{\mathrm{c}}^{+}(\mathcal{X}).$$

PROOF. Since $f^{\#}$ commutes with open base change (Theorem 8.1) we may assume $\mathcal{Y}$ to be affine, say $\mathcal{Y} = \mathrm{Spf}(A)$. Since f is of pseudo-finite type, every point of $\mathcal{Y}$ has an open neighborhood $\mathcal{U}$ such that $f|_{\mathcal{U}}$ factors as

$$\mathcal{U} \overset{i}{\hookrightarrow} \mathrm{Spf}(B) \xrightarrow{p} \mathrm{Spf}(A) = \mathcal{Y}$$

where B is the completion of a polynomial ring $P := A[T_0, T_1, \dots, T_n]$ with respect to an ideal I whose intersection with A is open, i is a closed immersion, and p corresponds to the obvious continuous ring homomorphism $A \to B$ (see footnote in Section 1.2.2). This $\mathrm{Spf}(B)$ is an open subscheme of the completion $\mathcal{P}$ of the projective space $\mathbf{P}_A^n$ along the closure of its subscheme $\mathrm{Spec}(P/I)$. Thus by Proposition 8.3.1 and item (c) in its proof, we can replace $\mathcal{X}$ by a closed formal subscheme of $\mathcal{P}$ having $\mathcal{U}$ as an open subscheme. In other words, we may assume that f factors as $\mathcal{X} \overset{i_1}{\hookrightarrow} \mathcal{P} \xrightarrow{p_1} \mathrm{Spf}(A) = \mathcal{Y}$ with i_1 a closed immersion and p_1 the natural map. Then $f^{\#} = i_1^{\#} p_1^{\#}$, and we need only consider the two cases (a) $f = p_1$ and (b) $f = i_1$.

Case (a) is given by Corollary 6.2.3. In case (b) we see as in example 6.1.3(4) that for $\mathcal{F} \in \mathbf{D}_{\mathrm{c}}^{+}(\mathcal{Y})$ we have $f^{\natural}\mathcal{F} \in \mathbf{D}_{\mathrm{c}}^{+}(\mathcal{X})$ and

$$f^{\#}\mathcal{F} = \mathbf{\Lambda}_{\mathcal{X}} \mathbf{R}\varGamma'_{\mathcal{X}} f^{\natural}\mathcal{F} \underset{\text{or } 5.2.3}{\overset{6.3.1(c)}{=\!=}} \mathbf{\Lambda}_{\mathcal{X}} f^{\natural}\mathcal{F} \overset{6.2.1}{=\!=} f^{\natural}\mathcal{F} \in \mathbf{D}_{\mathrm{c}}^{+}(\mathcal{X}). \qquad \square$$

COROLLARY 8.3.3. *For all $\mathcal{F} \in \mathbf{D}_{\mathrm{c}}^{+}(\mathcal{Y})$ the base-change map $\beta_{\mathcal{F}}^{\#}$ of Theorem* 8.1 *is an* isomorphism

$$\beta_{\mathcal{F}}^{\#}\colon v^* f^{\#}\mathcal{F} \xrightarrow{\sim} g^{\#} u^* \mathcal{F}.$$

PROOF. Proposition 6.2.1 gives an isomorphism $v^* f^{\#}\mathcal{F} \xrightarrow{\sim} \mathbf{\Lambda}_{\mathcal{V}} v^* f^{\#}\mathcal{F}$. $\square$

We have now the following duality theorem for proper maps and bounded-below complexes with coherent homology.

THEOREM 8.4. *Let $f\colon \mathcal{X} \to \mathcal{Y}$ be a proper map of noetherian formal schemes, so that $\mathbf{R}f_*(\mathbf{D}_{\mathrm{c}}^{+}(\mathcal{X})) \subset \mathbf{D}_{\mathrm{c}}^{+}(\mathcal{Y})$ and $f^{\#}(\mathbf{D}_{\mathrm{c}}^{+}(\mathcal{Y})) \subset \mathbf{D}_{\mathrm{c}}^{+}(\mathcal{X})$ (see Propositions* 3.5.1 *and* 8.3.2). *Then for $\mathcal{G} \in \mathbf{D}_{\mathrm{c}}^{+}(\mathcal{X})$ and $\mathcal{F} \in \mathbf{D}_{\mathrm{c}}^{+}(\mathcal{Y})$ there are functorial* isomorphisms

$$\mathbf{R}f_*\mathbf{R}\mathcal{H}om^{\bullet}(\mathcal{G}, f^{\#}\mathcal{F}) \underset{8.2}{\xrightarrow{\sim}} \mathbf{R}\mathcal{H}om^{\bullet}(\mathbf{R}f_*\mathbf{R}\varGamma'_{\mathcal{X}}\mathcal{G}, \mathcal{F})$$
$$\underset{5.2.11(d)}{\xrightarrow{\sim}} \mathbf{R}\mathcal{H}om^{\bullet}(\mathbf{R}\varGamma'_{\mathcal{Y}}\mathbf{R}f_*\mathcal{G}, \mathcal{F}) \underset{6.2.1}{\xrightarrow{\sim}} \mathbf{R}\mathcal{H}om^{\bullet}(\mathbf{R}f_*\mathcal{G}, \mathcal{F}).$$

In particular, $f^{\#}\colon \mathbf{D}_{\mathrm{c}}^{+}(\mathcal{Y}) \to \mathbf{D}_{\mathrm{c}}^{+}(\mathcal{X})$ is right-adjoint to $\mathbf{R}f_*\colon \mathbf{D}_{\mathrm{c}}^{+}(\mathcal{X}) \to \mathbf{D}_{\mathrm{c}}^{+}(\mathcal{Y})$.

If $\mathcal{X}$ is properly algebraic we can replace $f^{\#}$ by the functor $f^{\times}$ of Corollary 4.1.1.

PROOF. Left to reader. (For the last assertion see Corollaries 6.2.2 and 3.3.4.) $\square$

References

[AJL] L. Alonso Tarrío, A. Jeremías López and J. Lipman, Local homology and cohomology on schemes, *Ann. Scient. Éc. Norm. Sup.* **30** (1997), 1–39.

[AJL′] ———, Greenlees-May duality on formal schemes, this volume.

[AJS] L. Alonso Tarrío, A. Jeremías López and M. J. Souto Salorio, Localization in categories of complexes and unbounded resolutions, *Canadian Math. J.*, to appear.

[AHK] A. Altman, R. Hoobler and S. L. Kleiman, A note on the base change map for cohomology, *Compositio Math.* **27** (1973), 25–37.

[B] N. Bourbaki, *Algèbre Commutative,* Actualités Sci. et Industrielles, no. 1293, Hermann, Paris, 1961.

[BN] M. Bökstedt and A. Neeman, Homotopy limits in triangulated categories, *Compositio Math.* **86** (1993), 209–234.

[De] P. Deligne, Cohomologie à supports propres, in *Théorie des topos et Cohomologie Étale des Schémes (SGA 4) Tome 3,* Lecture Notes in Math., no. **305**, Springer-Verlag, New York, 1973, pp. 250–461.

[F] P. Freyd, *Abelian Categories,* Harper and Row, New York, 1964.

[Fe] J. Franke, Brown representability for triangulated categories, *Topology,* to appear.

[EGA] A. Grothendieck and J. Dieudonné, *Élements de Géométrie Algébrique* **III**, Publications Math. IHES **11**, Paris, 1961.

[GD] ———, *Élements de Géométrie Algébrique* **I**, Springer-Verlag, New York, 1971.

[H1] R. Hartshorne, *Residues and Duality,* Lecture Notes in Math., no. **20**, Springer-Verlag, New York, 1966.

[H2] ———, On the De Rham cohomology of algebraic varieties, *Publications Math. IHES* **45** (1976), 5–99.

[HüK] R. Hübl and E. Kunz, Integration of differential forms on schemes, *J. Reine u. Angew. Math.* **410** (1990), 53–83.

[HüS] R. Hübl and P. Sastry, Regular differential forms and relative duality, *American J. Math.* **115** (1993), 749–787.

[I] L. Illusie, Existence de résolutions globales, in *Théorie des Intersections et Théorème de Riemann-Roch (SGA 6),* Lecture Notes in Math., no. **225**, Springer-Verlag, New York, 1971, pp. 160-222.

[Ke] G. R. Kempf, Some elementary proofs of basic theorems in the cohomology of quasi-coherent sheaves, *Rocky Mountain J. Math.* **10** (1980), 637–645.

[L1] J. Lipman, *Dualizing sheaves, differentials and residues on algebraic varieties,* Astérisque, vol. **117**, Soc. Math. France, Paris, 1984.

[L2] ———, *Residues and traces of differentials forms via Hochschild homology,* Contemporary Mathematics, vol. **61**, AMS, Providence, 1987.

[L3] ———, Desingularization of two-dimensional schemes, *Annals of Math.* **107** (1978), 151–207.

[L4] ———, *Notes on Derived Categories,* preprint, at `www.math.purdue.edu/~lipman/`

[L5] ———, Non-noetherian Grothendieck duality, this volume.

[Lü] W. Lütkebohmert, On compactification of schemes, *Manuscripta Math.* **80** (1993), 95–111.

[M1] S. Mac Lane, *Categories for the Working Mathematician,* Springer-Verlag, New York, 1971.

[M2] ———, *Homology,* Springer-Verlag, New York, 1975.

[N1] A. Neeman, The Grothendieck duality theorem via Bousfield's techniques and Brown representability, *J. Amer. Math. Soc.* **9** (1996), 205–236.

[N2] ———, *Triangulated Categories,* preprint.

[Sp] N. Spaltenstein, Resolution of unbounded complexes, *Compositio Math.* **65** (1988), 121–154.

[St] B. Stenström, *Rings of Quotients,* Grundlehren der Math. Wiss., bd. 217, Springer-Verlag, New York, 1975.

[V] J. L. Verdier, Base change for twisted inverse image of coherent sheaves, in *Algebraic Geometry,* Oxford Univ. press, 1969, pp. 393–408.

[Y] A. Yekutieli, Smooth formal embeddings and the residue complex, *Canadian J. Math.* **50** (1998), 863–896.

Universidade de Santiago de Compostela, Facultade de Matemáticas, E-15771 Santiago de Compostela, SPAIN
E-mail address: `lalonso@zmat.usc.es`
URL: `http://web.usc.es/~lalonso/`

Universidade de Santiago de Compostela, Facultade de Matemáticas, E-15771 Santiago de Compostela, SPAIN
E-mail address: `jeremias@zmat.usc.es`

Dept. of Mathematics, Purdue University, W. Lafayette IN 47907, USA
E-mail address: `lipman@math.purdue.edu`
URL: `www.math.purdue.edu/~lipman/`

Part 2

Greenlees-May Duality on Formal Schemes

Contemporary Mathematics
Volume **244**, 1999

Greenlees-May Duality on Formal Schemes

Leovigildo Alonso Tarrío, Ana Jeremías López, and Joseph Lipman

Abstract. For a closed subscheme Z of a noetherian separated scheme X, let $\varGamma_Z$ be the functor of sections with support in Z, taking $\mathcal{O}_X$-modules to $\mathcal{O}_X$-modules. Inspired by a theorem of Greenlees and May [**GM**] about duality between local cohomology and local homology for modules over a commutative ring, we gave in [**AJL**] the result that on quasi-coherent complexes $\mathcal{F}$ the "homology localization" functor $\mathbf{R}\mathcal{H}om^\bullet(\mathbf{R}\varGamma_Z\mathcal{O}_X, \mathcal{F})$ is a left-derived functor of $\Lambda_Z :=$ completion along Z, the corresponding map to $\Lambda_Z\mathcal{F}$ being such that its composition with the natural map $\mathcal{F} \to \mathbf{R}\mathcal{H}om^\bullet(\mathbf{R}\varGamma_Z\mathcal{O}_X, \mathcal{F})$ is the completion map $\mathcal{F} \to \Lambda_Z\mathcal{F}$. We also showed how this unifies and generalizes several other recorded duality theorems. Here we extend the result to an arbitrary noetherian separated formal scheme $\mathcal{X}$, with "quasi-coherent" replaced by "$\varinjlim$ of coherent."

Given an open $\mathcal{O}_\mathcal{X}$-ideal $\mathcal{I}$, with $\kappa\colon \hat{\mathcal{X}}_\mathcal{I} \to \mathcal{X}$ the completion by $\mathcal{I}$ and with $\varGamma_\mathcal{I}$ the functor $\varinjlim \mathcal{H}om^\bullet(\mathcal{O}_\mathcal{X}/\mathcal{I}^n, -)$, we deduce for any $\mathcal{O}_\mathcal{X}$-complex $\mathcal{F}$ having coherent homology sheaves a canonical duality isomorphism

$$\mathbf{R}\mathcal{H}om^\bullet(\mathcal{E}, \kappa_*\kappa^*\mathcal{F}) \xrightarrow{\ \sim\ } \mathbf{R}\mathcal{H}om^\bullet(\mathbf{R}\varGamma_\mathcal{I}\mathcal{E}, \mathcal{F}) \qquad (\mathcal{E} \in \mathbf{D}(\mathcal{X})).$$

The special case where $\mathcal{I}$ is an ideal of definition—so that $\kappa =$ identity—plays an important part in the duality theory of coherent sheaves on formal schemes.

Introduction

We use the terminology of [**DFS**, Section 1], and basic facts about formal schemes found in [**GD**, Chapter 1, §10]. In particular, we assume familiarity with the notion of triangle-preserving functors (Δ-functors) between triangulated categories (Δ categories), and of (Δ-functorial) maps between such functors.

Throughout, $\mathcal{X}$ is a separated noetherian formal scheme, $\mathcal{A}$ is the category $\mathcal{A}(\mathcal{X})$ of $\mathcal{O}_\mathcal{X}$-modules, $\mathbf{K}$ is the homotopy category of $\mathcal{O}_\mathcal{X}$-complexes, $\mathbf{D}$ is its derived category (i.e., its localization with respect to quasi-isomorphisms [**Hrt**, pp. 28–35]), and $q\colon \mathbf{K} \to \mathbf{D}$ is the canonical functor. $\mathcal{A}_{\vec{c}} \subset \mathcal{A}$ is the plump subcategory (see [**DFS**, Section 1]) whose objects are $\varinjlim$'s of coherent $\mathcal{O}_\mathcal{X}$-modules. The derived category $\mathbf{D}_{\vec{c}}$ of the homotopy category $\mathbf{K}_{\vec{c}}$ of $\mathcal{A}$-complexes whose homology sheaves are all in $\mathcal{A}_{\vec{c}}$ is a Δ-subcategory of $\mathbf{D}$ ([**Hrt**, p. 50, Example 3]).

We work with *affine-acyclic* complexes, $\mathcal{O}_\mathcal{X}$-complexes $\mathcal{E}$ such that for each affine open $\mathcal{U} \subset \mathcal{X}$ the natural map $\Gamma(\mathcal{U}, \mathcal{E}) \to \mathbf{R}\Gamma(\mathcal{U}, \mathcal{E})$ is an isomorphism in the derived category of abelian groups, i.e., such that if $\mathcal{E} \to \mathcal{J}$ is a quasi-isomorphism from $\mathcal{E}$ to a K-injective complex $\mathcal{J}$—a K-injective resolution [**Spn**, 4.5])—then $\Gamma(\mathcal{U}, \mathcal{E}) \to \Gamma(\mathcal{U}, \mathcal{J})$ is a quasi-isomorphism. (Observe, using the functor "extension by 0," that the restriction of a K-injective complex to an open subset remains K-injective; and that $\mathbf{R}\Gamma$ can be realized via K-injective resolutions [**Spn**, 5.12, 6.4].)

First two authors partially supported by Xunta de Galicia research project XUGA20702B98 and Spain's DGES grant PB97-0530. They also thank the Mathematics Department of Purdue University for its hospitality, help and support.

Third author partially supported by the National Security Agency.

For example, every $\mathcal{A}_{\vec{c}}$-complex is affine-acyclic (Example 2.2.1(c)).

The homotopy category $\mathbf{K}^{\mathrm{a}}_{\vec{c}}$ of affine-acyclic complexes with $\mathcal{A}_{\vec{c}}$-homology is a Δ-subcategory of $\mathbf{K}$, see Remark 2.1.1(4). Since every complex admits a quasi-isomorphism into a K-injective (clearly affine-acyclic) one, the derived category $\mathbf{D}^{\mathrm{a}}_{\vec{c}}$ of $\mathbf{K}^{\mathrm{a}}_{\vec{c}}$ can be identified with a Δ-subcategory of $\mathbf{D}_{\vec{c}}$, whose inclusion $\mathbf{D}^{\mathrm{a}}_{\vec{c}} \hookrightarrow \mathbf{D}_{\vec{c}}$ is an *equivalence of Δ-categories.* The canonical functor $q^{\mathrm{a}}\colon \mathbf{K}^{\mathrm{a}}_{\vec{c}} \to \mathbf{D}^{\mathrm{a}}_{\vec{c}}$ is then the restriction of q.

Fix a coherent $\mathcal{O}_{\mathfrak{X}}$-ideal $\mathcal{I}$. The *torsion functor* $\varGamma_{\mathcal{I}}\colon \mathbf{K} \to \mathbf{K}$ is described by

$$\varGamma_{\mathcal{I}}\mathcal{G} := \varinjlim_{n>0} \mathcal{H}om^{\bullet}(\mathcal{O}_{\mathfrak{X}}/\mathcal{I}^n, \mathcal{G}) \qquad (\mathcal{G} \in \mathbf{K}),$$

(see [**DFS**, Section 1.2.1]). The *completion functor* $\Lambda_{\mathcal{I}}\colon \mathbf{K} \to \mathbf{K}$ is described by

$$\Lambda_{\mathcal{I}}(\mathcal{G}) := \varprojlim_{n>0} \big((\mathcal{O}_{\mathfrak{X}}/\mathcal{I}^n) \otimes \mathcal{G}\big) \qquad (\mathcal{G} \in \mathbf{K}).$$

By the universal property of right-derived functors, there is a unique functorial map $\mathbf{R}\varGamma_{\mathcal{I}} \to \mathbf{1}$ whose composition with the canonical map $q\varGamma_{\mathcal{I}} \to \mathbf{R}\varGamma_{\mathcal{I}}$ is the inclusion map $q\varGamma_{\mathcal{I}} \hookrightarrow \mathbf{1}$. For any K-injective resolution $\mathcal{E} \to \mathcal{J}$ as above, $\mathbf{R}\varGamma_{\mathcal{I}}\mathcal{E} \to \mathcal{E}$ is naturally isomorphic to the canonical $\mathbf{D}$-morphism $\varGamma_{\mathcal{I}}\mathcal{J} \to \mathcal{J}$.

The functor $\mathbf{R}\mathcal{H}om^{\bullet}(\mathbf{R}\varGamma_{\mathcal{I}}\mathcal{O}_{\mathfrak{X}}, -)$ is right-adjoint to $\mathbf{R}\varGamma_{\mathcal{I}}$ (via (4.1.1.2), (4.1.1.1) below). So the following **main result** establishes a *duality* (= adjunction) between the right-derived torsion and the left-derived completion functors associated to $\mathcal{I}$.

THEOREM 0.1. *There exists a unique Δ-functorial $\mathbf{D}$-morphism*

$$\zeta_{\mathcal{F}}\colon \mathbf{R}\mathcal{H}om^{\bullet}(\mathbf{R}\varGamma_{\mathcal{I}}\mathcal{O}_{\mathfrak{X}}, \mathcal{F}) \to \Lambda_{\mathcal{I}}\mathcal{F} \qquad (\mathcal{F} \in \mathbf{K}^{\mathrm{a}}_{\vec{c}})$$

such that:

(i) *the pair* $(\mathbf{R}\mathcal{H}om^{\bullet}(\mathbf{R}\varGamma_{\mathcal{I}}\mathcal{O}_{\mathfrak{X}}, -), \zeta)$ *is a* left-derived functor *of* $q\Lambda_{\mathcal{I}}|_{\mathbf{K}^{\mathrm{a}}_{\vec{c}}}$ *(i.e., a final object in the category of pairs (S, ϱ) with $S\colon \mathbf{D}^{\mathrm{a}}_{\vec{c}} \to \mathbf{D}$ a Δ-functor and $\varrho\colon Sq^{\mathrm{a}} \to q\Lambda_{\mathcal{I}}|_{\mathbf{K}^{\mathrm{a}}_{\vec{c}}}$ a map of Δ-functors), and*

(ii) *the composition of $\zeta_{\mathcal{F}}$ with the natural map*

$$\rho_{\mathcal{F}}\colon \mathcal{F} \cong \mathbf{R}\mathcal{H}om^{\bullet}(\mathcal{O}_{\mathfrak{X}}, \mathcal{F}) \to \mathbf{R}\mathcal{H}om^{\bullet}(\mathbf{R}\varGamma_{\mathcal{I}}\mathcal{O}_{\mathfrak{X}}, \mathcal{F})$$

is the canonical completion map $\mathcal{F} \to \Lambda_{\mathcal{I}}\mathcal{F}$.

Moreover, this $\zeta_{\mathcal{F}}$ is an isomorphism whenever $\mathcal{F}$ is a K-flat *$\mathcal{A}_{\vec{c}}$-complex.*[1]

The uniqueness of ζ is shown as follows. It results from (i) and (ii) that any two choices of ζ are the same modulo a functorial automorphism $\theta_{\mathcal{F}}$ of $\mathbf{\Lambda}_{\mathcal{I}}\mathcal{F} := \mathbf{R}\mathcal{H}om^{\bullet}(\mathbf{R}\varGamma_{\mathcal{I}}\mathcal{O}_{\mathfrak{X}}, \mathcal{F})$ respecting $\rho_{\mathcal{F}}$. But just as in setting up the isomorphism (B) in [**DFS**, Remark 6.3.1](1), we can see that composition with $\rho_{\mathcal{F}}$ is an *isomorphism* $\mathrm{Hom}(\mathbf{\Lambda}_{\mathcal{I}}\mathcal{F}, \mathbf{\Lambda}_{\mathcal{I}}\mathcal{F}) \xrightarrow{\sim} \mathrm{Hom}(\mathcal{F}, \mathbf{\Lambda}_{\mathcal{I}}\mathcal{F})$, whence $\theta_{\mathcal{F}} =$ identity.

0.2. Existence of ζ in Theorem 0.1 is proved along lines similar to those in [**AJL**]—which deals with ordinary schemes. Here there are new technical problems. For example, the canonical functor $\mathbf{D}(\mathcal{A}_{\vec{c}}) \to \mathbf{D}_{\vec{c}}$ might not be an equivalence (as it is for separated quasi-compact ordinary schemes). Another—related—problem is that for an open immersion $i\colon \mathfrak{U} \hookrightarrow \mathfrak{X}$, the functor i_* doesn't necessarily take $\mathcal{A}_{\vec{c}}(\mathfrak{U})$ into $\mathcal{A}_{\vec{c}}$ (as it does for separated quasi-compact ordinary schemes.) To get around such obstacles we need properties of sheaves on formal schemes established in [**DFS**].

[1] An $\mathcal{O}_{\mathfrak{X}}$-complex is *K-flat* if its tensor product with every exact $\mathcal{O}_{\mathfrak{X}}$-complex remains exact, cf. [**Spn**, pp. 139–140]. For example, any $\varinjlim$ of bounded-above flat complexes is K-flat (cf. e.g., [**Lpm**, Example (2.5.4)]).

In outline, we proceed as follows. After setting up some preliminaries on Čech functors in Section 1, we show in Section 2 that the completion functor $\Lambda_{\mathcal{I}}|_{\mathbf{K}^{\mathrm{a}}_{\vec{c}}}$ has a left-derived functor, for which the property corresponding to the last assertion in Theorem 0.1 holds. Then in Section 3 we identify this left-derived functor with a pair $(\mathbf{R}\mathcal{H}om^\bullet(\mathbf{R}\varGamma_{\mathcal{I}}\mathcal{O}_{\mathfrak{X}}, -), \zeta)$ such that Theorem 0.1(ii) is satisfied. In so doing we describe only the modifications needed in the corresponding proof in [**AJL**], referring to *loc. cit.* for the remaining details.

It must be noted that the main results in [**AJL**] are inadequately packaged, at least for non-affine $\mathfrak{X}$. There $\mathfrak{X}$ is an ordinary scheme. The completion functor $\Lambda_{\mathcal{I}}$ is said to have a left-derived functor (denoted $\mathbf{L}\Lambda_Z$, where $Z = \mathrm{Supp}(\mathcal{O}_{\mathfrak{X}}/\mathcal{I})$) on $\mathbf{D}_{\vec{c}}(\mathfrak{X}) = \mathbf{D}_{\mathrm{qc}}(\mathfrak{X})$. However, that holds only on some *equivalent subcategory* $\mathbf{D}^{\mathrm{a}}_{\mathrm{qc}}(\mathfrak{X})$, among whose objects are all the quasi-coherent complexes. (The error lies in the unduly facile paragraph preceding (1.2) on page 10 of [**AJL**]. To fill the gap we have had to work with the functor $\check{c}^\infty$ of Section 1.2 below.) Thus the map

$$\mathcal{F} \to \mathbf{L}\Lambda_Z\mathcal{F} \qquad (\mathcal{F} \in \mathbf{D}_{\mathrm{qc}}(\mathfrak{X}),\ \kappa\colon \mathfrak{X}_{/Z} \to \mathfrak{X} \text{ the completion map})$$

and its factorization via $\mathcal{F} \to \kappa_*\kappa^*\mathcal{F}$, which play an important role in Theorem (0.3) and Proposition (0.4.1) of *loc. cit.*, make sense, as defined there, only on $\mathbf{D}^{\mathrm{a}}_{\mathrm{qc}}(\mathfrak{X})$. But they *can* be extended to all of $\mathbf{D}_{\mathrm{qc}}(\mathfrak{X})$, by means of its equivalence with $\mathbf{D}^{\mathrm{a}}_{\mathrm{qc}}(\mathfrak{X})$ or (better) as described above in Theorem 0.1, and then the rest is okay.

0.3. The oft-used Proposition 6.2.1 in [**DFS**] follows from Theorem 0.1. In fact we will use Theorem 0.1 to show that a certain Δ-bifunctorial map, more general than the one in *loc. cit.*, is an isomorphism. To wit:

Suppose $\mathcal{I}$ is open, i.e., contains an ideal of definition of $\mathfrak{X}$. Let

$$\kappa = \kappa_{\mathcal{I}}\colon \hat{\mathfrak{X}}_{\mathcal{I}} \to \mathfrak{X}$$

be *completion by* $\mathcal{I}$. Thus $\hat{\mathfrak{X}}_{\mathcal{I}}$ is the topological space $\mathrm{Supp}(\mathcal{O}_{\mathfrak{X}}/\mathcal{I})$ together with the sheaf of topological rings $\varprojlim \mathcal{O}_{\mathfrak{X}}/\mathcal{I}^n$; and κ is the obvious ringed-space map. By [**GD**, p. 412, Proposition (10.6.3)], $\hat{\mathfrak{X}}_{\mathcal{I}}$ is a formal scheme, and by [**GD**, p. 422, Corollaire (10.8.9)], [**DFS**, Lemma 7.1.1], and [**Brb**, p. 103, Corollaire], the map κ is *flat*. (Both assertions are local, so need only be verified for affine $\mathfrak{X}$.)

In particular, if $\mathcal{I}$ itself is an ideal of definition then $\hat{\mathfrak{X}}_{\mathcal{I}} = \mathfrak{X}$ and κ is the identity.

Denote by $\mathbf{D}_{\mathrm{c}} \subset \mathbf{D}_{\vec{c}}$ the Δ-subcategory having as objects the complexes whose homology sheaves are all coherent. The following proposition is proved in Section 4.

PROPOSITION 0.3.1. (i) *For each pair* $\mathcal{E}, \mathcal{F} \in \mathbf{D}$ *the map induced by the canonical map* $\mathbf{R}\varGamma_{\mathcal{I}}\mathcal{E} \to \mathcal{E}$ *factors uniquely as*

$$\mathbf{R}\mathcal{H}om^\bullet(\mathcal{E},\mathcal{F}) \xrightarrow{\text{natural}} \mathbf{R}\mathcal{H}om^\bullet(\mathcal{E},\kappa_*\kappa^*\mathcal{F}) \xrightarrow{\lambda(\mathcal{E},\mathcal{F})} \mathbf{R}\mathcal{H}om^\bullet(\mathbf{R}\varGamma_{\mathcal{I}}\mathcal{E},\mathcal{F}).$$

(ii) *If* $\mathcal{F} \in \mathbf{D}_{\mathrm{c}}$ *then* $\lambda(\mathcal{E},\mathcal{F})$ *is an* isomorphism.

Remark. To be explicit, $\lambda(\mathcal{E},\mathcal{F})$ is the composition $s^{-1}t$ in the following natural commutative diagram, where the isomorphisms p and p' can be established by imitating [**DFS**, Corollary 5.2.3], and r (hence s) is an isomorphism by Lemma 4.1:

$$\begin{array}{ccccc}
\mathbf{R}\mathcal{H}om^\bullet(\mathcal{E},\kappa_*\kappa^*\mathcal{F}) & \xrightarrow[t]{} & \mathbf{R}\mathcal{H}om^\bullet(\mathbf{R}\varGamma_{\mathcal{I}}\mathcal{E},\kappa_*\kappa^*\mathcal{F}) & \xleftarrow[p]{\simeq} & \mathbf{R}\mathcal{H}om^\bullet(\mathbf{R}\varGamma_{\mathcal{I}}\mathcal{E},\mathbf{R}\varGamma_{\mathcal{I}}\kappa_*\kappa^*\mathcal{F}) \\
\uparrow & & s\uparrow\simeq & & r\uparrow\simeq \\
\mathbf{R}\mathcal{H}om^\bullet(\mathcal{E},\mathcal{F}) & \longrightarrow & \mathbf{R}\mathcal{H}om^\bullet(\mathbf{R}\varGamma_{\mathcal{I}}\mathcal{E},\mathcal{F}) & \xleftarrow[p']{\simeq} & \mathbf{R}\mathcal{H}om^\bullet(\mathbf{R}\varGamma_{\mathcal{I}}\mathcal{E},\mathbf{R}\varGamma_{\mathcal{I}}\mathcal{F})
\end{array}$$

This description makes it clear that $\lambda(\mathcal{E},\mathcal{F})$ is Δ-bifunctorial.

1. Preliminaries on Čech functors and left-derivability

1.1. Let $\mathcal{U} = (\mathfrak{U}_\alpha)_{1\le\alpha\le t}$ be an open cover of the separated formal scheme $\mathfrak{X}$. Let $\mathfrak{P}_t$ be the set of subsets of $\{1,2,\dots,t\}$. For $i \in \mathfrak{P}_t$ set

$$\mathfrak{U}_i := \bigcap_{\alpha\in i} \mathfrak{U}_\alpha, \quad \text{and} \quad \mathcal{O}_i := \mathcal{O}_{\mathfrak{U}_i}.$$

For $i \supset j$ in $\mathfrak{P}_t$ let $\lambda_{ij}\colon \mathfrak{U}_i \hookrightarrow \mathfrak{U}_j$ be the inclusion. Set $\lambda_i := \lambda_{i\phi}\colon \mathfrak{U}_i \hookrightarrow \mathfrak{X}$.

A *$\mathcal{U}$-module* is a family $\mathcal{F} = (\mathcal{F}_i)_{i\in\mathfrak{P}_t}$ such that $\mathcal{F}_i$ is an $\mathcal{O}_i$-module, together with a family of sheaf homomorphisms

$$\varphi_{jk}\colon \lambda_{jk}^*\mathcal{F}_k \to \mathcal{F}_j \qquad (j \supset k)$$

such that φ_{jk} is an $\mathcal{O}_j$-homomorphism, φ_{jj} is the identity map of $\mathcal{F}_j$, and whenever $i \supset j \supset k$ we have the transitivity relation $\varphi_{ik} = \varphi_{ij} \circ (\varphi_{jk}|_{\mathfrak{U}_i})$, i.e., φ_{ik} factors as

$$\lambda_{ik}^*\mathcal{F}_k = \lambda_{ij}^*\lambda_{jk}^*\mathcal{F}_k \xrightarrow{\lambda_{ij}^*(\varphi_{jk})} \lambda_{ij}^*\mathcal{F}_j \xrightarrow{\varphi_{ij}} \mathcal{F}_i.$$

We say that the $\mathcal{U}$-module $\mathcal{F}$ is *quasi-coherent* (resp. *flat*, resp. ...) if each one of the $\mathcal{O}_i$-modules $\mathcal{F}_i$ is such. So, for example, $\mathcal{A}_{\vec{c}}(\mathcal{U})$ denotes the category of $\mathcal{U}$-modules $\mathcal{F}$ such that $\mathcal{F}_i \in \mathcal{A}_{\vec{c}}(\mathfrak{U}_i)$ for all $i \in \mathfrak{P}_t$.

The $\mathcal{U}$-modules and their morphisms (defined in the obvious manner) form an abelian category $\mathcal{A}(\mathcal{U})$ having $\varinjlim$'s and $\varprojlim$'s. For instance one checks that the $\varinjlim$ of a direct system $(\mathcal{F}^m, \varphi_{jk}^m)_{m\in I}$ in $\mathcal{A}(\mathcal{U})$ is $(\varinjlim_m \mathcal{F}_i^m, \varinjlim_m \varphi_{jk}^m)$.

The *Čech functor* $\check{C}^\bullet$ from the homotopy category $\mathbf{K}(\mathcal{U})$ of $\mathcal{U}$-complexes to $\mathbf{K} := \mathbf{K}(\mathfrak{X})$ is defined as follows. With

$$|i| := (\text{cardinality of } i) - 1 \qquad (i \in \mathfrak{P}_t),$$

set, for $\mathcal{F} \in \mathcal{A}(\mathcal{U})$,

$$\check{C}^s(\mathcal{F}) := \bigoplus_{|i|=s} \lambda_{i*}\mathcal{F}_i \ \text{ if } 0 \le s < t, \text{ and } 0 \text{ otherwise.}$$

Whenever k is obtained from $j = \{j_0 < j_1 < \cdots < j_{s+1}\} \in \mathfrak{P}_t$ by removing a single element j_a, set $\epsilon_{jk} := (-1)^a$; and otherwise set $\epsilon_{jk} := 0$. (Thus if k is fixed then $\epsilon_{jk} = 0$ for all but finitely many j.) We specify the differential $\delta^s\colon \check{C}^s(\mathcal{F}) \to \check{C}^{s+1}(\mathcal{F})$ by requiring its restriction to $\lambda_{k*}\mathcal{F}_k$ $(|k| = s)$ to be the natural composition

$$\lambda_{k*}\mathcal{F}_k \longrightarrow \lambda_{k*}\lambda_{jk*}\lambda_{jk}^*\mathcal{F}_k = \lambda_{j*}\lambda_{jk}^*\mathcal{F}_k \xrightarrow{\oplus\lambda_{j*}(\epsilon_{jk}\varphi_{jk})} \bigoplus_{|j|=s+1} \lambda_{j*}\mathcal{F}_j.$$

Then $\delta^{s+1}\circ\delta^s = 0$ for all s, and so we get a functor $\check{C}^\bullet$ from $\mathcal{U}$-modules to $\mathcal{O}_{\mathfrak{X}}$-complexes. Next, for a complex $\mathcal{F}^\bullet \in \mathbf{K}(\mathcal{U})$, $\check{C}^\bullet(\mathcal{F}^\bullet)$ is defined to be the total complex associated to the double complex $\check{C}^p(\mathcal{F}^q)$:

$$\check{C}^s(\mathcal{F}^\bullet) := \oplus_{p+q=s}\check{C}^p(\mathcal{F}^q),$$

and the differential $\check{C}^s(\mathcal{F}^\bullet) \to \check{C}^{s+1}(\mathcal{F}^\bullet)$ restricts on $\check{C}^p(\mathcal{F}^q)$ to

$$(\check{C}^p(d^q) \oplus (-1)^q\check{d}^p)\colon \check{C}^p(\mathcal{F}^q) \longrightarrow (\check{C}^p(\mathcal{F}^{q+1}) \oplus \check{C}^{p+1}(\mathcal{F}^q))$$

where d^q (resp. $\check{d}^p$) is the differential in the complex $\mathcal{F}^\bullet$ (resp. $\check{C}^\bullet(\mathcal{F}^q)$). One checks that $\check{C}^\bullet$ commutes with translation and with forming mapping cones, so that $\check{C}^\bullet$ can (and will) be regarded as a Δ-functor from $\mathbf{K}(\mathcal{U})$ to $\mathbf{K}$.

For instance, if $\mathcal{M}$ is an $\mathcal{O}_{\mathfrak{X}}$-module and $\mathcal{M}^*$ is the "pulled-back" $\mathcal{U}$-module such that $\mathcal{M}_i^* := \lambda_i^*\mathcal{M}$ and φ_{jk} is the identity map of $\mathcal{M}_j^* = \lambda_{jk}^*\mathcal{M}_k^*$ for all $j \supset k$, then $\check{C}^\bullet(\mathcal{M}^*)$ is the usual $\mathcal{U}$-Čech resolution of $\mathcal{M}$ [**Gdm**, p. 206, Théorème 5.2.1]. More generally, composing $\check{C}^\bullet$ with the pullback functor $\mathbf{K} \to \mathbf{K}(\mathcal{U})$ gives the functor

$$\check{c} = \check{c}_{\mathcal{U}} \colon \mathbf{K} \to \mathbf{K}$$

sending each $\mathcal{O}_{\mathfrak{X}}$-complex $\mathcal{E}$ to its $\mathcal{U}$-Čech resolution $\check{c}\mathcal{E}$.

1.2. There is a functorial quasi-isomorphism $\chi_{\mathcal{E}} \colon \mathcal{E} \to \check{c}\mathcal{E}$ $(\mathcal{E} \in \mathbf{K})$, giving rise to an inductive system of functors $\left(\chi_{\check{c}^m} \colon \check{c}^m \to \check{c}^{m+1}\right)_{m>0}$. Set

$$\check{c}^\infty := \varinjlim_m \check{c}^m.$$

There is then a functorial quasi-isomorphism

$$\chi_{\mathcal{E}}^\infty \colon \mathcal{E} \to \check{c}^\infty\mathcal{E} \qquad (\mathcal{E} \in \mathbf{K}).$$

Since $\mathfrak{X}$ is noetherian, the functor $\check{c}^\infty$ commutes with $\varinjlim$, and hence is *idempotent*—$\chi_{\check{c}^\infty\mathcal{E}}^\infty \colon \check{c}^\infty\mathcal{E} \to \check{c}^\infty\check{c}^\infty\mathcal{E}$ is an isomorphism of complexes for all $\mathcal{E}$.

LEMMA 1.2.1. *$\Lambda_{\mathcal{I}}(\chi_{\mathcal{E}}^\infty)$ is a quasi-isomorphism. In other words, $q(\Lambda_{\mathcal{I}}(\chi_{\mathcal{E}}^\infty))$ is a* **D**-*isomorphism.*

PROOF. The question is readily seen to be local; and over any member $\mathfrak{U}$ of the open covering $\mathcal{U}$, $\chi_{\mathcal{E}}^\infty$ is, by the following Corollary 1.2.3, a *homotopy isomorphism,* whence so is $\Lambda_{\mathcal{I}}(\chi_{\mathcal{E}}^\infty)$. (Observe that $(\check{c}\mathcal{E})|_{\mathfrak{U}}$ is the Čech resolution of $\mathcal{E}|_{\mathfrak{U}}$ with respect to the cover $(\mathfrak{U}\cap\mathfrak{U}_\alpha)_{1\le\alpha\le t}$ of $\mathfrak{U}$, a cover of which $\mathfrak{U}$ itself is a member). □

LEMMA 1.2.2. *Let $\mathcal{U} = (\mathfrak{U}_1, \mathfrak{U}_2, \dots, \mathfrak{U}_t)$ be an open cover of $\mathfrak{X}$, with $\mathfrak{U}_1 = \mathfrak{X}$. Then for any $\mathcal{O}_{\mathfrak{X}}$-complex $\mathcal{E}$, the natural map $\chi = \chi_{\mathcal{E}} \colon \mathcal{E} \to \check{c}\mathcal{E}$ has a left inverse whose kernel is homotopically trivial (i.e., its identity map is homotopic to 0).*

PROOF. Let $\mathcal{E}^*$ be the pullback of $\mathcal{E}$ to a $\mathcal{U}$-complex, so that $\check{c}\mathcal{E} = \check{C}^\bullet\mathcal{E}^*$. Then

$$\check{C}_0^\bullet\mathcal{E}^* := \mathcal{E} \oplus \lambda_{2*}\lambda_2^*\mathcal{E} \oplus \cdots \oplus \lambda_{t*}\lambda_t^*\mathcal{E} \cong \check{C}^\bullet\mathcal{E}^*/\check{C}_{>0}^\bullet\mathcal{E}^*$$

where $\check{C}_{>0}^\bullet\mathcal{E}^*$ is the total complex associated to the double complex $\check{C}^p(\mathcal{E}^{*q})_{p>0;\, q\in\mathbb{Z}}$. Composing the natural map $\check{C}^\bullet\mathcal{E}^* \twoheadrightarrow \check{C}_0^\bullet\mathcal{E}^*$ and the projection $\check{C}_0^\bullet\mathcal{E}^* \twoheadrightarrow \mathcal{E}$ we get a left inverse $\pi = \pi_{\mathcal{E}}$ of χ.

Furthermore, π is a right *homotopy*-inverse of χ. Indeed, if $h \colon \check{c}\mathcal{E} \to \check{c}\mathcal{E}[-1]$ is the map of graded sheaves defined on the summand $\lambda_{i*}\lambda_i^*\mathcal{E}^p$ $(i \in \mathfrak{P}_t,\ p \in \mathbb{Z})$ by:

if $1 \in i$, and $i' := i - \{1\}$, then the restriction of h to $\lambda_{i*}\lambda_i^*\mathcal{E}^p$ is

$$\lambda_{i*}\lambda_i^*\mathcal{E}^p \xrightarrow{(-1)^p} \lambda_{i*}\lambda_i^*\mathcal{E}^p = \lambda_{i'*}\lambda_{i'}^*\mathcal{E}^p,$$

and otherwise the restriction is 0,

then one calculates, with d the differential in $\check{c}\mathcal{E}$, that $hd + dh = 1 - \chi\pi$.

As $\pi\chi$ is the identity map of $\mathcal{E}$, we have a functorial direct-sum decomposition

$$\check{c}\mathcal{E} = \mathcal{E} \oplus \check{c}_0\mathcal{E} \tag{1.2.2.1}$$

where $\check{c}_0\mathcal{E}$ is the kernel of $\pi_{\mathcal{E}}$. If $\nu \colon \check{c}_0\mathcal{E} \hookrightarrow \check{c}\mathcal{E}$ is the inclusion, then in $\mathbf{K}$,

$$\nu = (\chi\pi)\circ\nu = \chi\circ(\pi\nu) = 0,$$

and since ν has a left inverse therefore $\check{c}_0\mathcal{E}$ is homotopically trivial. □

COROLLARY 1.2.3. *With $\mathcal{U}$ as in Lemma 1.2.2 the natural map of complexes*

$$\chi_{\mathcal{E}}^{\infty}\colon \mathcal{E} \to \check{c}^{\infty}\mathcal{E} := \varinjlim_{m} \check{c}^{m}\mathcal{E}$$

has a left inverse whose kernel is homotopically trivial.

PROOF. From (1.2.2.1) we deduce via induction that

$$\check{c}^{\infty}\mathcal{E} \cong \mathcal{E} \oplus \bigoplus_{m=0}^{\infty} \check{c}_0(\check{c}^{m}\mathcal{E}).$$

Then we need only note that any direct sum of homotopically trivial complexes is homotopically trivial. □

1.3. The category $\mathbf{K}^{\mathrm{a}}_{\vec{c}}$ in Theorem 0.1 satisfies $\check{c}^{\infty}\mathbf{K}^{\mathrm{a}}_{\vec{c}} \subset \mathbf{K}^{\mathrm{a}}_{\vec{c}}$ (Corollary 2.2.5).

Let $\check{\mathbf{K}}^{\mathrm{a}}_{\vec{c}}$ be the essential image of $\check{c}^{\infty}|_{\mathbf{K}^{\mathrm{a}}_{\vec{c}}}$, i.e., the full subcategory of $\mathbf{K}^{\mathrm{a}}_{\vec{c}}$ whose objects are those complexes which are isomorphic to one of the form $\check{c}^{\infty}\mathcal{E}$ ($\mathcal{E} \in \mathbf{K}^{\mathrm{a}}_{\vec{c}}$). Then $\check{\mathbf{K}}^{\mathrm{a}}_{\vec{c}}$ *is a Δ-subcategory of* $\mathbf{K}_{\vec{c}}$. For if T is a $\mathbf{K}_{\vec{c}}$-triangle, with summit $\mathcal{E}$, whose base is a map $\mathcal{E}_1 \to \mathcal{E}_2$ with $\mathcal{E}_i \in \check{\mathbf{K}}^{\mathrm{a}}_{\vec{c}}$, then $\mathcal{E} \in \mathbf{K}^{\mathrm{a}}_{\vec{c}}$ (Remark 2.1.1(4)); and considering the map of triangles $\chi_T^{\infty}\colon T \to \check{c}^{\infty}T$ in light of the idempotence of $\chi^{\infty}_{\mathcal{E}_i}$ (see §1.2), we find that $\chi_{\mathcal{E}}^{\infty}$ is a $\mathbf{K}$-isomorphism, whence $\mathcal{E} \cong \check{c}^{\infty}\mathcal{E} \in \check{\mathbf{K}}^{\mathrm{a}}_{\vec{c}}$.

The functorial quasi-isomorphism $\chi^{\infty}\colon \mathbf{1} \to \check{c}^{\infty}$ shows that the additive functor $\check{c}^{\infty}\colon \mathbf{K}^{\mathrm{a}}_{\vec{c}} \to \check{\mathbf{K}}^{\mathrm{a}}_{\vec{c}}$ takes quasi-isomorphisms to quasi-isomorphisms, so induces a Δ-functor $\boldsymbol{\rho}\colon \mathbf{D}^{\mathrm{a}}_{\vec{c}} \to \check{\mathbf{D}}^{\mathrm{a}}_{\vec{c}}$ between the corresponding derived categories. Let $\boldsymbol{j}\colon \check{\mathbf{D}}^{\mathrm{a}}_{\vec{c}} \to \mathbf{D}^{\mathrm{a}}_{\vec{c}}$ be the natural Δ-functor. Then χ^{∞} induces a functorial isomorphism from the identity of $\mathbf{D}^{\mathrm{a}}_{\vec{c}}$ to $\boldsymbol{j\rho}$, and from the identity of $\check{\mathbf{D}}^{\mathrm{a}}_{\vec{c}}$ to $\boldsymbol{\rho j}$ (see [**Hrt**, p. 33, Proposition 3.4]), so that $\boldsymbol{\rho}$ *and* $\boldsymbol{j}$ *are quasi-inverse equivalences of Δ-categories.*

We have then the natural commutative diagram of functorial maps:

(1.3.1)
$$\begin{array}{ccccc}
\check{\mathbf{D}}^{\mathrm{a}}_{\vec{c}} & \xrightarrow[\approx]{\ \boldsymbol{j}\ } & \mathbf{D}^{\mathrm{a}}_{\vec{c}} & \xrightarrow[\approx]{\ \boldsymbol{\rho}\ } & \check{\mathbf{D}}^{\mathrm{a}}_{\vec{c}} \\
{\scriptstyle \check{q}}\uparrow & & \uparrow{\scriptstyle q^{\mathrm{a}}} & & \uparrow{\scriptstyle \check{q}} \\
\check{\mathbf{K}}^{\mathrm{a}}_{\vec{c}} & \xrightarrow[j]{\qquad} & \mathbf{K}^{\mathrm{a}}_{\vec{c}} & \xrightarrow[r:=\check{c}^{\infty}]{\qquad} & \check{\mathbf{K}}^{\mathrm{a}}_{\vec{c}}
\end{array}$$

1.3.2. In the next section, we will identify a Δ-subcategory $\check{\mathbf{L}}^{\mathrm{a}}_{\vec{c}} \subset \check{\mathbf{K}}^{\mathrm{a}}_{\vec{c}}$ such that:

(a) for every exact complex $\mathcal{P}$ in $\check{\mathbf{L}}^{\mathrm{a}}_{\vec{c}}$, the complex $\Lambda_{\mathcal{I}}\mathcal{P}$ is exact (Lemma 2.1.2),

(b) every complex $\mathcal{E}$ in $\check{\mathbf{K}}^{\mathrm{a}}_{\vec{c}}$ is the target of a quasi-isomorphism with source in $\check{\mathbf{L}}^{\mathrm{a}}_{\vec{c}}$ (Proposition 2.1.3), and

(c) if $\mathcal{E}$ is a K-flat $\mathcal{A}_{\vec{c}}$-complex then $\check{c}^{\infty}\mathcal{E} \in \check{\mathbf{L}}^{\mathrm{a}}_{\vec{c}}$ (Corollary 2.3.5).

From (a) and (b) it follows that $\check{\Lambda} := \Lambda_{\mathcal{I}}|_{\check{\mathbf{K}}^{\mathrm{a}}_{\vec{c}}}$ has a left-derived functor $\mathbf{L}\check{\Lambda}\colon \check{\mathbf{D}}^{\mathrm{a}}_{\vec{c}} \to \mathbf{D}$ such that the canonical map $\mathbf{L}\check{\Lambda}\mathcal{E} \to q\check{\Lambda}\mathcal{E}$ is an isomorphism for all $\mathcal{E} \in \check{\mathbf{L}}^{\mathrm{a}}_{\vec{c}}$ [**Hrt**, p. 53, Theorem 5.1]. In view of (c), the following formal lemma shows then[2] that $\Lambda := \Lambda_{\mathcal{I}}|_{\mathbf{K}^{\mathrm{a}}_{\vec{c}}}$ *has a left-derived functor* $\mathbf{L}\Lambda\colon \mathbf{D}^{\mathrm{a}}_{\vec{c}} \to \mathbf{D}$ *such that the canonical map* $\mathbf{L}\Lambda\mathcal{F} \to q\Lambda\mathcal{F}$ *is an isomorphism for every K-flat $\mathcal{A}_{\vec{c}}$-complex $\mathcal{F}$*, thereby completing the first main step in the proof of Theorem 0.1.

---

[2]This point was overlooked in [**AJL**] (p. 10, just before §1.2). The resulting deficiency can be corrected by the arguments in this paper, which apply *mutatis mutandis* to any separated quasi-compact scheme.

LEMMA 1.3.3. *With reference to diagram* (1.3.1), *if the restriction* $\check\Lambda := \Lambda_{\mathcal{I}}|_{\check{\mathbf{K}}^{\mathrm{a}}_{\vec{c}}}$ *has a left-derived functor* $(\mathbf{L}\check\Lambda, \check\xi)$ *then the functor* $(\mathbf{L}\check\Lambda)\boldsymbol{\rho}\colon \mathbf{D}^{\mathrm{a}}_{\vec{c}} \to \mathbf{D}$ *together with the* Δ*-functorial map*

$$\xi\colon (\mathbf{L}\check\Lambda)\boldsymbol{\rho}q^{\mathrm{a}} = (\mathbf{L}\check\Lambda)\check q r \xrightarrow{\check\xi} q\check\Lambda r \xrightarrow[(1.2.1)]{q(\Lambda_{\mathcal{I}}(\chi^{\infty}))^{-1}} q\Lambda_{\mathcal{I}}|_{\mathbf{K}^{\mathrm{a}}_{\vec{c}}}$$

is a left-derived functor of $\Lambda_{\mathcal{I}}|_{\mathbf{K}^{\mathrm{a}}_{\vec{c}}}$.

PROOF. The assertion is that for any Δ-functor $S\colon \mathbf{D}^{\mathrm{a}}_{\vec{c}} \to \mathbf{D}$, with "Hom" denoting the group of morphisms of Δ-functors, and with $\Lambda := \Lambda_{\mathcal{I}}|_{\mathbf{K}^{\mathrm{a}}_{\vec{c}}}$, the composed map

$$\alpha\colon \mathrm{Hom}(S, (\mathbf{L}\check\Lambda)\boldsymbol{\rho}) \xrightarrow{\text{natural}} \mathrm{Hom}(Sq^{\mathrm{a}}, (\mathbf{L}\check\Lambda)\boldsymbol{\rho}q^{\mathrm{a}}) \xrightarrow{\text{via } \xi} \mathrm{Hom}(Sq^{\mathrm{a}}, q\Lambda)$$

is *bijective.*

In fact, $(\mathbf{L}\check\Lambda, \check\xi)$ being a left-derived functor of $\check\Lambda$, the composition

$$\gamma\colon \mathrm{Hom}(S\boldsymbol{j}, \mathbf{L}\check\Lambda) \xrightarrow{\text{natural}} \mathrm{Hom}(S\boldsymbol{j}\check q, (\mathbf{L}\check\Lambda)\check q) \xrightarrow{\text{via } \check\xi} \mathrm{Hom}(S\boldsymbol{j}\check q, q\check\Lambda)$$

is bijective; and we show now that the following natural composition, denoted β, is an inverse of α:

$$\begin{aligned}\mathrm{Hom}(Sq^{\mathrm{a}}, q\Lambda) &\longrightarrow \mathrm{Hom}(Sq^{\mathrm{a}}j, q\Lambda j)\\ &=\!= \mathrm{Hom}(S\boldsymbol{j}\check q, q\check\Lambda)\\ &\underset{\gamma^{-1}}{\xrightarrow{\sim}} \mathrm{Hom}(S\boldsymbol{j}, \mathbf{L}\check\Lambda) \xrightarrow{\sim} \mathrm{Hom}(S, (\mathbf{L}\check\Lambda)\boldsymbol{\rho}).\end{aligned}$$

(The last isomorphism reflects the fact that $\boldsymbol{j}$ and $\boldsymbol{\rho}$ are quasi-inverse equivalences.)

First, we claim, $\beta\alpha = $ identity, i.e., the following natural diagram commutes:

$$\begin{array}{ccccc}\mathrm{Hom}(S, (\mathbf{L}\check\Lambda)\boldsymbol{\rho}) & \xrightarrow{\ \alpha\ } & \mathrm{Hom}(Sq^{\mathrm{a}}, q\Lambda) & \longrightarrow & \mathrm{Hom}(Sq^{\mathrm{a}}j, q\Lambda j)\\ \simeq\downarrow & & & & \Vert\\ \mathrm{Hom}(S\boldsymbol{j}, \mathbf{L}\check\Lambda) & & \xrightarrow[\gamma]{\ \sim\ } & & \mathrm{Hom}(S\boldsymbol{j}\check q, q\check\Lambda)\end{array}$$

Indeed, going from the top left corner to the bottom right corner either clockwise or counterclockwise has the same effect on any $\varrho \in \mathrm{Hom}(S, (\mathbf{L}\check\Lambda)\boldsymbol{\rho})$, as can be verified in a mechanical way with the aid of the commutative diagram:

$$\begin{array}{ccccccc}Sq^{\mathrm{a}}j & \xrightarrow{\text{via } \varrho} & (\mathbf{L}\check\Lambda)\boldsymbol{\rho}q^{\mathrm{a}}j & =\!=\!= & (\mathbf{L}\check\Lambda)\check q r j & \xrightarrow{\text{via } \check\xi} & q\check\Lambda r j\\ \Vert & & \Vert & & {\scriptstyle \check q(\chi^{\infty})^{-1}}\downarrow & & \downarrow{\scriptstyle q\check\Lambda(\chi^{\infty})^{-1}}\\ S\boldsymbol{j}\check q & \xrightarrow[\text{via } \varrho]{} & (\mathbf{L}\check\Lambda)\boldsymbol{\rho}\boldsymbol{j}\check q & \longrightarrow & (\mathbf{L}\check\Lambda)\check q & \xrightarrow[\text{via } \check\xi]{} & q\check\Lambda\end{array}$$

To see now that β and α are inverse to each other we need only check that *β is injective.* But this injectivity clearly follows from the fact that *any functorial map* $\varepsilon\colon S' \to q\Lambda$ *which vanishes on* $\check{\mathbf{K}}^{\mathrm{a}}_{\vec{c}}$ *must itself vanish,* a fact resulting from the commutative diagram (where $q\Lambda(\chi_{\mathcal{E}}^{\infty})$ is an isomorphism, see Lemma 1.2.1):

$$\begin{array}{ccc} S'\mathcal{E} & \xrightarrow{\varepsilon(\mathcal{E})} & q\Lambda\mathcal{E} \\ {\scriptstyle S'(\chi_{\mathcal{E}}^{\infty})}\big\downarrow & & \simeq\big\downarrow{\scriptstyle q\Lambda(\chi_{\mathcal{E}}^{\infty})} \\ S'r\mathcal{E} & \xrightarrow[\varepsilon(r\mathcal{E})=0]{} & q\Lambda r\mathcal{E} \end{array} \qquad (\mathcal{E}\in\mathbf{K}_{\vec{c}}^{\mathrm{a}}).$$

□

2. Affine-acyclicity

The purpose of this section is to establish that $\Lambda_{\mathcal{I}}|_{\mathbf{K}_{\vec{c}}^{\mathrm{a}}}$ has a left-derived functor, by showing that the full subcategory $\check{\mathbf{L}}_{\vec{c}}^{\mathrm{a}} \subset \check{\mathbf{K}}_{\vec{c}}^{\mathrm{a}}$ whose objects are the K-flat, locally $\mathcal{I}$-acyclic complexes in $\check{\mathbf{K}}_{\vec{c}}^{\mathrm{a}}$ (Definition 2.1) is a Δ-subcategory (Remark 2.1.1(4)) satisfying the conditions in Section 1.3.2.

Notation remains as before. Denote by $\mathsf{q}\colon \mathbf{K}(\mathfrak{Ab}) \to \mathbf{D}(\mathfrak{Ab})$ the natural functor from the homotopy category of complexes of abelian groups to its derived category.

DEFINITION 2.1. An $\mathcal{O}_{\mathfrak{X}}$-complex $\mathcal{E}$ is *affine-acyclic* if for each affine open $\mathfrak{U} \subset \mathfrak{X}$ the natural map is a $\mathbf{D}(\mathfrak{Ab})$-isomorphism

$$\mathsf{q}\Gamma(\mathfrak{U},\mathcal{E}) \xrightarrow{\ \sim\ } \mathbf{R}\Gamma(\mathfrak{U},\mathcal{E}).$$

$\mathcal{E}$ is *affine-$\mathcal{I}$-acyclic* if for all $n>0$, $\mathcal{E}_n := \mathcal{E}/\mathcal{I}^n\mathcal{E}$ is affine-acyclic and for any affine open $\mathfrak{U} \subset \mathfrak{X}$ the natural map is a surjection

$$\Gamma(\mathfrak{U},\mathcal{E}_n) \twoheadrightarrow \Gamma(\mathfrak{U},\mathcal{E}_{n-1}).$$

$\mathcal{E}$ is *locally $\mathcal{I}$-acyclic* if $\mathfrak{X}$ has an open cover $\mathcal{U}=(\mathfrak{U}_\alpha)$ such that the restriction $\mathcal{E}|_{\mathfrak{U}_\alpha}$ is an affine-$\mathcal{I}|_{\mathfrak{U}_\alpha}$-acyclic $\mathcal{O}_{\mathfrak{U}_\alpha}$-complex for every α.

REMARKS 2.1.1. (1) Affine-acyclicity is affine-$\mathcal{I}$-acyclicity for $\mathcal{I}=(0)$.

(2) If $\mathcal{E}$ and $\mathcal{F}$ are $\mathbf{K}$-isomorphic complexes and $\mathcal{E}$ is affine-acyclic then so is $\mathcal{F}$.

(3) An $\mathcal{O}_{\mathfrak{X}}$-module $\mathcal{E}$ (considered as an $\mathcal{O}_{\mathfrak{X}}$-complex concentrated in degree 0) is affine-acyclic iff for each affine open $\mathfrak{U} \subset \mathfrak{X}$, $\mathrm{H}^i(\mathfrak{U},\mathcal{E})=0$ for all $i>0$. In particular, any flasque $\mathcal{O}_{\mathfrak{X}}$-module is affine-acyclic.

(4) The cone $\mathcal{C}_u$ of any map $u\colon \mathcal{E}\to\mathcal{E}'$ of locally $\mathcal{I}$-acyclic complexes is itself locally $\mathcal{I}$-acyclic—an appropriate cover consisting of all intersections $\mathfrak{U}_\alpha\cap\mathfrak{U}'_\beta$ where $(\mathfrak{U}_\alpha)$ (resp. $(\mathfrak{U}'_\beta)$) is a cover over which $\mathcal{E}$ (resp. $\mathcal{E}'$) is affine-$\mathcal{I}$-acyclic.

(*Proof.* Let $u_n\colon \mathcal{E}_n \to \mathcal{E}'_n$ be the map induced by u, and consider the map of $\mathbf{D}(\mathfrak{Ab})$-triangles obtained by applying $\mathsf{q}\Gamma(\mathfrak{U},-)\to\mathbf{R}\Gamma(\mathfrak{U},-)$ $(\mathfrak{U}\subset\mathfrak{U}_\alpha\cap\mathfrak{U}'_\beta)$ to the standard $\mathbf{K}$-triangle $\mathcal{E}_n \xrightarrow{u_n} \mathcal{E}'_n \to \mathcal{C}_u/\mathcal{I}^n\mathcal{C}_u \to \mathcal{E}_n[1], \dots$)

Hence if $\mathbf{K}_0$ is a Δ-subcategory of $\mathbf{K}$, then the full subcategory of locally $\mathcal{I}$-acyclic complexes in $\mathbf{K}_0$ is also a Δ-subcategory of $\mathbf{K}$.

Similarly, the full subcategory $\mathbf{K}_0^{\mathrm{a}} \subset \mathbf{K}_0$ whose objects are the affine-acyclic complexes is a Δ-subcategory of $\mathbf{K}$. For example, as was noted in the Introduction, *the inclusion $\mathbf{K}_{\vec{c}}^{\mathrm{a}} \hookrightarrow \mathbf{K}_{\vec{c}}$ induces an equivalence of derived categories $\mathbf{D}_{\vec{c}}^{\mathrm{a}} \xrightarrow{\approx} \mathbf{D}_{\vec{c}}$.*

LEMMA 2.1.2. *If $\mathcal{E}$ is an exact K-flat locally $\mathcal{I}$-acyclic complex then $\Lambda_{\mathcal{I}}\mathcal{E}$ is exact.*

PROOF. Since $\mathcal{E}$ is K-flat the complexes $\mathcal{E}_n := \mathcal{E}/\mathcal{I}^n\mathcal{E}$ $(n\ge 0)$ are all exact ([**Spn**, p. 140, Proposition 5.7]), and being affine-acyclic, remain exact after application of any functor $\Gamma(\mathfrak{U},-)$ with $\mathfrak{U}$ an affine open subset of any $\mathfrak{U}_\alpha$ as in Definition 2.1. Moreover, the natural maps $\Gamma(\mathfrak{U},\mathcal{E}_n)\to\Gamma(\mathfrak{U},\mathcal{E}_{n-1})$ are surjections.

So by [**EGA**, p. 66, (13.2.3)], the complex

$$\Gamma(\mathfrak{U}, \Lambda\mathcal{E}) = \varprojlim_{n} \Gamma(\mathfrak{U}, \mathcal{E}_n)$$

is exact for any such $\mathfrak{U}$, whence the assertion. □

This takes care of condition (a) in Section 1.3.2. Condition (b) is given by the following proposition.

PROPOSITION 2.1.3. *Any $\mathcal{E} \in \check{\mathbf{K}}^{\mathrm{a}}_{\vec{\mathrm{c}}}$ admits a quasi-isomorphism $\mathcal{P}_{\mathcal{E}} \to \mathcal{E}$ where $\mathcal{P}_{\mathcal{E}} \in \check{\mathbf{K}}^{\mathrm{a}}_{\vec{\mathrm{c}}}$ is flat, K-flat and locally $\mathcal{I}$-acyclic.*

The *proof* will occupy the rest of section 2.

2.2. We need various examples of affine-acyclic and affine-$\mathcal{I}$-acyclic complexes.

EXAMPLES 2.2.1. (a) If $\mathcal{E} \in \mathcal{A}_{\vec{c}}$, then $\mathcal{I}^n\mathcal{E} \in \mathcal{A}_{\vec{c}}$ for all $n \geq 0$: this results from [**DFS**, Proposition 3.2.2], because $\mathcal{I}^n\mathcal{E}$ is the image of the natural map $\mathcal{I}^n \otimes \mathcal{E} \to \mathcal{E}$ whose source and target are both in $\mathcal{A}_{\vec{c}}$. [**DFS**, Proposition 3.2.2] shows further that $\mathcal{I}^m\mathcal{E}/\mathcal{I}^n\mathcal{E} \in \mathcal{A}_{\vec{c}}$ whenever $0 \leq m \leq n$. It follows then from [**DFS**, Corollary 3.1.8] that *$\mathcal{E}$ is affine-$\mathcal{I}$-acyclic.*

(b) *Any $\varinjlim$ of affine-acyclic complexes with $\mathcal{A}_{\vec{c}}$-homology is affine-acyclic and, of course, has $\mathcal{A}_{\vec{c}}$-homology.*

For, by [**DFS**, Proposition 3.2.2], $\mathcal{A}_{\vec{c}}$ is plump in $\mathcal{A}$, and by [**DFS**, Proposition 3.4.3], the functor $\mathbf{R}\Gamma(\mathfrak{U}, -)$ is bounded-above on $\mathbf{D}_{\vec{c}}(\mathfrak{U})$ for any open $\mathfrak{U} \subset \mathfrak{X}$. One can, then, adapt the proof of [**Lpm**, Corollary (3.9.3.2)], *mutatis mutandis,* to show that for any small directed system $\mathcal{E}_\eta$ of $\mathcal{O}_{\mathfrak{X}}$-complexes with $\mathcal{A}_{\vec{c}}$-homology, and each $i \in \mathbb{Z}$, the bottom row of the natural commutative diagram

$$\begin{array}{ccc} \varinjlim_{\eta} \mathrm{H}^i\Gamma(\mathfrak{U}, \mathcal{E}_\eta) & \xrightarrow{\ \sim\ } & \mathrm{H}^i\Gamma(\mathfrak{U}, \varinjlim_{\eta} \mathcal{E}_\eta) \\ {\scriptstyle \mu_i}\downarrow & & \downarrow{\scriptstyle \nu_i} \\ \varinjlim_{\eta} \mathrm{H}^i\mathbf{R}\Gamma(\mathfrak{U}, \mathcal{E}_\eta) & \longrightarrow & \mathrm{H}^i\mathbf{R}\Gamma(\mathfrak{U}, \varinjlim_{\eta} \mathcal{E}_\eta) \end{array}$$

is an *isomorphism.*[3] Thus if every $\mathcal{E}_\eta$ is affine-acyclic, so that each μ_i is an isomorphism, then each ν_i is an isomorphism, and so $\varinjlim \mathcal{E}_\eta$ is affine-acyclic, as asserted.

(c) *Any complex $\mathcal{E}$ in $\mathcal{A}_{\vec{c}}$ is affine-$\mathcal{I}$-acyclic.*

(In particular, since quasi-coherent $\mathcal{O}_{\mathfrak{X}}$-modules are locally in $\mathcal{A}_{\vec{c}}$ ([**DFS**, Corollary 3.1.4]), *any bounded quasi-coherent $\mathcal{O}_{\mathfrak{X}}$-complex is locally $\mathcal{I}$-acyclic.*)

Indeed, for all $n > 0$ and affine open $\mathfrak{U} \subset \mathfrak{X}$ one sees as in (a) above that $\mathrm{H}^1(\mathfrak{U}, \mathcal{I}^{n-1}\mathcal{E}/\mathcal{I}^n\mathcal{E}) = 0$, whence the natural map is a surjection

$$\Gamma(\mathfrak{U}, \mathcal{E}_n) \twoheadrightarrow \Gamma(\mathfrak{U}, \mathcal{E}_{n-1}) \qquad (\mathcal{E}_n := \mathcal{E}/\mathcal{I}^n\mathcal{E});$$

and that each component $\mathcal{E}_n^m$ ($m \in \mathbb{Z}$) of $\mathcal{E}_n$ is affine-acyclic. To see that $\mathcal{E}_n$ is affine-acyclic when $\mathcal{E}$ is bounded below, represent the complex $\mathbf{R}\Gamma(\mathfrak{U}, \mathcal{E}_n)$ as $\Gamma(\mathfrak{U}, \mathfrak{T}_n)$, where $\mathfrak{T}_n$ is the total complex associated to a Cartan-Eilenberg resolution of $\mathcal{E}_n$; and use standard arguments to deduce from affine-acyclicity of $\mathcal{E}_n^m$ that the natural map $\Gamma(\mathfrak{U}, \mathcal{E}_n) \to \Gamma(\mathfrak{U}, \mathfrak{T}_n)$ is a quasi-isomorphism. Then for the unbounded case, let $\mathcal{E}_{n,\eta}$ be the complex $\cdots \to 0 \to 0 \to \mathcal{E}_n^{-\eta} \to \mathcal{E}_n^{-\eta+1} \to \cdots$ ($\eta \in \mathbb{N}$) and apply (b) to $\mathcal{E}_n = \varinjlim_{\eta} \mathcal{E}_{n,\eta}$.

[3] Roughly, boundedness of $\mathbf{R}\Gamma(\mathfrak{U}, -)$ enables reduction to where the $\mathcal{E}_\eta$ are uniformly bounded below, and then $\mathbf{R}\Gamma(\mathfrak{U}, \varinjlim \mathcal{E}_\eta)$ can be represented via $\varinjlim$'s of canonical flasque resolutions ...

The next intermediate goal is Lemma 2.2.7, a generalization of Example (c).

LEMMA 2.2.2. *Let $\nu\colon \mathcal{V} \hookrightarrow \mathcal{X}$ be an open immersion with $\mathcal{V}$ affine. Then for any affine-acyclic $\mathcal{O}_{\mathcal{V}}$-complex $\mathcal{F}$ the natural map is a* **D***-isomorphism.*

$$\theta\colon \nu_*\mathcal{F} \xrightarrow{\;\sim\;} \mathbf{R}\nu_*\mathcal{F}.$$

PROOF. Let $\rho\colon \mathcal{F} \to \mathcal{J}$ be a K-injective resolution. Then θ can be identified with the map $\nu_*\rho\colon \nu_*\mathcal{F} \to \nu_*\mathcal{J}$. A **D**-map being an isomorphism if and only if it induces homology isomorphisms, Lemma 2.2.2 just asserts that for all $i \in \mathbb{Z}$, the resulting map Θ^i of presheaves

$$\begin{aligned}\Theta^i(\mathcal{U})\colon \mathrm{H}^i\Gamma(\nu^{-1}\mathcal{U},\mathcal{F}) &= \mathrm{H}^i\Gamma(\mathcal{U},\nu_*\mathcal{F})\\ &\to \mathrm{H}^i\Gamma(\mathcal{U},\nu_*\mathcal{J}) = \mathrm{H}^i\Gamma(\nu^{-1}\mathcal{U},\mathcal{J}) \qquad (\mathcal{U}\text{ open in }\mathcal{X})\end{aligned}$$

induces an isomorphism of the associated sheaves. When $\mathcal{U}$ is affine, $\nu^{-1}\mathcal{U}$ is affine (since $\mathcal{X}$ is separated), so by Definition 2.1 the natural map is a **D**-isomorphism

$$\Gamma(\nu^{-1}\mathcal{U},\mathcal{F}) \xrightarrow{\;\sim\;} \mathbf{R}\Gamma(\nu^{-1}\mathcal{U},\mathcal{F}) = \Gamma(\nu^{-1}\mathcal{U},\mathcal{J}),$$

i.e., $\Theta^i(\mathcal{U})$ itself is an isomorphism, whence the conclusion. □

COROLLARY 2.2.3. *If $\nu\colon \mathcal{V} \hookrightarrow \mathcal{X}$ is an open immersion with $\mathcal{V}$ affine, then for any affine-acyclic $\mathcal{O}_{\mathcal{V}}$-complex $\mathcal{F}$, $\nu_*\mathcal{F}$ is also affine-acyclic.*

PROOF. For the inclusion $\lambda\colon \mathcal{U} \hookrightarrow \mathcal{X}$ of an affine open subset, we have a commutative diagram of open immersions where, $\mathcal{X}$ being separated, $\nu^{-1}\mathcal{U}$ is affine:

$$\begin{array}{ccc}\nu^{-1}\mathcal{U} & \xrightarrow{\;\lambda'\;} & \mathcal{V}\\ {\scriptstyle\nu'}\downarrow & & \downarrow{\scriptstyle\nu}\\ \mathcal{U} & \xrightarrow[\;\lambda\;]{} & \mathcal{X}\end{array}$$

Let $\rho\colon \mathcal{F} \to \mathcal{J}$ be a K-injective resolution. Since ν_* has an exact left adjoint (namely ν^*), therefore $\nu_*\mathcal{J}$ is K-injective; and so by Lemma 2.2.2, $\nu_*\rho\colon \nu_*\mathcal{F} \to \nu_*\mathcal{J}$ is a K-injective resolution. As $\mathcal{F}$ is affine-acyclic, the map

$$\Gamma(\mathcal{U},\nu_*\mathcal{F}) = \Gamma(\nu^{-1}\mathcal{U},\mathcal{F}) \xrightarrow{\text{via }\rho} \Gamma(\nu^{-1}\mathcal{U},\mathcal{J}) = \Gamma(\mathcal{U},\nu_*\mathcal{J})$$

is a quasi-isomorphism; and the conclusion follows. □

LEMMA 2.2.4. *Let $\mathcal{U} = (\mathcal{U}_\alpha)_{1\le\alpha\le t}$ be a finite affine open cover of $\mathcal{X}$, and let $\check{C}^\bullet$ be the corresponding Čech functor from $\mathcal{U}$-complexes to $\mathcal{O}_{\mathcal{X}}$-complexes. If $\mathcal{F}$ is a $\mathcal{U}$-complex such that $\mathcal{F}_j$ is an affine-acyclic $\mathcal{O}_{\mathcal{U}_j}$-complex for all $j \in \mathfrak{P}_t$ then $\check{C}^\bullet\mathcal{F}$ is an affine–acyclic $\mathcal{O}_{\mathcal{X}}$-complex. Moreover, if $\mathcal{F}$ is exact then so is $\check{C}^\bullet\mathcal{F}$.*

PROOF. The complex $\check{C}^\bullet\mathcal{F}$ has a finite filtration

$$\check{C}^\bullet\mathcal{F} = C_0 \supset C_1 \supset C_2 \cdots \supset C_t = 0,$$

with C_s the total complex associated to the double complex $\check{C}^p(\mathcal{F}^q)_{p\ge s;\, q\in\mathbb{Z}}$. Since C_{s+1} is, as graded module, a direct summand of C_s, we have triangles in $\mathbf{K}(\mathcal{X})$

$$C_{s+1} \longrightarrow C_s \longrightarrow C_s/C_{s+1} = \oplus_{|j|=s}\lambda_{j*}\mathcal{F}_j \to C_{s+1}[1] \qquad (0 \le s < t)$$

(see e.g., [**Lpm**, Example (1.4.3)]).

Since $\lambda_{j*}\mathcal{F}_j$ is affine-acyclic (Corollary 2.2.3), descending induction on s, starting with $s = t-1$, yields that each C_s is affine-acyclic. Thus $\check{C}^\bullet\mathcal{F}$ $(= C_0)$ is affine-acyclic.

If $\mathcal{F}$ is exact then for $k \in \mathfrak{P}_t$, Lemma 2.2.2 shows that $\lambda_{k*}\mathcal{F}_k$ is exact. Hence for $p \geq 0$, the complex $\check{C}^p\mathcal{F}$ is exact. The family $(\check{C}^p\mathcal{F})_{0\leq p<t}$ contains all the non-vanishing columns of the double complex of which $\check{C}^\bullet\mathcal{F}$ is the total complex, and the assertion results. $\square$

COROLLARY 2.2.5. *Let $\mathcal{U} = (\mathfrak{U}_\alpha)_{1\leq\alpha\leq t}$ be a finite affine open cover of $\mathfrak{X}$. Let $\check{c} := \check{c}_{\mathcal{U}}$ and $\check{c}^\infty := \varinjlim \check{c}^m$ be the corresponding functors (see §1.2). If $\mathcal{E} \in \mathbf{K}_{\vec{c}}(\mathfrak{X})$ is an affine-acyclic $\mathcal{O}_{\mathfrak{X}}$-complex, then so are $\check{c}\mathcal{E}$ and $\check{c}^\infty\mathcal{E}$.*

PROOF. Affine-acyclicity of $\check{c}\mathcal{E}$ is given by Lemma 2.2.4, applied to the $\mathcal{U}$-pullback $\mathcal{F}$ of $\mathcal{E}$ ($\mathcal{F}_j := \lambda_j^*\mathcal{E}$ for all $j \in \mathfrak{P}_t \ldots$; and $\check{c}\mathcal{E} = \check{C}^\bullet\mathcal{F}$).

By induction, $\check{c}^m\mathcal{E}$ is affine-acyclic for all $m > 0$. Being the target of a quasi-isomorphism from $\mathcal{E}$, $\check{c}^m\mathcal{E}$ is in $\mathbf{K}_{\vec{c}}(\mathfrak{X})$. It results then from Example 2.2.1(b) that $\check{c}^\infty\mathcal{E}$ is affine-acyclic. $\square$

LEMMA 2.2.6. *Let $\lambda\colon \mathfrak{U} \to \mathfrak{X}$ be an affine map of locally noetherian formal schemes. Then for any $\mathcal{F} \in \mathcal{A}_{\vec{c}}(\mathfrak{U})$ the natural map is an isomorphism*

$$\pi_{\mathcal{G}}\colon \lambda_*\mathcal{F} \otimes_{\mathcal{O}_{\mathfrak{X}}} \mathcal{G} \xrightarrow{\ \sim\ } \lambda_*(\mathcal{F} \otimes_{\mathcal{O}_{\mathfrak{U}}} \lambda^*\mathcal{G}) \qquad (\mathcal{G} \in \mathcal{A}_{\mathrm{qc}}(\mathfrak{X})).$$

PROOF. The map $\pi_{\mathcal{G}}$ is, by definition, adjoint to the natural map

$$\lambda^*(\lambda_*\mathcal{F} \otimes_{\mathcal{O}_{\mathfrak{X}}} \mathcal{G}) \xrightarrow{\ \sim\ } \lambda^*\lambda_*\mathcal{F} \otimes_{\mathcal{O}_{\mathfrak{U}}} \lambda^*\mathcal{G} \to \mathcal{F} \otimes_{\mathcal{O}_{\mathfrak{U}}} \lambda^*\mathcal{G}.$$

It follows that $\pi_{\mathcal{G}}$ "commutes" (in the obvious sense) with open immersions into $\mathfrak{X}$. Hence we may assume that $\mathfrak{X}$ and $\mathfrak{U}$ are affine, and that there is an exact sequence

$$\mathcal{G}_2 \to \mathcal{G}_1 \to \mathcal{G} \to 0$$

with $\mathcal{G}_2$ and $\mathcal{G}_1$ direct sums of copies of $\mathcal{O}_{\mathfrak{X}}$.

Then both $\mathcal{F} \otimes \lambda^*\mathcal{G}_1$ and $\mathcal{F} \otimes \lambda^*\mathcal{G}_2$, being direct sums of copies of $\mathcal{F}$, are in $\mathcal{A}_{\vec{c}}(\mathfrak{U})$. So [**DFS**, Proposition 3.2.2] yields that the kernels of the maps a and b in the natural exact sequence

$$\mathcal{F} \otimes \lambda^*\mathcal{G}_2 \xrightarrow{a} \mathcal{F} \otimes \lambda^*\mathcal{G}_1 \xrightarrow{b} \mathcal{F} \otimes \lambda^*\mathcal{G} \to 0$$

are in $\mathcal{A}_{\vec{c}}(\mathfrak{U})$. From [**DFS**, Lemma 3.4.2] it follows then that the bottom row in the natural commutative diagram

$$\begin{array}{ccccccc}
\lambda_*\mathcal{F} \otimes \mathcal{G}_2 & \longrightarrow & \lambda_*\mathcal{F} \otimes \mathcal{G}_1 & \longrightarrow & \lambda_*\mathcal{F} \otimes \mathcal{G} & \longrightarrow & 0 \\
\downarrow{\scriptstyle \pi_{\mathcal{G}_2}} & & \downarrow{\scriptstyle \pi_{\mathcal{G}_1}} & & \downarrow{\scriptstyle \pi_{\mathcal{G}}} & & \\
\lambda_*(\mathcal{F} \otimes \lambda^*\mathcal{G}_2) & \longrightarrow & \lambda_*(\mathcal{F} \otimes \lambda^*\mathcal{G}_1) & \longrightarrow & \lambda_*(\mathcal{F} \otimes \lambda^*\mathcal{G}) & \longrightarrow & 0
\end{array}$$

is exact, and of course the top row is exact.

Since $\mathfrak{U}$ is noetherian, the functor λ_* commutes with direct sums, whence $\pi_{\mathcal{G}_2}$ and $\pi_{\mathcal{G}_1}$ are isomorphisms; so $\pi_{\mathcal{G}}$ is an isomorphism too. $\square$

LEMMA 2.2.7. *Let $\mathcal{U} = (\mathfrak{U}_\alpha)_{1\leq\alpha\leq t}$ be a finite affine open cover of $\mathfrak{X}$, and let $\check{C}^\bullet$ be the corresponding Čech functor from $\mathcal{U}$-complexes to $\mathcal{O}_{\mathfrak{X}}$-complexes. Let $\mathcal{F}$ be a $\mathcal{U}$-complex such that for every $j \in \mathfrak{P}_t$, $\mathcal{F}_j$ is a direct sum of complexes of the form $\lambda_{ij*}\mathcal{F}_{ij}$ $(i \supset j)$ with $\mathcal{F}_{ij}$ an $\mathcal{A}_{\vec{c}}(\mathfrak{U}_i)$-complex. Then $\check{C}^\bullet\mathcal{F}$ is affine-$\mathcal{I}$-acyclic; and $\check{c}^\infty\check{C}^\bullet\mathcal{F}$ is locally $\mathcal{I}$-acyclic.*

PROOF. Lemma 2.2.6 (with $\lambda = \lambda_{ij}, \mathcal{F} = \mathcal{F}_{ij}$ and $\mathcal{G} = \mathcal{O}_{\mathfrak{U}_j}/\mathcal{I}^n\mathcal{O}_{\mathfrak{U}_j}$) provides an isomorphism

$$\lambda_{ij*}\mathcal{F}_{ij}/\mathcal{I}^n\lambda_{ij*}\mathcal{F}_{ij} \xrightarrow{\sim} \lambda_{ij*}(\mathcal{F}_{ij}/\mathcal{I}^n\mathcal{F}_{ij}).$$

Hence $\mathcal{F}_j/\mathcal{I}^n\mathcal{F}_j \cong \oplus_{i\supset j}\lambda_{ij*}(\mathcal{F}_{ij}/\mathcal{I}^n\mathcal{F}_{ij})$. As in Example 2.2.1(a), $\mathcal{F}_{ij}/\mathcal{I}^n\mathcal{F}_{ij}$ is an $\mathcal{A}_{\vec{c}}(\mathfrak{U}_i)$-complex for all $n > 0$, so that by Example 2.2.1(c), $\mathcal{F}_{ij}/\mathcal{I}^n\mathcal{F}_{ij}$ is affine-acyclic. By Corollary 2.2.3, then, $\mathcal{F}_j/\mathcal{I}^n\mathcal{F}_j$ is an affine-acyclic $\mathcal{O}_{U_j}$-complex; and so by Lemma 2.2.4, $\check{C}^\bullet(\mathcal{F}/\mathcal{I}^n\mathcal{F})$ is an affine-acyclic $\mathcal{O}_{\mathfrak{X}}$-complex.

Now the natural map $\check{C}^\bullet\mathcal{F}/(\mathcal{I}^n\check{C}^\bullet\mathcal{F}) \to \check{C}^\bullet(\mathcal{F}/\mathcal{I}^n\mathcal{F})$ is an isomorphism, since it is a direct sum of maps of the form

$$\begin{aligned}\lambda_{j*}\lambda_{ij*}\mathcal{F}_{ij}/\mathcal{I}^n\lambda_{j*}\lambda_{ij*}\mathcal{F}_{ij} &= \lambda_{i*}\mathcal{F}_{ij}/\mathcal{I}^n\lambda_{i*}\mathcal{F}_{ij}\\ &\underset{2.2.6}{\xrightarrow{\sim}} \lambda_{i*}(\mathcal{F}_{ij}/\mathcal{I}^n\mathcal{F}_{ij}) = \lambda_{j*}\lambda_{ij*}(\mathcal{F}_{ij}/\mathcal{I}^n\mathcal{F}_{ij}).\end{aligned}$$

So $\check{C}^\bullet\mathcal{F}/(\mathcal{I}^n\check{C}^\bullet\mathcal{F})$ is affine-acyclic.

Also, for any affine open $\mathfrak{U} \subset \mathfrak{X}$ and $n > 0$, the natural map is a *surjection*

$$\Gamma(\mathfrak{U}, \check{C}^\bullet\mathcal{F}/(\mathcal{I}^{n+1}\check{C}^\bullet\mathcal{F})) \twoheadrightarrow \Gamma(\mathfrak{U}, \check{C}^\bullet\mathcal{F}/(\mathcal{I}^n\check{C}^\bullet\mathcal{F})),$$

because it is isomorphic to a direct sum of surjective maps of the form

$$\begin{aligned}\Gamma(\mathfrak{U}, \lambda_{i*}(\mathcal{F}_{ij}/\mathcal{I}^{n+1}\mathcal{F}_{ij})) &= \Gamma(\lambda_i^{-1}\mathfrak{U}, \mathcal{F}_{ij}/\mathcal{I}^{n+1}\mathcal{F}_{ij})\\ &\twoheadrightarrow \Gamma(\lambda_i^{-1}\mathfrak{U}, \mathcal{F}_{ij}/\mathcal{I}^n\mathcal{F}_{ij}) = \Gamma(\mathfrak{U}, \lambda_{i*}(\mathcal{F}_{ij}/\mathcal{I}^n\mathcal{F}_{ij})),\end{aligned}$$

surjectivity holding by [**DFS**, Lemma 3.1.8], since $\mathcal{I}^n\mathcal{F}_{ij}/\mathcal{I}^{n+1}\mathcal{F}_{ij} \in \mathcal{A}_{\vec{c}}(\mathfrak{U}_i)$ (see Example 2.2.1(a)). Thus $\check{C}^\bullet\mathcal{F}$ is indeed affine-$\mathcal{I}$-acyclic.

Next, the pullback of $\check{C}^\bullet\mathcal{F}$ to a $\mathcal{U}$-module $\mathcal{F}^*$ is such that for every $\ell \in \mathfrak{P}_t$, $\mathcal{F}^*_\ell$ is a direct sum of complexes of the form $\lambda_{k\ell*}\mathcal{F}^*_{k\ell}$ ($k \supset \ell$) with $\mathcal{F}^*_{k\ell}$ an $\mathcal{A}_{\vec{c}}(\mathfrak{U}_k)$-complex: that follows from the relation

$$\lambda^*_\ell\lambda_{j*}\lambda_{ij*}\mathcal{F}_{ij} = \lambda_{(i\cup\ell)\ell*}\lambda^*_{(i\cup\ell)i}\mathcal{F}_{ij} \qquad (i \supset j \in \mathfrak{P}_t,\ \ell \in \mathfrak{P}_t).$$

Hence $\check{c}\check{C}^\bullet\mathcal{F} = \check{C}^\bullet\mathcal{F}^*$ is again affine-$\mathcal{I}$-acyclic. And as above, the natural map is an isomorphism

$$\check{c}\check{C}^\bullet\mathcal{F}/(\mathcal{I}^n\check{c}\check{C}^\bullet\mathcal{F}) = \check{C}^\bullet\mathcal{F}^*/(\mathcal{I}^n\check{C}^\bullet\mathcal{F}^*) \xrightarrow{\sim} \check{C}^\bullet(\mathcal{F}^*/\mathcal{I}^n\mathcal{F}^*) = \check{c}(\check{C}^\bullet\mathcal{F}/\mathcal{I}^n\check{C}^\bullet\mathcal{F}).$$

By induction we find that $\check{c}^m\check{C}^\bullet\mathcal{F} = \check{c}^{m-1}\check{C}^\bullet\mathcal{F}^*$ is affine-$\mathcal{I}$-acyclic for all $m > 0$. So for all $n > 0$ and any affine open $\mathfrak{U} \subset \mathfrak{X}$, the natural map is a surjection

$$\Gamma(\mathfrak{U}, \check{c}^m\check{C}^\bullet\mathcal{F}/\mathcal{I}^{n+1}\check{c}^m\check{C}^\bullet\mathcal{F}) \twoheadrightarrow \Gamma(\mathfrak{U}, \check{c}^m\check{C}^\bullet\mathcal{F}/\mathcal{I}^n\check{c}^m\check{C}^\bullet\mathcal{F});$$

and applying $\varinjlim_m$, we get the same with ∞ in place of m.

By induction again, there is a natural isomorphism

$$\begin{aligned}\check{c}^m\check{C}^\bullet\mathcal{F}/(\mathcal{I}^n\check{c}^m\check{C}^\bullet\mathcal{F}) &= \check{c}^{m-1}\check{C}^\bullet\mathcal{F}^*/(\mathcal{I}^n\check{c}^{m-1}\check{C}^\bullet\mathcal{F}^*)\\ &\xrightarrow{\sim} \check{c}^{m-1}(\check{C}^\bullet\mathcal{F}^*/\mathcal{I}^n\check{C}^\bullet\mathcal{F}^*) \xrightarrow{\sim} \check{c}^m(\check{C}^\bullet\mathcal{F}/\mathcal{I}^n\check{C}^\bullet\mathcal{F}).\end{aligned}$$

So for all $n > 0$, the restriction to any $\mathfrak{U}_\alpha$ of

$$\check{c}^\infty\check{C}^\bullet\mathcal{F}/(\mathcal{I}^n\check{c}^\infty\check{C}^\bullet\mathcal{F}) = \varinjlim_m \check{c}^m\check{C}^\bullet\mathcal{F}/(\mathcal{I}^n\check{c}^m\check{C}^\bullet\mathcal{F}) \cong \varinjlim_m \check{c}^m(\check{C}^\bullet\mathcal{F}/\mathcal{I}^n\check{C}^\bullet\mathcal{F})$$

is homotopy-isomorphic to the restriction of $\check{C}^\bullet\mathcal{F}/\mathcal{I}^n\check{C}^\bullet\mathcal{F}$ (cf. end of Section 1.2), so that $(\check{c}^\infty\check{C}^\bullet\mathcal{F}/\mathcal{I}^n\check{c}^\infty\check{C}^\bullet\mathcal{F})|_{\mathfrak{U}_\alpha}$ is affine-acyclic (Remark 2.1.1(2)).

Thus $\check{c}^\infty\check{C}^\bullet\mathcal{F}$ is locally $\mathcal{I}$-acyclic (over the cover $\mathcal{U}$), as asserted. □

2.3. We also need a few results concerning flatness of complexes.

LEMMA 2.3.1. *Let $\lambda\colon \mathfrak{U} \to \mathfrak{X}$ be a flat affine map of locally noetherian formal schemes. If $\mathcal{F} \in \mathcal{A}_{\vec{c}}(\mathfrak{U})$ is $\mathcal{O}_{\mathfrak{U}}$-flat, then $\lambda_*\mathcal{F}$ is $\mathcal{O}_{\mathfrak{X}}$-flat.*

PROOF. The question being local, we may assume that $\mathfrak{U} = \mathrm{Spf}(B)$ and $\mathfrak{X} = \mathrm{Spf}(A)$ where A and B are noetherian adic rings, and that $\lambda = \mathrm{Spf}(\phi)$ where $\phi\colon A \to B$ is a continuous ring homomorphism. Then by [**DFS**, Proposition 3.1.1], there is a B-module F such that $\mathcal{F} \cong \kappa^*\widetilde{F}$, where $\kappa\colon \mathrm{Spf}(B) \to \mathrm{Spec}(B)$ is the canonical map and $\widetilde{F}$ is the quasi-coherent sheaf on $\mathrm{Spec}(B)$ corresponding to F.

We claim that F *is B-flat.* Indeed, if J is any B-ideal and N is the kernel of the natural map $\tau_J\colon J \otimes_B F \to F$, then $\kappa^*\widetilde{N}$ is the kernel of the induced map $\kappa^*\widetilde{J} \otimes_{\mathcal{O}_{\mathfrak{X}}} \kappa^*\widetilde{F} \to \kappa^*\widetilde{F}$, a map which is injective since $\kappa^*\widetilde{F} \cong \mathcal{F}$ is flat, so that $\kappa^*\widetilde{N} = 0$, whence by [**DFS**, Proposition 3.1.1], $\widetilde{N} = 0$, and so $N = 0$. Thus τ_J is injective for all J, i.e., F is B-flat.

Now for each $f \in A$, with $\kappa_f\colon \mathrm{Spf}(B_{\{f\}}) \to \mathrm{Spec}(B_{\{f\}})$ the natural map, and for any $B_{\{f\}}$-module M with corresponding quasi-coherent $\mathcal{O}_{\mathrm{Spec}(B_{\{f\}})}$-module $M^{\approx}$, it follows from [**DFS**, Proposition 3.1.1] (as the "quasi-coherator" $Q = Q_{\mathrm{Spec}(B_{\{f\}})}$ can be taken to be $\Gamma(\mathrm{Spec}(B_{\{f\}}), -)^{\sim}$) that

$$\Gamma(\mathrm{Spf}(B_{\{f\}}), \kappa_f^*(M^{\approx})) = \Gamma(\mathrm{Spec}(B_{\{f\}}), Q\kappa_{f*}\kappa_f^*(M^{\approx})) \cong \Gamma(\mathrm{Spec}(B_{\{f\}}), M^{\approx}) \cong M.$$

Taking $M = F \otimes_B B_{\{f\}}$—so that with $i_f\colon \mathrm{Spec}(B_{\{f\}}) \to \mathrm{Spec}(B)$ the natural map, $M^{\approx} = i_f^*\widetilde{F}$—we conclude that

$$\Gamma(\mathrm{Spf}(A_{\{f\}}), \lambda_*\kappa^*\widetilde{F}) = \Gamma(\mathrm{Spf}(B_{\{f\}}), \kappa^*\widetilde{F}) = \Gamma(\mathrm{Spf}(B_{\{f\}}), \kappa_f^*(M^{\approx})) \cong F \otimes_B B_{\{f\}}.$$

Hence, since $B_{\{f\}}$ is $A_{\{f\}}$-flat ([**DFS**, Lemma 7.1.1]), $\Gamma(\mathrm{Spf}(A_{\{f\}}), \lambda_*\kappa^*\widetilde{F})$ is flat over $A_{\{f\}} = \Gamma(\mathrm{Spf}(A_{\{f\}}), \mathcal{O}_{\mathfrak{X}})$ for any $f \in A$; and so $\lambda_*\kappa^*\widetilde{F} \cong \lambda_*\mathcal{F}$ is $\mathcal{O}_{\mathfrak{X}}$-flat. □

From the definitions of $\check{C}^\bullet$ and $\check{c}^m$ it follows now that:

COROLLARY 2.3.2. *If $\mathcal{U}$ is a finite affine open cover of $\mathfrak{X}$ and $\mathcal{P}$ is a flat $\mathcal{A}_{\vec{c}}(\mathcal{U})$-complex, then $\check{c}^m\check{C}^\bullet\mathcal{P}$ $(0 \le m \le \infty)$ is a flat $\mathcal{O}_{\mathfrak{X}}$-complex.*

If $\lambda\colon \mathfrak{U} \hookrightarrow \mathfrak{X}$ is the inclusion of an open subset then since $\mathfrak{X}$ is noetherian, λ_* preserves $\varinjlim$. Hence $\check{C}^\bullet$ and $\check{c}^m$ $(0 \le m \le \infty)$ preserve $\varinjlim$, and so:

COROLLARY 2.3.3. *If $\mathcal{U}$ is a finite affine open cover of $\mathfrak{X}$ and $\mathcal{P}$ is a $\varinjlim$ of bounded-above flat $\mathcal{A}_{\vec{c}}(\mathcal{U})$-complexes, then $\check{c}^m\check{C}^\bullet\mathcal{P}$ $(0 \le m \le \infty)$ is a $\varinjlim$ of bounded-above flat $\mathcal{O}_{\mathfrak{X}}$-complexes, and hence is flat and K-flat.*

COROLLARY 2.3.4. (i) *If an $\mathcal{O}_{\mathfrak{X}}$-complex $\mathcal{E}$ is K-flat then so is $\check{c}^m\mathcal{E}$ $(0 \le m \le \infty)$.*
(ii) *If an $\mathcal{A}_{\vec{c}}(\mathfrak{X})$-complex $\mathcal{E}$ is flat then so is $\check{c}^m\mathcal{E}$ $(0 \le m \le \infty)$.*

PROOF. (i) The restriction of a K-flat complex to an open subset is still K-flat, as one sees using "extension by 0." Thus for K-flatness the question is local, and we need only recall that $\check{c}^m\mathcal{E}$ is locally homotopy-isomorphic to $\mathcal{E}$ (cf. end of Section 1.2).

(ii) Apply Corollary 2.3.2 with $\mathcal{P}$ the pullback to $\mathcal{U}$ of $\mathcal{E}$. □

Applying Lemma 2.2.7 to the $\mathcal{U}$-pullback $\mathcal{F}$ of $\mathcal{E}$, we get (c) in Section 1.3.2:

COROLLARY 2.3.5. *For any K-flat $\mathcal{A}_{\vec{c}}$-complex $\mathcal{E}$, $\check{c}^{\infty}\mathcal{E}$ is K-flat and locally $\mathcal{I}$-acyclic.*

LEMMA 2.4.1. *Let $Q_{\mathfrak{X}}$ be right-adjoint to the inclusion functor $\mathcal{A}_{\vec{c}} \hookrightarrow \mathcal{A}$ (see* [**DFS**, Proposition 3.2.3]*). If $\mathfrak{X}$ is affine then for any affine-acyclic $\mathcal{E} \in \mathbf{K}_{\vec{c}}$ the canonical map $Q_{\mathfrak{X}}\mathcal{E} \to \mathcal{E}$ is a quasi-isomorphism.*

PROOF. Let A be a noetherian adic ring such that $\mathfrak{X} \cong \mathrm{Spf}(A)$ (for example, $A := \Gamma(\mathfrak{X}, \mathcal{O}_{\mathfrak{X}})$), and let $\kappa\colon \mathfrak{X} \to X := \mathrm{Spec}(A)$ be the canonical map. Then with $^{\sim}$ denoting the usual functor from A-modules to $\mathcal{O}_X$-modules, [**DFS**, Proposition 3.2.3] yields

$$Q_{\mathfrak{X}}\mathcal{E} \cong \kappa^*(Q_X \kappa_*\mathcal{E}) = \kappa^*(\Gamma(X, \kappa_*\mathcal{E})^{\sim}) = \kappa^*(\Gamma(\mathfrak{X}, \mathcal{E})^{\sim}).$$

By assumption, any K-injective resolution $i\colon \mathcal{E} \to \mathcal{J}$ induces a quasi-isomorphism $\Gamma(\mathfrak{X}, \mathcal{E}) \to \Gamma(\mathfrak{X}, \mathcal{J})$; and so since κ^* and $^{\sim}$ are exact, $Q_{\mathfrak{X}}(i)\colon Q_{\mathfrak{X}}\mathcal{E} \to Q_{\mathfrak{X}}\mathcal{J}$ is a quasi-isomorphism.

Furthermore, [**DFS**, Corollary 3.3.4] implies that the natural map $Q_{\mathfrak{X}}\mathcal{J} \to \mathcal{J}$ is a quasi-isomorphism. Conclude via the natural commutative diagram

$$\begin{array}{ccc} Q_{\mathfrak{X}}\mathcal{E} & \longrightarrow & \mathcal{E} \\ {\scriptstyle Q_{\mathfrak{X}}(i)}\downarrow & & \downarrow{\scriptstyle i} \\ Q_{\mathfrak{X}}\mathcal{J} & \longrightarrow & \mathcal{J} \end{array}$$

□

PROPOSITION 2.4.2. *Let $\mathcal{U} = (\mathfrak{U}_\alpha)_{1\le\alpha\le t}$ be an affine open cover of $\mathfrak{X}$. Let $\mathcal{G}$ be a $\mathcal{U}$-complex with $\mathcal{A}_{\vec{c}}(\mathcal{U})$-homology such that for each $k \in \mathfrak{P}_t$, $\mathcal{G}_k$ is affine-acyclic. Then $\mathcal{G}$ receives a quasi-isomorphism from a $\varinjlim$ of bounded-above flat complexes of $\mathcal{A}_{\vec{c}}(\mathcal{U})$-modules.*

PROOF. If $i \supset j$ are in $\mathfrak{P}_t$, then $\lambda_{ij}^* Q_{\mathfrak{U}_j}\mathcal{G}_j$ is an $\mathcal{A}_{\vec{c}}(\mathfrak{U}_i)$-complex, and so the natural composition

$$\lambda_{ij}^* Q_{\mathfrak{U}_j}\mathcal{G}_j \to \lambda_{ij}^*\mathcal{G}_j \to \mathcal{G}_i$$

factors naturally as

$$\lambda_{ij}^* Q_{\mathfrak{U}_j}\mathcal{G}_j \xrightarrow{\ \phi_{ij}\ } Q_{\mathfrak{U}_i}\mathcal{G}_i \to \mathcal{G}_i\,.$$

The maps ϕ_{ij} make the family $(Q_{\mathfrak{U}_k}\mathcal{G}_k)_{k\in\mathfrak{P}_t}$ into an $\mathcal{A}_{\vec{c}}(\mathcal{U})$-complex. Lemma 2.4.1 lets us replace $\mathcal{G}$ by this complex, i.e., we may assume that $\mathcal{G}$ is an $\mathcal{A}_{\vec{c}}(\mathcal{U})$-complex.

After noting that if $\mathfrak{U}$ is an affine noetherian formal scheme then, by [**DFS**, Corollary 3.1.4], any $\mathcal{A}_{\vec{c}}(\mathfrak{U})$-module is a homomorphic image of a free $\mathcal{O}_{\mathfrak{U}}$-module, we can conclude as in the proof of [**AJL**, p. 11, Corollary 1.2.2]. □

2.5. We can now complete the proof of Proposition 2.1.3.

Let $\mathcal{U} = (\mathfrak{U}_\alpha)_{1\le\alpha\le t}$ be an affine open cover of $\mathfrak{X}$. For any $\mathcal{E} \in \mathbf{K}_{\vec{c}}^{\mathrm{a}}$, the pullback is a $\mathcal{U}$-module $\mathcal{G}$ satisfying the hypotheses of Proposition 2.4.2; so there is a quasi-isomorphism $\mathcal{P} \to \mathcal{G}$ with $\mathcal{P}$ a $\varinjlim$ of bounded-above flat complexes of $\mathcal{A}_{\vec{c}}(\mathcal{U})$-modules. Since $\mathcal{A}_{\vec{c}}$-complexes are affine-acyclic (Example 2.2.1(c) and Remark 2.1.1(1)), the (exact) mapping cone $\mathcal{F}$ of this quasi-isomorphism satisfies the hypotheses of Lemma 2.2.4, so $\check{C}^\bullet\mathcal{F}$ is exact, and hence the induced map $\check{C}^\bullet\mathcal{P} \to \check{C}^\bullet\mathcal{G} = \check{c}\mathcal{E}$ is a quasi-isomorphism, as is the resulting map

$$\mathcal{P}_{\mathcal{E}} := \check{c}^{\infty}\check{C}^\bullet\mathcal{P} \to \check{c}^{\infty}\check{c}\mathcal{E} \cong \check{c}^{\infty}\mathcal{E}.$$

Since $\check{C}^\bullet\mathcal{P}$, $\check{c}\mathcal{E}$, and $\mathcal{E}$ have isomorphic homology, therefore $\check{C}^\bullet\mathcal{P} \in \mathbf{K}_{\vec{c}}(\mathfrak{X})$; and $\check{C}^\bullet\mathcal{P}$ is affine-acyclic (Lemma 2.2.7, with $\mathcal{I}$=0), so $\mathcal{P}_\mathcal{E} \in \check{\mathbf{K}}^{\mathrm{a}}_{\vec{c}}$. (Look at the actual construction of $\mathcal{P}$—as given in [**AJL**, p. 11, Corollary 1.2.2]—to see that the hypotheses of Lemma 2.2.7 are satisfied.) By Lemma 2.2.7, furthermore, $\mathcal{P}_\mathcal{E}$ is locally $\mathcal{I}$-acyclic. From Corollary 2.3.3, it follows that $\mathcal{P}_\mathcal{E}$ is flat and K-flat.

Finally, since $\mathcal{E} \in \check{\mathbf{K}}_{\vec{c}}(\mathfrak{X})$ the natural map of complexes $\mathcal{E} \to \check{c}^\infty\mathcal{E}$ has an inverse, which can be composed with the preceding quasi-isomorphism $\mathcal{P}_\mathcal{E} \to \check{c}^\infty\mathcal{E}$ to give the desired quasi-isomorphism $\mathcal{P}_\mathcal{E} \to \mathcal{E}$. □

3. A functorial isomorphism

Denote by Λ the restricted completion functor $\Lambda_\mathcal{I}|_{\check{\mathbf{K}}^{\mathrm{a}}_{\vec{c}}}$, and by $\mathbf{L}\Lambda\colon \mathbf{D}^{\mathrm{a}}_{\vec{c}} \to \mathbf{D}$ its left-derived functor, whose existence has now been proved according to Section 1.3.2. The proof shows we may assume $\mathbf{L}\Lambda\mathcal{P} = \Lambda\mathcal{P}$ for any K-flat locally $\mathcal{I}$-acyclic $\mathcal{P} \in \check{\mathbf{K}}^{\mathrm{a}}_{\vec{c}}$.

3.1. We first construct a Δ-functorial map

$$\Phi\colon \mathbf{L}\Lambda\mathcal{E} \to \mathbf{R}\mathcal{H}om^\bullet(\mathbf{R}\varGamma_\mathcal{I}\mathcal{O}_\mathfrak{X}, \mathcal{E}) \qquad (\mathcal{E} \in \mathbf{D}^{\mathrm{a}}_{\vec{c}}).$$

Using the equivalence $\rho\colon \mathbf{D}^{\mathrm{a}}_{\vec{c}} \to \check{\mathbf{D}}^{\mathrm{a}}_{\vec{c}}$ (Section 1.3), we can replace $\mathbf{D}^{\mathrm{a}}_{\vec{c}}$ by its Δ-subcategory $\check{\mathbf{D}}^{\mathrm{a}}_{\vec{c}}$. Because of Proposition 2.1.3, we can further restrict to the equivalent Δ-subcategory of $\check{\mathbf{D}}^{\mathrm{a}}_{\vec{c}}$ whose objects are the flat, K-flat, locally $\mathcal{I}$-acyclic complexes in $\check{\mathbf{K}}_{\vec{c}}(\mathfrak{X})$—this is the derived category of the homotopy category $\mathbf{H}$ of such complexes. Thus, with $q\colon \mathbf{K} \to \mathbf{D}$ the natural functor and q' the natural functor from $\mathbf{H}$ to its derived category, [**Hrt**, p. 33, Proposition 3.4] shows it sufficient to construct a map of functors

$$(\mathbf{L}\Lambda)q'\mathcal{P} = q\Lambda\mathcal{P} \to \mathbf{R}\mathcal{H}om^\bullet(\mathbf{R}\varGamma_\mathcal{I}\mathcal{O}_\mathfrak{X}, q'\mathcal{P}) \qquad (\mathcal{P} \in \mathbf{H}).$$

Hence if $\mathcal{O}_\mathfrak{X} \to \mathcal{R}$ is a K-injective resolution, it will suffice to construct a map

$$\Lambda\mathcal{P} \to \mathcal{H}om^\bullet(\varGamma_\mathcal{I}\mathcal{R}, \mathcal{P} \otimes \mathcal{R}) \qquad (\mathcal{P} \in \mathbf{H}) \tag{3.1.1}$$

between functors from $\mathbf{H}$ to $\mathbf{K}$.

For any $\mathcal{O}_\mathfrak{X}$-complexes $\mathcal{P}$, $\mathcal{Q}$, $\mathcal{R}$, the natural map

$$(\mathcal{P} \otimes \mathcal{Q}) \otimes (\mathcal{H}om^\bullet(\mathcal{Q}, \mathcal{R})) \cong \mathcal{P} \otimes (\mathcal{Q} \otimes \mathcal{H}om^\bullet(\mathcal{Q}, \mathcal{R})) \to \mathcal{P} \otimes \mathcal{R}$$

induces (via $\otimes$–$\mathcal{H}om$ adjunction) a functorial map

$$\mathcal{P} \otimes \mathcal{Q} \to \mathcal{H}om^\bullet(\mathcal{H}om^\bullet(\mathcal{Q}, \mathcal{R}), \mathcal{P} \otimes \mathcal{R}).$$

Letting $\mathcal{Q}$ run through the inverse system $\mathcal{O}_\mathfrak{X}/\mathcal{I}^n$ $(n > 0)$ one gets the desired natural map

$$\begin{aligned}\Lambda\mathcal{P} = \varprojlim\, (\mathcal{P} \otimes \mathcal{O}_\mathfrak{X}/\mathcal{I}^n) \to \varprojlim\, &\mathcal{H}om^\bullet(\mathcal{H}om^\bullet(\mathcal{O}_\mathfrak{X}/\mathcal{I}^n, \mathcal{R}), \mathcal{P} \otimes \mathcal{R})\\ &\cong \mathcal{H}om^\bullet(\varinjlim\, \mathcal{H}om^\bullet(\mathcal{O}_\mathfrak{X}/\mathcal{I}^n, \mathcal{R}), \mathcal{P} \otimes \mathcal{R})\\ &\cong \mathcal{H}om^\bullet(\varGamma_\mathcal{I}\mathcal{R}, \mathcal{P} \otimes \mathcal{R}).\end{aligned}$$

Now comes the principal result.

PROPOSITION 3.2. *The Δ-functorial map Φ is an isomorphism.*

Before describing its proof, we note that Proposition 3.2 allows us to complete the proof of Theorem 0.1, as follows.

There is a canonical functorial map $\xi\colon \mathbf{L}\Lambda \to q\Lambda$. With $\zeta := \xi\Phi^{-1}$, property (i) in Theorem 0.1 results. The functorial map $\mathcal{F} \to q\Lambda\mathcal{F}$ ($\mathcal{F} \in \mathbf{D}_{\vec{c}}^{\mathrm{a}}$) factors uniquely as $\mathcal{F} \xrightarrow{\iota_{\mathcal{F}}} \mathbf{L}\Lambda\mathcal{F} \xrightarrow{\xi_{\mathcal{F}}} q\Lambda\mathcal{F}$ (thanks to the defining property of "left-derived functor"), and by Theorem 0.1(ii), $\Phi_{\mathcal{F}} \circ \iota(\mathcal{F})$ is the natural map $\mathcal{F} \to \mathbf{R}\mathcal{H}om^{\bullet}(\mathbf{R}\varGamma_{\mathcal{I}}\mathcal{O}_{\mathfrak{X}}, \mathcal{F})$. One checks that this comes down to the straightforward verification (left to the reader) that the following natural diagram commutes for all $\mathcal{P} \in \mathbf{H}$:

$$\begin{array}{ccc} \mathcal{P} & \longrightarrow & \Lambda\mathcal{P} \\ \big\downarrow & & \big\downarrow{\scriptstyle (3.1.1)} \\ \mathcal{H}om^{\bullet}(\mathcal{R}, \mathcal{P}\otimes\mathcal{R}) & \longrightarrow & \mathcal{H}om^{\bullet}(\varGamma_{\mathcal{I}}\mathcal{R}, \mathcal{P}\otimes\mathcal{R}) \end{array}$$

Finally, for the last assertion in Theorem 0.1 see Section 1.3.2.

Proof of Proposition 3.2. The proof is similar to that of the main Theorem (0.3) in [**AJL**]. We just indicate the non-trivial modifications.

Everything in sight commutes with restriction to open subsets, so we may assume that $\mathfrak{X}$ is affine and that $\mathcal{I}$ is generated by a finite number of global sections $t_1, \dots, t_m$. Recalling that every sheaf in $\mathcal{A}_{\vec{c}}$ is then a homomorphic image of a free $\mathcal{O}_{\mathfrak{X}}$-module ([**DFS**, Corollary 3.1.4]), we have a "way-out" reduction, just as in [**AJL**, p. 15], to the case where $\mathcal{E}$ is a single flat $\mathcal{O}_{\mathfrak{X}}$-module, denoted $\mathcal{P}$, in $\mathcal{A}_{\vec{c}}$.

This case is dealt with as in [**AJL**, §4]. Let R be the adic ring $\Gamma(\mathfrak{X}, \mathcal{O}_{\mathfrak{X}})$ (topologized by the global sections of a defining ideal of $\mathfrak{X}$), so that $\mathfrak{X} \cong \mathrm{Spf}(R)$. Since R is noetherian, the sequence $\mathbf{t} := (t_1, \dots, t_m)$ in R is *proregular* (see [**AJL**, p. 16, Example (a) following Definition (3.0.1)]). Throughout §4 of [**AJL**], replace the scheme $(X, \mathcal{O}_X)$ by the noetherian affine formal scheme $(\mathfrak{X}, \mathcal{O}_{\mathfrak{X}})$, and interpret "quasi-coherent" to mean "$\varinjlim$ of coherent." Also, replace $\mathbf{R}\varGamma_{Z}\mathcal{O}_X$ by $\mathbf{R}\varGamma_{\mathcal{I}}\mathcal{O}_{\mathfrak{X}}$, which is $\mathbf{D}(\mathfrak{X})$-isomorphic to the complex

$$\mathcal{K}^{\bullet}_{\infty}(\mathbf{t}) := \varinjlim_{r>0} \mathcal{K}^{\bullet}(\mathbf{t}^r)$$

where $\mathcal{K}^{\bullet}(\mathbf{t}^r)$ is the Koszul $\mathcal{O}_{\mathfrak{X}}$-complex determined by the sequence $(t_1^r, \dots, t_m^r)$, see [**AJL**, p. 18, Lemma (3.1.1)].

To validate the sequence of isomorphisms (4.1.3) of [**AJL**, p. 28] in the present context, use the following observations:

(i) If $\mathcal{P} \to \mathcal{J}$ is an injective resolution, the resulting map $\Gamma(\mathfrak{X}, \mathcal{P}) \to \Gamma(\mathfrak{X}, \mathcal{J})$ is a *quasi-isomorphism.*

This is because $\mathcal{P}$ is affine-acyclic (Example 2.2.1(a)).

(ii) If $\mathcal{E}$ is any complex in $\mathcal{A}_{\vec{c}}$, then the natural maps

$$\mathrm{H}^i\Gamma(\mathfrak{X}, \mathcal{E}) \to \Gamma(\mathfrak{X}, H^i\mathcal{E}) \qquad (i \in \mathbb{Z})$$

are all isomorphisms.

This is because the functor $\Gamma := \Gamma(\mathfrak{X}, -)$ on the abelian category $\mathcal{A}_{\vec{c}}$ is *exact* ([**DFS**, Proposition 3.2.2], [**DFS**, Corollary 3.1.8]). In particular, since $\mathcal{K}^{\bullet}(\mathbf{t}^r)$ is a free, finite-rank $\mathcal{O}_{\mathfrak{X}}$-complex and $R = \Gamma\mathcal{O}_{\mathfrak{X}}$, we have natural isomorphisms

$$\mathrm{H}^i\mathrm{Hom}^{\bullet}_R(\Gamma\mathcal{K}^{\bullet}(\mathbf{t}^r), \Gamma\mathcal{P}) \xrightarrow{\ \sim\ } \mathrm{H}^i\mathrm{Hom}^{\bullet}_{\mathcal{O}_{\mathfrak{X}}}(\mathcal{K}^{\bullet}(\mathbf{t}^r), \mathcal{P}) \xrightarrow{\ \sim\ } \Gamma H^i\mathcal{H}om^{\bullet}_{\mathcal{O}_{\mathfrak{X}}}(\mathcal{K}^{\bullet}(\mathbf{t}^r), \mathcal{P}).$$

The rest of the translation of [**AJL**, §4] to the present context is straightforward, except for the last half of the proof of Lemma (4.3) in *loc. cit.* For that, one needs [**DFS**, Corollary 3.3.4] (to replace the reference to [**BN**]), and [**DFS**, Corollary 3.1.8] (for imitating the proof of [**Spn**, p. 134, Proposition 3.13]); and one needs to keep in mind that $\mathcal{A}_{\vec{c}}$-complexes are affine-acyclic (Example 2.2.1(c)). □

4. Proof of Proposition 0.3.1.

Refer to Section 0.3 for notation, and for the statement of Proposition 0.3.1.

LEMMA 4.1. *The canonical map* $\mathcal{F} \to \kappa_*\kappa^*\mathcal{F}$ *induces an isomorphism*

$$\mathbf{R}\varGamma_{\mathcal{I}}\mathcal{F} \xrightarrow{\sim} \mathbf{R}\varGamma_{\mathcal{I}}\kappa_*\kappa^*\mathcal{F} \qquad (\mathcal{F} \in \mathbf{D}).$$

PROOF. The question being local, we may assume that $\mathfrak{X}$ is the completion of an affine scheme $X = \mathrm{Spec}(A)$ along a closed subscheme Z, and that $\hat{\mathfrak{X}}_{\mathcal{I}}$ is the completion of X along a closed subscheme $Y \subset Z$, so that there is a natural factorization of the completion map $\kappa_Y \colon \hat{\mathfrak{X}}_{\mathcal{I}} \to X$ as

$$\hat{\mathfrak{X}}_{\mathcal{I}} \xrightarrow{\kappa} \mathfrak{X} \xrightarrow{\kappa_Z} X.$$

Now let $\mathcal{I}' \subset \mathcal{I}$ be an ideal of definition of $\mathfrak{X}$. Then for any $\mathcal{O}_{\mathfrak{X}}$-module $\mathcal{E}$ such that $\mathbf{R}\varGamma_{\mathcal{I}'}\mathcal{E} = \mathcal{E}$, we have $\mathcal{E} \cong \kappa_Z^*E$ for some $\mathcal{O}_X$-module E [**DFS**, Proposition 5.2.4]. In particular, since $\mathbf{R}\varGamma_{\mathcal{I}'}\mathbf{R}\varGamma_{\mathcal{I}}\mathcal{F} = \mathbf{R}\varGamma_{\mathcal{I}}\mathcal{F}$ (cf. [**AJL**, p. 20, Corollary 3.1.3]), therefore $\mathbf{R}\varGamma_{\mathcal{I}}\mathcal{F} \cong \kappa_Z^*F$ for some $\mathcal{O}_X$-module F.

By [**DFS**, Proposition 5.2.4] again, the natural maps are isomorphisms

$$\mathrm{Hom}_X(E,F) \xrightarrow{\sim} \mathrm{Hom}_{\mathfrak{X}}(\kappa_Z^*E, \kappa_Z^*F) = \mathrm{Hom}_{\mathfrak{X}}(\mathcal{E}, \mathbf{R}\varGamma_{\mathcal{I}}\mathcal{F}),$$
$$\mathrm{Hom}_X(E,F) \xrightarrow{\sim} \mathrm{Hom}_{\hat{\mathfrak{X}}_{\mathcal{I}}}(\kappa_Y^*E, \kappa_Y^*F) = \mathrm{Hom}_{\hat{\mathfrak{X}}_{\mathcal{I}}}(\kappa^*\mathcal{E}, \kappa^*\mathbf{R}\varGamma_{\mathcal{I}}\mathcal{F}).$$

Furthermore, arguing as in the proof of [**DFS**, Proposition 5.2.4] we get a natural isomorphism

$$\kappa_*\kappa^*\mathbf{R}\varGamma_{\mathcal{I}}\mathcal{F} \xrightarrow{\sim} \mathbf{R}\varGamma_{\mathcal{I}}\kappa_*\kappa^*\mathcal{F}$$

whose composition with the natural map $\mathbf{R}\varGamma_{\mathcal{I}}\mathcal{F} \to \kappa_*\kappa^*\mathbf{R}\varGamma_{\mathcal{I}}\mathcal{F}$ is the natural map $\mathbf{R}\varGamma_{\mathcal{I}}\mathcal{F} \to \mathbf{R}\varGamma_{\mathcal{I}}\kappa_*\kappa^*\mathcal{F}$.

Thus for any $\mathcal{E}$ such that $\mathbf{R}\varGamma_{\mathcal{I}'}\mathcal{E} = \mathcal{E}$, we have a composed isomorphism

$$\begin{aligned}\mathrm{Hom}_{\mathfrak{X}}(\mathcal{E}, \mathbf{R}\varGamma_{\mathcal{I}}\mathcal{F}) \xrightarrow{\sim} \mathrm{Hom}_X(E,F) &\xrightarrow{\sim} \mathrm{Hom}_{\hat{\mathfrak{X}}_{\mathcal{I}}}(\kappa^*\mathcal{E}, \kappa^*\mathbf{R}\varGamma_{\mathcal{I}}\mathcal{F}) \\ &\xrightarrow{\sim} \mathrm{Hom}_{\mathfrak{X}}(\mathcal{E}, \kappa_*\kappa^*\mathbf{R}\varGamma_{\mathcal{I}}\mathcal{F}) \xrightarrow{\sim} \mathrm{Hom}_{\mathfrak{X}}(\mathcal{E}, \mathbf{R}\varGamma_{\mathcal{I}}\kappa_*\kappa^*\mathcal{F})\end{aligned}$$

which (one checks) is induced by $\mathcal{F} \to \kappa_*\kappa^*\mathcal{F}$, whence the conclusion. □

COROLLARY 4.1.1. *The canonical map* $\mathcal{F} \to \kappa_*\kappa^*\mathcal{F}$ *induces an isomorphism*

$$\mathrm{Hom}(\kappa_*\kappa^*\mathcal{F}, \mathbf{R}\mathcal{H}om^\bullet(\mathbf{R}\varGamma_{\mathcal{I}}\mathcal{O}_{\mathfrak{X}}, \mathcal{F})) \xrightarrow{\sim} \mathrm{Hom}(\mathcal{F}, \mathbf{R}\mathcal{H}om^\bullet(\mathbf{R}\varGamma_{\mathcal{I}}\mathcal{O}_{\mathfrak{X}}, \mathcal{F})).$$

PROOF. Using that $\mathbf{R}\varGamma_{\mathcal{I}}$ is right-adjoint to the inclusion into $\mathbf{D}$ of the derived category of the category of $\mathcal{I}$-torsion sheaves ([**DFS**, Lemma 5.2.2]), one finds that there is a unique functorial map

$$(4.1.1.1) \qquad \psi_{\mathcal{E}}\colon \mathcal{E} \underset{=}{\otimes} \mathbf{R}\varGamma_{\mathcal{I}}\mathcal{O}_{\mathfrak{X}} \to \mathbf{R}\varGamma_{\mathcal{I}}\mathcal{E} \qquad (\mathcal{E} \in \mathbf{D})$$

making the following natural diagram commute:

$$\begin{array}{ccc}
\mathcal{E} \underset{=}{\otimes} \mathbf{R}\varGamma_{\mathcal{I}}\mathcal{O}_{\mathfrak{X}} & \xrightarrow{\ \psi_{\mathcal{E}}\ } & \mathbf{R}\varGamma_{\mathcal{I}}\mathcal{E} \\
\Big\downarrow & & \Big\downarrow \\
\mathcal{E} \underset{=}{\otimes} \mathcal{O}_{\mathfrak{X}} & \xrightarrow{\ \sim\ } & \mathcal{E}
\end{array}$$

It follows from [**AJL**, p. 20, Corollary (3.1.2)] that $\psi_{\mathcal{E}}$ is an *isomorphism.*

Let us also recall the basic *adjoint associativity* isomorphism (see, e.g., [**Lpm**, Proposition (2.6.1)*]):

$$(4.1.1.2) \quad \mathbf{R}\mathcal{H}om^{\bullet}(\mathcal{D} \underset{=}{\otimes} \mathcal{E}, \mathcal{F}) \xrightarrow{\ \sim\ } \mathbf{R}\mathcal{H}om^{\bullet}(\mathcal{D}, \mathbf{R}\mathcal{H}om^{\bullet}(\mathcal{E}, \mathcal{F})) \quad (\mathcal{D}, \mathcal{E}, \mathcal{F} \in \mathbf{D}).$$

The isomorphism also holds with the outer $\mathbf{R}\mathcal{H}om^{\bullet}$'s replaced by $\mathrm{Hom} := \mathrm{Hom}_{\mathbf{D}}$, as one sees by applying the functor $\mathrm{H}^0\mathbf{R}\Gamma(\mathfrak{X}, -)$.

Corollary 4.1.1 results then from the natural commutative diagram, in which all the vertical arrows are isomorphisms, as is the bottom horizontal one:

$$\begin{array}{ccc}
\mathrm{Hom}(\kappa_*\kappa^*\mathcal{F}, \mathbf{R}\mathcal{H}om^{\bullet}(\mathbf{R}\varGamma_{\mathcal{I}}\mathcal{O}_{\mathfrak{X}}, \mathcal{F})) & \longrightarrow & \mathrm{Hom}(\mathcal{F}, \mathbf{R}\mathcal{H}om^{\bullet}(\mathbf{R}\varGamma_{\mathcal{I}}\mathcal{O}_{\mathfrak{X}}, \mathcal{F})) \\
\simeq\Big\downarrow & & \Big\downarrow\simeq \\
\mathrm{Hom}(\kappa_*\kappa^*\mathcal{F} \underset{=}{\otimes} \mathbf{R}\varGamma_{\mathcal{I}}\mathcal{O}_{\mathfrak{X}}, \mathcal{F}) & \longrightarrow & \mathrm{Hom}(\mathcal{F} \underset{=}{\otimes} \mathbf{R}\varGamma_{\mathcal{I}}\mathcal{O}_{\mathfrak{X}}, \mathcal{F}) \\
\simeq\Big\uparrow & & \Big\uparrow\simeq \\
\mathrm{Hom}(\mathbf{R}\varGamma_{\mathcal{I}}\kappa_*\kappa^*\mathcal{F}, \mathcal{F}) & \underset{\text{Lemma 4.1}}{\xrightarrow{\ \sim\ }} & \mathrm{Hom}(\mathbf{R}\varGamma_{\mathcal{I}}\mathcal{F}, \mathcal{F})
\end{array}$$

□

In the following diagram, where "H" stands for "Hom," the isomorphisms a and a^* are obtained by adjoint associativity and multiple use of ψ (see (4.1.1.1)), and the isomorphisms b and b^* are obtained by imitating [**DFS**, Remark 5.2.10](4). The diagram commutes, so c is an isomorphism.

Assertion (i) in Proposition 0.3.1 results.

$$\begin{array}{ccc}
\mathrm{H}(\mathbf{R}\mathcal{H}om^{\bullet}(\mathcal{E}, \kappa_*\kappa^*\mathcal{F}), \mathbf{R}\mathcal{H}om^{\bullet}(\mathbf{R}\varGamma_{\mathcal{I}}\mathcal{E}, \mathcal{F})) & \xrightarrow{\ c\ } & \mathrm{H}(\mathbf{R}\mathcal{H}om^{\bullet}(\mathcal{E}, \mathcal{F}), \mathbf{R}\mathcal{H}om^{\bullet}(\mathbf{R}\varGamma_{\mathcal{I}}\mathcal{E}, \mathcal{F})) \\
a^*\Big\downarrow\simeq & & \simeq\Big\downarrow a \\
\mathrm{H}(\mathbf{R}\varGamma_{\mathcal{I}}\mathbf{R}\mathcal{H}om^{\bullet}(\mathcal{E}, \kappa_*\kappa^*\mathcal{F}) \underset{=}{\otimes} \mathcal{E}, \mathcal{F})) & \underset{\text{natural}}{\longrightarrow} & \mathrm{H}(\mathbf{R}\varGamma_{\mathcal{I}}\mathbf{R}\mathcal{H}om^{\bullet}(\mathcal{E}, \mathcal{F}) \underset{=}{\otimes} \mathcal{E}, \mathcal{F})) \\
b^*\Big\downarrow\simeq & & \simeq\Big\downarrow b \\
\mathrm{H}(\mathbf{R}\varGamma_{\mathcal{I}}\mathbf{R}\mathcal{H}om^{\bullet}(\mathcal{E}, \mathbf{R}\varGamma_{\mathcal{I}}\kappa_*\kappa^*\mathcal{F}) \underset{=}{\otimes} \mathcal{E}, \mathcal{F})) & \underset{4.1.1}{\xrightarrow{\ \sim\ }} & \mathrm{H}(\mathbf{R}\varGamma_{\mathcal{I}}\mathbf{R}\mathcal{H}om^{\bullet}(\mathcal{E}, \mathbf{R}\varGamma_{\mathcal{I}}\mathcal{F}) \underset{=}{\otimes} \mathcal{E}, \mathcal{F}))
\end{array}$$

4.2. It remains to be shown that $\lambda(\mathcal{E}, \mathcal{F})$ is an isomorphism for all $\mathcal{F} \in \mathbf{D}_{\mathrm{c}}$.

We can reduce the problem to the case $\mathcal{E} = \mathcal{O}_{\mathfrak{X}}$ by factoring $\lambda(\mathcal{E}, \mathcal{F})$ as

$$\mathbf{R}\mathcal{H}om^{\bullet}(\mathcal{E}, \kappa_*\kappa^*\mathcal{F}) \xrightarrow{\text{via } \lambda(\mathcal{O}_{\mathfrak{X}}, \mathcal{F})} \mathbf{R}\mathcal{H}om^{\bullet}(\mathcal{E}, \mathbf{R}\mathcal{H}om^{\bullet}(\mathbf{R}\varGamma_{\mathcal{I}}\mathcal{O}_{\mathfrak{X}}, \mathcal{F}))$$

$$\underset{(4.1.1.2)}{\xrightarrow{\ \sim\ }} \mathbf{R}\mathcal{H}om^{\bullet}(\mathcal{E} \underset{=}{\otimes} \mathbf{R}\varGamma_{\mathcal{I}}\mathcal{O}_{\mathfrak{X}}, \mathcal{F}) \underset{(4.1.1.1)}{\xrightarrow{\ \sim\ }} \mathbf{R}\mathcal{H}om^{\bullet}(\mathbf{R}\varGamma_{\mathcal{I}}\mathcal{E}, \mathcal{F}).$$

To check that this composition is in fact $\lambda(\mathcal{E},\mathcal{F})$, using the characterization in Proposition 0.3.1(i), consider the following natural diagram:

$$
\begin{array}{ccc}
\mathbf{R}\mathcal{H}om^\bullet(\mathcal{E},\mathcal{F}) & \longrightarrow & \mathbf{R}\mathcal{H}om^\bullet(\mathcal{E},\kappa_*\kappa^*\mathcal{F}) \\
\Big\downarrow{\scriptstyle \simeq} & & \Big\downarrow{\scriptstyle \text{via } \lambda(\mathcal{O}_\mathfrak{X},\mathcal{F})} \\
\mathbf{R}\mathcal{H}om^\bullet(\mathcal{E},\mathbf{R}\mathcal{H}om^\bullet(\mathcal{O}_\mathfrak{X},\mathcal{F})) & \longrightarrow & \mathbf{R}\mathcal{H}om^\bullet(\mathcal{E},\mathbf{R}\mathcal{H}om^\bullet(\mathbf{R}\varGamma_{\mathcal{I}}\mathcal{O}_\mathfrak{X},\mathcal{F})) \\
{\scriptstyle (4.1.1.2)}\Big\downarrow{\scriptstyle \simeq} & & {\scriptstyle \simeq}\Big\downarrow{\scriptstyle (4.1.1.2)} \\
\mathbf{R}\mathcal{H}om^\bullet(\mathcal{E}\underset{=}{\otimes}\mathcal{O}_\mathfrak{X},\mathcal{F}) & \longrightarrow & \mathbf{R}\mathcal{H}om^\bullet(\mathcal{E}\underset{=}{\otimes}\mathbf{R}\varGamma_{\mathcal{I}}\mathcal{O}_\mathfrak{X},\mathcal{F}) \\
\Big\downarrow{\scriptstyle \simeq} & & {\scriptstyle \simeq}\Big\downarrow{\scriptstyle (4.1.1.1)} \\
\mathbf{R}\mathcal{H}om^\bullet(\mathcal{E},\mathcal{F}) & \longrightarrow & \mathbf{R}\mathcal{H}om^\bullet(\mathbf{R}\varGamma_{\mathcal{I}}\mathcal{E},\mathcal{F})
\end{array}
$$

The top rectangle commutes by the characterization of $\lambda(\mathcal{O}_\mathfrak{X},\mathcal{F})$, the middle one commutes by functoriality, and commutativity of the bottom one is given by the commutative diagram following (4.1.1.1). It suffices then to verify that the composition of the vertical maps on the left is the identity, which is easy to do after assuming (as one may) that $\mathcal{F}$ is K-injective, so that each $\mathbf{R}\mathcal{H}om^\bullet$ can be replaced by $\mathcal{H}om^\bullet$.

To show that $\lambda(\mathcal{O}_\mathfrak{X},\mathcal{F})$ is an isomorphism when $\mathcal{F}\in\mathbf{D}_{\mathrm{c}}$ we may assume, after replacing $\mathcal{F}$ by a K-injective complex into which $\mathcal{F}$ maps quasi-isomorphically, that $\mathcal{F}\in\mathbf{D}_{\vec{\mathrm{c}}}^{\mathrm{a}}$. For such an $\mathcal{F}$ we will now give another construction of $\lambda(\mathcal{O}_\mathfrak{X},\mathcal{F})$, adapted to application of Theorem 0.1.

For any $\mathcal{O}_\mathfrak{X}$-complex $\mathcal{F}$, we have the canonical completion map

$$\mathcal{F}\to\Lambda_{\mathcal{I}}\mathcal{F}=\kappa_*\varprojlim_n\big((\mathcal{F}/\mathcal{I}^n\mathcal{F})|_{\hat{\mathfrak{X}}_{\mathcal{I}}}\big),$$

whence a natural factorization $\mathcal{F}\longrightarrow\kappa_*\kappa^*\mathcal{F}\xrightarrow{\gamma_\mathcal{F}}\Lambda_{\mathcal{I}}\mathcal{F}$. The exact functor $\kappa_*\kappa^*$ induces a functor $\mathbf{D}_{\vec{\mathrm{c}}}^{\mathrm{a}}\to\mathbf{D}$, so for $\mathcal{F}\in\mathbf{D}_{\vec{\mathrm{c}}}^{\mathrm{a}}$ Theorem 0.1(i) yields a factorization $\gamma_\mathcal{F}=\zeta_\mathcal{F}\circ\lambda_0(\mathcal{F})$, as depicted:

$$\mathcal{F}\longrightarrow\kappa_*\kappa^*\mathcal{F}\xrightarrow{\lambda_0(\mathcal{F})}\mathbf{R}\mathcal{H}om^\bullet(\mathbf{R}\varGamma_{\mathcal{I}}\mathcal{O}_\mathfrak{X},\mathcal{F})\xrightarrow{\zeta_\mathcal{F}}\Lambda_{\mathcal{I}}\mathcal{F}\qquad(\mathcal{F}\in\mathbf{D}_{\vec{\mathrm{c}}}^{\mathrm{a}}).$$

In view of Theorem 0.1, (i) and (ii), *the composition of* $\lambda_0(\mathcal{F})$ *with the canonical map* $\mathcal{F}\to\kappa_*\kappa^*\mathcal{F}$ *is the natural map* $\mathcal{F}\to\mathbf{R}\mathcal{H}om^\bullet(\mathbf{R}\varGamma_{\mathcal{I}}\mathcal{O}_\mathfrak{X},\mathcal{F})$, since both this composition and the natural map give the same result when composed with $\zeta_\mathcal{F}$. Proposition 0.3.1(i) shows then that $\lambda_0(\mathcal{F})=\lambda(\mathcal{O}_\mathfrak{X},\mathcal{F})$.

The question of whether $\lambda(\mathcal{O}_\mathfrak{X},\mathcal{F})$ is an isomorphism is local. One can therefore assume that $\mathfrak{X}$ is affine, so that every coherent $\mathcal{O}_\mathfrak{X}$-module is a homomorphic image of a finite-rank free one [**GD**, p. 427, Théorème (10.10.2)]. Now the exact functor $\kappa_*\kappa^*$ is bounded on $\mathbf{D}_{\vec{\mathrm{c}}}$, and so is the functor $\mathbf{R}\mathcal{H}om^\bullet(\mathbf{R}\varGamma_{\mathcal{I}}\mathcal{O}_\mathfrak{X},-)$, by [**AJL**, p. 30, Lemma (4.3)], suitably modified (see discussion at the end of Section 3 above). Hence we have a "way-out" reduction [**Hrt**, p. 68, Proposition 7.1] to the case where $\mathcal{F}$ is a single finite-rank free $\mathcal{O}_\mathfrak{X}$-module. In this case it is straightforward to see that $\gamma_\mathcal{F}=\zeta_\mathcal{F}\lambda_0(\mathcal{F})$ is an isomorphism; and by the last assertion in Theorem 0.1, so is $\zeta_\mathcal{F}$. Thus $\lambda(\mathcal{O}_\mathfrak{X},\mathcal{F})=\lambda_0(\mathcal{F})$ *is* an isomorphism, and Proposition 0.3.1 is proved.

References

[AJL] L. Alonso Tarrío, A. Jeremías López and J. Lipman, Local homology and cohomology on schemes, *Ann. Scient. Éc. Norm. Sup.* **30** (1997), 1–39.

[DFS] ———, Duality and flat base change on formal schemes, this volume.

[BN] M. Bökstedt and A. Neeman, Homotopy limits in triangulated categories, *Compositio Math.* **86** (1993), 209–234.

[Brb] N. Bourbaki, *Algèbre Commutative,* Actualités Sci. et Industrielles, no. 1293, Hermann, Paris, 1961.

[Gdm] R. Godement, *Théorie des Faisceaux,* Act. Sci. et Industrielles 1252, Herrmann, Paris, 1964.

[EGA] A. Grothendieck and J. Dieudonné, *Élements de Géométrie Algébrique* **III**, Publications Math. IHES **11**, Paris, 1961.

[GD] ———, *Élements de Géométrie Algébrique* **I**, Springer-Verlag, New York, 1971.

[GM] J. P. C. Greenlees and J. P. May, Derived functors of I-adic completion and local homology *J. Algebra* **149** (1992), 438–453

[Hrt] R. Hartshorne, *Residues and Duality,* Lecture Notes in Math., no. **20**, Springer-Verlag, New York, 1966.

[Lpm] J. Lipman, *Notes on Derived Categories,* preprint, at `www.math.purdue.edu/~lipman/`

[Kmp] G. R. Kempf, Some elementary proofs of basic theorems in the cohomology of quasi-coherent sheaves, *Rocky Mountain J. Math.* **10** (1980), 637–645.

[Spn] N. Spaltenstein, Resolution of unbounded complexes, *Compositio Math.* **65** (1988), 121–154.

Universidade de Santiago de Compostela, Facultade de Matemáticas, E-15771 Santiago de Compostela, SPAIN

E-mail address: `lalonso@zmat.usc.es`

URL: `http://web.usc.es/~lalonso/`

Universidade de Santiago de Compostela, Facultade de Matemáticas, E-15771 Santiago de Compostela, SPAIN

E-mail address: `jeremias@zmat.usc.es`

Dept. of Mathematics, Purdue University, W. Lafayette IN 47907, USA

E-mail address: `lipman@math.purdue.edu`

URL: `www.math.purdue.edu/~lipman/`

Part 3

Non-noetherian Grothendieck Duality

Contemporary Mathematics
Volume **244**, 1999

Non-noetherian Grothendieck Duality

Joseph Lipman

ABSTRACT. For any separated map $f : X \to Y$ of quasi-compact quasi-separated schemes, $\mathbf{R}f_*\colon \mathbf{D}^+_{\mathrm{qc}}(X) \to \mathbf{D}^+(Y)$ has a right adjoint $f^!$. If f is proper and pseudo-coherent (e.g., finitely-presented and flat) then Duality and tor-independent Base Change hold for $f^!$.

Preface

This is a research summary written early in 1991, concerning results obtained by the author during a stay at MSRI in Berkeley during 1989–90. The intention then—and now—is to include details in the notes [**Li**] (whose completion has been delayed by the development of the other papers in this volume). The methods still seem relevant, though in the meantime fresh ideas have been brought to the subject by Neeman [**N**], who has given, in particular, new proofs of existence and sheafification (= open base change) for the Grothendieck duality functor.

Introduction

The fundamental results of scheme-theoretic Grothendieck Duality were first treated systematically in [**H**] (see also [**Co**]). Another approach is indicated in [**V**]. In these sources, heavy use is made of noetherian hypotheses. (See, e.g., the discussion on p. 12 of [**H**].) Also, in [**H**] and [**V**] (though not in [**Co**]) short shrift is made of "compatibilities," i.e., the commutativity of certain functorial diagrams, see e,g, [**H,** p. 118].

In the intervening years several developments have made possible substantial improvements in this theory, notably with regard to *elimination of noetherian hypotheses.* Also, many of the definitions and supporting lemmas can now be more simply formulated in the language of categories, thus acquiring meaning in larger contexts and revealing more of their interrelationships. In particular, the compatibilities problem can be better understood in terms of *coherence in categories.*

In §1, we review the intervening developments just alluded to. The category-theoretic formalism is sketched in §2. Some preliminary isomorphisms appear in §3,

1991 *Mathematics Subject Classification.* Primary 14F99; Secondary 18D99, 18F99.

Supported by NSF through DMS-8803054 at Purdue University, and through MSRI during 1989-90.

while the main new results (Existence, Duality, and tor-independent Base Change, in a not-necessarily-noetherian setting) are described in §4.

It should be noted that we have not de-noetherianized all of the main results of duality theory. For example, we are as yet unable to show the existence and uniqueness of a pseudofunctor $f^!$ defined for *all* separated pseudo-coherent maps f of quasi-compact quasi-separated schemes (1.2), right adjoint to $\mathbf{R}f_*$ when f is proper, and isomorphic to f^* when f is étale. (Existence, at least, can be shown for Cohen-Macaulay maps, by descent to the noetherian case and the fact that they satisfy *universal* Base Change, i.e., (4.3) without the assumption that u has finite tor-dimension.)

1. Background

(1.1) On a quasi-compact separated scheme X, the natural functor from the derived category of (homologically) bounded-below quasi-coherent $\mathcal{O}_X$-complexes to the derived category $\mathbf{D}^+_{\mathrm{qc}}(X)$ of bounded-below complexes with quasi-coherent homology, is an *equivalence of categories.* (This is due to Verdier, [**I,** p. 189, Proposition 3.5]. The same holds true on any locally noetherian scheme [**H,** p. 133, 7.19].)

(1.2) *Pseudo-coherence* of complexes and maps will replace noetherianness of schemes in our principal results. On a scheme X, a bounded $\mathcal{O}_X$-complex F is pseudo-coherent if each $x \in X$ has a neighborhood over which F is isomorphic, in the derived category, to a bounded-above complex of *finite-rank* free $\mathcal{O}_X$-modules [**I,** p. 175, 2.2.10]. For unbounded complexes the idea is similar, but a little more involved [**I,** p. 98, Définition 2.3; p. 108, 2.15]. If $\mathcal{O}_X$ is coherent, pseudo-coherence means simply that F has coherent homology [**I,** p. 115, Corollary 3.5 b)].

A map $f\colon X \to Y$ is said to be pseudo-coherent if it factors locally as $f = p \circ i$ where $i\colon U \to Z$ (U open in X) is a closed immersion such that $i_*\mathcal{O}_U$ is pseudo-coherent on Z, and $p\colon Z \to Y$ is smooth [**I,** p. 228, Déf. 1.2]. For example, if Y is noetherian then any finite-type $f\colon X \to Y$ is pseudo-coherent. Pseudo-coherence of maps is preserved by tor-independent base change [**I,** p. 233, Cor. 1.10]. It follows, by reduction to the noetherian case [**GD,** (11.2.7), proof], that *any flat finitely-presentable scheme-map is pseudo-coherent.*

Kiehl's Finiteness Theorem [**K**][1] generalizes preservation of coherence by higher direct images under proper maps of noetherian schemes: *If $f\colon X \to Y$ is a proper pseudo-coherent map of arbitrary schemes, then $\mathbf{R}f_*$ preserves pseudo-coherence of complexes.*

(1.3) (Derived functors for unbounded complexes.) In [**Sp**], Spaltenstein has proved existence of the standard derived functors $\mathbf{R}f_*$, $\mathbf{L}f^*$, $\mathbf{R}\mathcal{H}om^\bullet$, and $\underline{\otimes}$, for *all* $\mathcal{O}_X$-complexes (X a scheme, or more generally any ringed space). This elimination of the usual boundedness conditions is a technical godsend. For instance, Spaltenstein establishes a natural functorial isomorphism

$$\mathrm{Hom}_{\mathbf{D}(X)}\big(A \underline{\otimes} B, C\big) \xrightarrow{\ \sim\ } \mathrm{Hom}_{\mathbf{D}(X)}\big(A, \mathbf{R}\mathcal{H}om^\bullet(B, C)\big)\,,$$

making the derived category $\mathbf{D}(X)$ of $\mathcal{O}_X$-modules into a *monoidal closed category,* so that all the facts about categories with that structure become available (cf. [**EK**], or [**So**] and its references). This would not be so under boundedness restrictions.

---

[1] conjectured by Illusie, and proved by him for projective maps [**I,** p. 236, Thm. 2.2]

Moreover, given a scheme-map $f\colon X \to Y$, the functors $\mathbf{L}f^*$ and $\mathbf{R}f_*$ are *adjoint;* and they are *monoidal* [**EK,** pp. 473 and 513]: there are natural functorial maps (corresponding by adjunction, the second an isomorphism)

$$\mathbf{R}f_*(A) \underline{\underline{\otimes}} \mathbf{R}f_*(B) \to \mathbf{R}f_*(A \underline{\underline{\otimes}} B), \qquad \mathbf{L}f^*(C \underline{\underline{\otimes}} D) \xrightarrow{\sim} \mathbf{L}f^*(C) \underline{\underline{\otimes}} \mathbf{L}f^*(D),$$

respecting the monoidal structures on $\mathbf{D}(X)$ and $\mathbf{D}(Y)$, in that certain obvious diagrams commute. We also have, for a pair of maps $X \xrightarrow{f} Y \xrightarrow{g} Z$, natural isomorphisms $\mathbf{R}(gf)_* \xrightarrow{\sim} \mathbf{R}g_*\mathbf{R}f_*$ and (by adjointness) $\mathbf{L}f^*\mathbf{L}g^* \xrightarrow{\sim} \mathbf{L}(gf)^*$, making derived direct and inverse image into adjoint pseudofunctors [**Li,** §3.6].

(1.4) Beginning in the 1960's, Mac Lane and others have studied the problem, in structured categories, of deducing commutativity of complicated functorial diagrams from simple ones (cf. [**So,** references]). This subject is called "Coherence in Categories" (where "coherence" has nothing to do with sheaves). Its relevance for duality theory will be illustrated in §2.

2. Formal definitions

To reduce clutter, we will use the following abbreviations, for a scheme-map h or a scheme Z:

$$h_* := \mathbf{R}h_*\,, \quad h^* := \mathbf{L}h^*\,, \quad \mathcal{H}_Z := \mathbf{R}\mathcal{H}om^{\bullet}{}_Z\,, \quad \otimes_Z := \underline{\underline{\otimes}}_Z\,.$$

(2.1) Except for the right adjoint $f^\times$ of f_*, whose existence is asserted in (4.0) below, all the functorial maps to be considered, and the relations among them, can be derived formally from the data in (1.3). The basic results in §§3–4 state that under suitable conditions certain such maps are isomorphisms.

For example, given $f\colon X \to Y$, the bifunctorial *projection map*

$$p = p(F,G)\colon F \otimes_Y f_*G \to f_*(f^*F \otimes_X G) \qquad \big(F \in \mathbf{D}(Y),\ G \in \mathbf{D}(X)\big)$$

is defined to be the natural composition

$$F \otimes_Y f_*G \xrightarrow{\eta\otimes 1} f_*f^*F \otimes_Y f_*G \longrightarrow f_*(f^*F \otimes_X G).$$

(Here $\eta\colon F \to f_*f^*F$ corresponds to the identity map of f^*F via adjointness of f_* and f^*.)

For fixed $F \in \mathbf{D}(Y)$, the functors (from $\mathbf{D}(Y)$ to $\mathbf{D}(X)$)

$$H \mapsto f^*(H \otimes_Y F) \quad \text{and} \quad H \mapsto f^*H \otimes_X f^*F$$

both have right adjoints, namely

$$G \mapsto \mathcal{H}_Y(F, f_*G) \quad \text{and} \quad G \mapsto f_*\mathcal{H}_X(f^*F, G).$$

Hence there is a functorial isomorphism

$$\alpha(F,G)\colon f_*\mathcal{H}_X(f^*F, G) \xrightarrow{\sim} \mathcal{H}_Y(F, f_*G) \qquad \big(F \in \mathbf{D}(Y),\ G \in \mathbf{D}(X)\big)$$

conjugate to the natural functorial isomorphism $f^*(H \otimes_Y F) \xrightarrow{\sim} f^*H \otimes_X f^*F$.

We deduce functorial maps

$$\begin{aligned}\nu(E,G)\colon f_*\mathcal{H}_X(E,\,G) &\longrightarrow f_*\mathcal{H}_X(f^*f_*E,\,G)\\ &\underset{\alpha}{\xrightarrow{\ \sim\ }} \mathcal{H}_Y(f_*E,\,f_*G) \qquad\qquad \big(E,G\in\mathbf{D}(X)\big)\end{aligned}$$

and

$$\begin{aligned}\rho(F,H)\colon f^*\mathcal{H}_Y(F,\,H) &\longrightarrow f^*\mathcal{H}_Y(F,\,f_*f^*H)\\ &\underset{\alpha^{-1}}{\xrightarrow{\ \sim\ }} f^*f_*\mathcal{H}_X(f^*F,\,f^*H)\longrightarrow\mathcal{H}_X(f^*F,\,f^*H)\\ &\qquad\qquad\qquad\qquad\qquad\qquad\qquad \big(F,H\in\mathbf{D}(Y)\big).\end{aligned}$$

To any commutative square

$$\text{(2.1.1)}\qquad \begin{array}{ccc} X' & \xrightarrow{\ v\ } & X\\ {\scriptstyle g}\big\downarrow & & \big\downarrow{\scriptstyle f}\\ Y' & \xrightarrow[\ u\ ]{} & Y\end{array}$$

there is associated a functorial map

$$\theta(G)\colon u^*f_*G\to g_*v^*G \qquad \big(G\in\mathbf{D}(X)\big)$$

which corresponds by adjunction to either of the functorial compositions

$$f_*\longrightarrow f_*v_*v^*\xrightarrow{\ \sim\ } u_*g_*v^*,\qquad g^*u^*f_*\xrightarrow{\ \sim\ } v^*f^*f_*\longrightarrow v_*\,.$$

(2.2) Such functorial maps are related by various commutative diagrams. Here are two useful examples, with the maps p,θ,ν,ρ as above. (Commutativity is proved formally, i.e., category-theoretically—see e.g., [**Li,** Proposition (3.7.3)].)

(2.2.1) In (2.1.1), set $h = fv = ug$. Then the following natural diagram commutes, for all $F\in\mathbf{D}(Y)$ and $G\in\mathbf{D}(X)$:

$$\begin{array}{ccccc} u^*F\otimes u^*f_*G & \xleftarrow{\ \sim\ } & u^*(F\otimes f_*G) & \xrightarrow{\ u^*(p)\ } & u^*f_*(f^*F\otimes G)\\ {\scriptstyle 1\otimes\theta}\big\downarrow & & & & \big\downarrow{\scriptstyle\theta}\\ u^*F\otimes g_*v^*G & & & & g_*v^*(f^*F\otimes G)\\ {\scriptstyle p}\big\downarrow & & & & \big\downarrow{\scriptstyle\simeq}\\ g_*(g^*u^*F\otimes v^*G) & \xrightarrow{\ \sim\ } & g_*(h^*F\otimes v^*G) & \xleftarrow{\ \sim\ } & g_*(v^*f^*F\otimes v^*G)\end{array}$$

(2.2.2) Given a commutative square (2.1.1), let $E \in \mathbf{D}(X)$, $G \in \mathbf{D}(X)$. Then the following diagram commutes:

$$
\begin{array}{ccccc}
u^*f_*\mathcal{H}_X(E,\,G) & & \xrightarrow{\quad\nu\quad} & & u^*\mathcal{H}_Y(f_*E,\,f_*G) \\
\Big\downarrow{\scriptstyle\theta} & & & & \Big\downarrow{\scriptstyle\rho} \\
g_*v^*\mathcal{H}_X(E,\,G) & & & & \mathcal{H}_{Y'}(u^*f_*E,\,u^*f_*G) \\
\Big\downarrow{\scriptstyle\rho} & & & & \Big\downarrow{\scriptstyle\theta} \\
g_*\mathcal{H}_{X'}(v^*E,\,v^*G) & \underset{\nu}{\longrightarrow} & \mathcal{H}_{Y'}(g_*v^*E,\,g_*v^*G) & \underset{\theta}{\longrightarrow} & \mathcal{H}_{Y'}(u^*f_*E,\,g_*v^*G)
\end{array}
$$

(2.3) Typically, one will want to prove that one part of a functorial diagram is an isomorphism, knowing that another part is. Thus the question will be whether the diagram commutes—in other words, can it be decomposed into the elementary diagrams expressing the axioms of monoidal closed categories, monoidal functors, and adjointness and pseudofunctoriality of f^* and f_*? Rather than attacking diagrams one by one, it would be better to prove theorems showing that whole classes of diagrams must commute. This is the stuff of "Coherence in Categories." Such theorems have been proved for situations in which there is a functor f_*, but without the adjoint f^*, cf. [**Le**]. A corresponding investigation including f^* would surely be useful for our present purposes. (To see why, write down a detailed proof of commutativity in (2.2.1) and (2.2.2)!)

3. Minor results

(3.1) Some conditions under which the projection map $p(F, G)$ in (2.1) is an isomorphism are given in [**I,** p. 247, Proposition 3.7] and [**Li,** Proposition (3.9.4)]. Usually, the complex F has to have quasi-coherent homology.

(3.2) The map $\rho(F, H)$ in (2.1) is an isomorphism if f is an open immersion; or if f has finite tor-dimension, F is pseudo-coherent, and H is bounded-below.

(3.3) (A kind of Künneth isomorphism, see [**Li,** Proposition (3.10.3)].) Suppose (2.1.1) is a fibre square (i.e., cartesian), with f, g quasi-compact quasi-separated, and u, v of finite tor-dimension. If f and u are *tor-independent,* i.e., if for all $y \in Y$, $y' \in Y'$, and $x \in X$ such that $y = u(y') = f(x)$, we have, with $R = \mathcal{O}_{Y,y}$, $R' = \mathcal{O}_{Y',y'}$, $S = \mathcal{O}_{X,x}$ that

$$\mathrm{Tor}_n^R(R', S) = 0 \quad \text{for all } n > 0,$$

then for any $G \in \mathbf{D}^+_{\mathrm{qc}}(X)$, the map $\theta(G)$ in (2.1) is an isomorphism; and the converse is true if the scheme X is quasi-separated.

4. Major results

Recall the abbreviations introduced in §2. All schemes are now assumed to be quasi-compact and quasi-separated, and all scheme-maps separated.

THEOREM (4.0) (Existence). *For any $f\colon X \to Y$ (X, Y quasi-compact quasi-separated, f separated), the functor $f_*\colon \mathbf{D}^+_{\mathrm{qc}}(X) \to \mathbf{D}^+(Y)$ has a right adjoint.*

The proof is based on ideas of Verdier [**G**] and Deligne [**H,** appendix], and also uses (1.1). (It is similar to the one in §4 of the paper "Duality ... on formal schemes" in this volume. See also [**Li,** Chapter 4].)

Denote the right adjoint in question by $f^\times$, or by $f^!$ when f is proper; and denote by $\tau\colon f_*f^\times \to \mathbf{1}$ the natural functorial map.

THEOREM (4.1). *Suppose* (2.1.1) *is a fibre square, with f proper and pseudo-coherent, u of finite tor-dimension, and f and u tor-independent, see* (3.3). *Then for any bounded $G' \in \mathbf{D}^+_{\mathrm{qc}}(X')$ and any $H \in \mathbf{D}^+_{\mathrm{qc}}(Y)$, the composition*

$$g_*\mathcal{H}_{X'}(G',\, v^*f^!H) \xrightarrow{\ \nu\ } \mathcal{H}_{Y'}(g_*G',\, g_*v^*f^!H) \xrightarrow[(3.3)]{\ \sim\ } \mathcal{H}_{Y'}(g_*G',\, u^*f_*f^!H) \xrightarrow[\tau]{} \mathcal{H}_{Y'}(g_*G',\, u^*H)$$

is an isomorphism.

In (4.1), if u and v are identity maps then so is the map labeled (3.3), and the resulting composition

$$\delta(G,H)\colon f_*\mathcal{H}_X(G,\, f^!H) \xrightarrow{\nu} \mathcal{H}_Y(f_*G,\, f_*f^!H) \xrightarrow{\tau} \mathcal{H}_Y(f_*G, H)$$

is called the *duality map*. Thus:

COROLLARY (4.2) (Duality). *Let $f\colon X \to Y$ be proper and pseudo-coherent. Then for any bounded $G \in \mathbf{D}^+_{\mathrm{qc}}(X)$ and any $H \in \mathbf{D}^+_{\mathrm{qc}}(Y)$, the duality map $\delta(G,H)$ is an isomorphism.*

Moreover:

COROLLARY (4.3) (Base Change). *With assumptions as in* (4.1), *the map of functors (from $\mathbf{D}^+_{\mathrm{qc}}(Y)$ to $\mathbf{D}^+_{\mathrm{qc}}(X)$) corresponding by* (4.0) *to the composition*

$$g_*v^*f^! \xrightarrow[(3.3)]{\ \sim\ } u^*f_*f^! \xrightarrow[\text{via } \tau]{} u^*$$

is an isomorphism

$$\beta\colon v^*f^! \xrightarrow{\sim} g^!u^*.$$

To deduce (4.3) from (4.1), consider the next diagram, whose commutativity follows from the definition of β:

$$\begin{array}{ccc} g_*\mathcal{H}_{X'}(G',\, v^*f^!H) & \xrightarrow{\ \beta\ } & g_*\mathcal{H}_{X'}(G',\, g^!u^*H) \\ \nu\big\downarrow & & \big\downarrow\nu \\ \mathcal{H}_{Y'}(g_*G',\, g_*v^*f^!H) & \xrightarrow{\ \beta\ } & \mathcal{H}_{Y'}(g_*G',\, g_*g^!u^*H) \\ (3.3)\big\downarrow\simeq & & \big\downarrow\tau \\ \mathcal{H}_{Y'}(g_*G',\, u^*f_*f^!H) & \xrightarrow[\tau]{} & \mathcal{H}_{Y'}(g_*G',\, u^*H) \end{array} \tag{4.3.1}$$

By (4.1), $\tau\circ(3.3)\circ\nu$ is an isomorphism; and by (4.2), the right column is an isomorphism too. It follows that the top row is an isomorphism, and applying the functor $H^0\mathbf{R}\Gamma_{Y'}$, we conclude that for all $G' \in \mathbf{D}^+_{\mathrm{qc}}(X')$,

$$\mathrm{Hom}_{\mathbf{D}(X')}(G',\, v^*f^!H) \xrightarrow{\text{via } \beta} \mathrm{Hom}_{\mathbf{D}(X')}(G',\, g^!u^*H)$$

is an isomorphism. Consequently β itself must be an isomorphism. □

REMARK (4.3.2). Conversely, the commutativity of (4.3.1) shows that (4.2) and (4.3) together imply (4.1).

In describing the organization of the *proof of* (4.1), we will attach symbols to labels of the form (4.x) to refer to special cases of (4.x):

$(4.1)^*_{\mathrm{pc}}$:= (4.1) with $G' = v^*G$, where $G \in \mathbf{D}(X)$ is pseudo-coherent.
$(4.2)_{\mathrm{pc}}$:= (4.2) with G pseudo-coherent := $(4.1)^*_{\mathrm{pc}}$ with $u = v = \mathrm{id}$.
$(4.3)^{\mathrm{o}}$:= (4.3) with the map u an open immersion.
$(4.3)^{\mathrm{af}}$:= (4.3) with the map u affine.

Our strategy is to prove $(4.2)_{\mathrm{pc}}$ and the chain of implications

$$(4.2)_{\mathrm{pc}} \Leftrightarrow (4.1)^*_{\mathrm{pc}} \Rightarrow \big((4.3)^{\mathrm{o}} + (4.3)^{\mathrm{af}}\big) \Rightarrow (4.3) \Rightarrow (4.3)^{\mathrm{o}} \Rightarrow (4.2).$$

By (4.3.2) then, (4.1) results. Here are some brief indications.

Proof of $(4.2)_{\mathrm{pc}}$ (with roots in [**V**, p. 404, proof of Proposition 3]): Formally, the source and target of the duality map δ are right-adjoint to the target and source of the projection map p, and δ is conjugate to p; so if p is an isomorphism .... This approach has pitfalls, because of the boundedness restriction in (4.0); but the idea can be pushed through, as follows. Let τ_m $(m \in \mathbb{Z})$ be the functor taking a complex $A^\bullet$ to the complex

$$\cdots \longrightarrow 0 \longrightarrow 0 \longrightarrow \operatorname{coker}(A^{m-1} \to A^m) \longrightarrow A^{m+1} \longrightarrow A^{m+2} \longrightarrow \cdots .$$

One shows, for any $F \in \mathbf{D}^-_{\mathrm{qc}}(Y)$, that for $m \ll 0$ the map

$$\mathrm{Hom}_{\mathbf{D}(Y)}\big(F,\, f_*\mathcal{H}_X(G,\, f^!H)\big) \xrightarrow{\text{via }\delta} \mathrm{Hom}_{\mathbf{D}(Y)}\big(F,\, \mathcal{H}_Y(f_*G,\, H)\big)$$

is naturally isomorphic to the map

$$\mathrm{Hom}_{\mathbf{D}(Y)}\big(f_*\tau_m(f^*F \otimes G),\, H\big) \xrightarrow{p_m} \mathrm{Hom}_{\mathbf{D}(Y)}\big(F \otimes f_*G,\, H\big)$$

derived from p and the natural map $(f^*F \otimes G) \to \tau_m(f^*F \otimes G)$. A variant of (3.1) yields that for $m \ll 0$, the latter map is an isomorphism, whence so is the former, and hence so is δ itself if its source and target have quasi-coherent homology. When G is pseudo-coherent, then it is easily seen that the source does indeed have quasi-coherent homology; and if further the map f is pseudo-coherent and proper, then it follows from the Finiteness Theorem in (1.2) that the target does too.

Similarly, for *any* $f\colon X \to Y$, and pseudo-coherent G, we have an *isomorphism*

$$\mathcal{H}_X(f^*G, f^\times H) \xrightarrow{\sim} f^\times \mathcal{H}_Y(G, H)$$

adjoint to the composition $f_*\mathcal{H}_X(f^*G, f^\times H) \underset{\alpha}{\longrightarrow} \mathcal{H}_X(f^*G, f_*f^\times H) \underset{\tau}{\longrightarrow} \mathcal{H}_Y(G, H)$.

$(4.2)_{\mathrm{pc}} \Leftrightarrow (4.1)^*_{\mathrm{pc}}$: The implication $\Leftarrow$ is obvious. Now in (2.2.2), replace E by G and G by $f^!H$, and derive the commutative diagram

(4.3.3)
$$\begin{array}{ccc} u^*f_*\mathcal{H}_X(G,\, f^!H) & \xrightarrow{u^*(\delta)} & u^*\mathcal{H}_Y(f_*G,\, H) \\ \big\downarrow & & \big\downarrow \\ g_*\mathcal{H}_{X'}(v^*G,\, v^*f^!H) & \underset{(4.1)}{\longrightarrow} & \mathcal{H}_{Y'}(g_*v^*G,\, u^*H) \end{array}$$

If G—and hence, by the Finiteness Theorem, f_*G—is pseudo-coherent, then by §3, the columns are isomorphisms, whence the implication $\Rightarrow$.

$(4.1)^*_{\mathrm{pc}} \Rightarrow (4.3)^{\mathrm{o}} + (4.3)^{\mathrm{af}}$: Since u (hence v) is an open immersion or affine, it suffices that $v_*\beta$ be an isomorphism. Let G, and hence v^*G, be pseudo-coherent. Using (4.3.1), we derive a commutative diagram

$$\begin{array}{ccc}
f_*\mathcal{H}_X(G,\, v_*v^*f^!H) & \xrightarrow{\;v_*\beta\;} & f_*\mathcal{H}_X(G,\, v_*g^!u^*H) \\
{\scriptstyle \alpha^{-1}}\big\downarrow{\scriptstyle \simeq} & & {\scriptstyle \simeq}\big\downarrow{\scriptstyle \alpha^{-1}} \\
f_*v_*\mathcal{H}_{X'}(v^*G,\, v^*f^!H) & \xrightarrow{\;\beta\;} & f_*v_*\mathcal{H}_{X'}(v^*G,\, g^!u^*H) \\
\big\downarrow{\scriptstyle \simeq} & & {\scriptstyle \simeq}\big\downarrow \\
u_*g_*\mathcal{H}_{X'}(v^*G,\, v^*f^!H) & \underset{(4.1)^*_{\mathrm{pc}}}{\xrightarrow{\;\sim\;}} & u_*\mathcal{H}_{Y'}(g_*v^*G,\, u^*H)
\end{array}$$

in which α is as in (2.1), and the right column is an isomorphism by $(4.2)_{\mathrm{pc}}$ (for g). The desired implication results then from the following:

Key Fact (4.4). *If $f\colon X \to Y$ is a finitely presented scheme-map, then a map ϕ in $\mathbf{D}^+_{\mathrm{qc}}(X)$ is an isomorphism iff so is the map $f_*\mathcal{H}_X(G,\,\phi)$ (in $\mathbf{D}(Y)$) for every pseudo-coherent $G \in \mathbf{D}(X)$.*

$(4.3)^{\mathrm{o}} + (4.3)^{\mathrm{af}} \Rightarrow (4.3)$: Follows from a simple "transitivity" property of θ (hence β) with respect to horizontal composition of fibre squares.

$(4.3)^{\mathrm{o}} \Rightarrow (4.2)$: When u is an open immersion, the columns of (4.3.3) are isomorphisms. And applying the functor $\mathbf{R}\Gamma_{Y'}$ to the bottom row of (4.3.3) produces an isomorphism: to see this, apply $\mathbf{R}\Gamma_{Y'}$ to (4.3.1), and note that then the right hand column induces homology isomorphisms (by definition of $g^!$), hence is itself an isomorphism, and that the top row is an isomorphism by assumption. We conclude that $\mathbf{R}\Gamma_{Y'}u^*(\delta)$ is an isomorphism for all open immersions $u\colon Y' \to Y$; and (4.2) follows. (Taking $G = v_*G'$, we get the converse implication.)

Remark (4.5). Let $F \in \mathbf{D}^+_{\mathrm{qc}}(Y)$ be isomorphic to a bounded flat complex, and let $H \in \mathbf{D}^+_{\mathrm{qc}}(Y)$. We have a map

(4.5.1) $$f^*F \otimes_X f^\times H \to f^\times(F \otimes_Y H)$$

adjoint to

$$f_*(f^*F \otimes_X f^\times H) \underset{(3.1)}{\xrightarrow{\;\sim\;}} F \otimes_Y f_*f^\times H \underset{\tau}{\longrightarrow} F \otimes_Y H\,.$$

If f is proper and pseudo-coherent, then (4.5.1) is an *isomorphism.* To see this, one reduces to where F is a single quasi-coherent flat $\mathcal{O}_Y$-module, and applies (4.3) with $Y' = \mathrm{Spec}(\mathcal{O}_Y \oplus F)$ (where $F^2 = 0$). Alternatively, one can deduce this isomorphism from (4.2) and (4.4), and use it to prove (4.3).

References

[Co] B. Conrad, *Base Change and Grothendieck Duality*, Springer-Verlag,, New York (to appear).

[GD] A. Grothendieck, J. Dieudonné, *Éléments de Géométrie Algébrique IV*, Publ. Math. IHES no. 28, 1966.

[G] P.-P. Grivel, *Une démonstration du théorème de dualité de Verdier*, L'Enseignement Math. **31** (1985), 227–247.

[EK] S. Eilenberg and G. M. Kelly, *Closed categories*, Proceedings of the Conference on Categorical Algebra, La Jolla, 1965, Springer-Verlag, New York, 1966, pp. 421–562.

[H] R. Hartshorne, *Residues and Duality*, Lecture Notes in Math., no. 20, Springer Verlag, New York, 1966.

[I] L. Illusie, *Généralités sur les conditions de finitude dans les catégories dérivées, etc.*, Théorie des Intersections et Théorème de Riemann-Roch (SGA 6), Lecture Notes in Math., no. 225, Springer Verlag, New York, 1971, pp. 78–273.

[K] R. Kiehl, *Ein "Descente"-Lemma und Grothendiecks Projektionssatz für nichtnoethersche Schemata*, Math. Ann. **198** (1972), 287–316.

[Le] G. Lewis, *Coherence for a closed functor*, Coherence in Categories, Lecture Notes in Math., no. 281, Springer Verlag, New York, 1972, pp. 148–195.

[Li] J. Lipman, *Notes on Derived Categories*, preprint, at `www.math.purdue.edu/~lipman/`.

[N] A. Neeman, *The Grothendieck duality theorem via Bousfield's techniques and Brown representability*, Jour. Amer. Math. Soc. **9** (1996), 205–236..

[So] S.V. Soloviev, *On the conditions of full coherence in closed categories*, J. Pure and Applied Algebra **69** (1990), 301–329.

[Sp] N. Spaltenstein, *Resolutions of unbounded complexes*, Compositio Math. **65** (1988), 121–154.

[V] J.-L. Verdier, *Base change for twisted inverse image of coherent sheaves*, Algebraic Geometry (Bombay, 1968), Oxford Univ. Press, London, 1969, pp. 393–408.

Department of Mathematics, Purdue University, W. Lafayette IN 47907
E-mail address: `lipman@math.purdue.edu`
URL: `www.math.purdue.edu/~lipman`

Index